WITHDRAWN
UTSA LIBRARIES

RENEWALS 458-4574
DATE DUE

Applied Mathematical Sciences
Volume 155

Editors
S.S. Antman J.E. Marsden L. Sirovich

Advisors
J.K. Hale P. Holmes J. Keener
J. Keller B.J. Matkowsky A. Mielke
C.S. Peskin K.R.S. Sreenivasan

Springer
New York
Berlin
Heidelberg
Hong Kong
London
Milan
Paris
Tokyo

Applied Mathematical Sciences

1. *John:* Partial Differential Equations, 4th ed.
2. *Sirovich:* Techniques of Asymptotic Analysis.
3. *Hale:* Theory of Functional Differential Equations, 2nd ed.
4. *Percus:* Combinatorial Methods.
5. *von Mises/Friedrichs:* Fluid Dynamics.
6. *Freiberger/Grenander:* A Short Course in Computational Probability and Statistics.
7. *Pipkin:* Lectures on Viscoelasticity Theory.
8. *Giacaglia:* Perturbation Methods in Non-linear Systems.
9. *Friedrichs:* Spectral Theory of Operators in Hilbert Space.
10. *Stroud:* Numerical Quadrature and Solution of Ordinary Differential Equations.
11. *Wolovich:* Linear Multivariable Systems.
12. *Berkovitz:* Optimal Control Theory.
13. *Bluman/Cole:* Similarity Methods for Differential Equations.
14. *Yoshizawa:* Stability Theory and the Existence of Periodic Solution and Almost Periodic Solutions.
15. *Braun:* Differential Equations and Their Applications, 3rd ed.
16. *Lefschetz:* Applications of Algebraic Topology.
17. *Collatz/Wetterling:* Optimization Problems.
18. *Grenander:* Pattern Synthesis: Lectures in Pattern Theory, Vol. I.
19. *Marsden/McCracken:* Hopf Bifurcation and Its Applications.
20. *Driver:* Ordinary and Delay Differential Equations.
21. *Courant/Friedrichs:* Supersonic Flow and Shock Waves.
22. *Rouche/Habets/Laloy:* Stability Theory by Liapunov's Direct Method.
23. *Lamperti:* Stochastic Processes: A Survey of the Mathematical Theory.
24. *Grenander:* Pattern Analysis: Lectures in Pattern Theory, Vol. II.
25. *Davies:* Integral Transforms and Their Applications, 2nd ed.
26. *Kushner/Clark:* Stochastic Approximation Methods for Constrained and Unconstrained Systems.
27. *de Boor:* A Practical Guide to Splines: Revised Edition.
28. *Keilson:* Markov Chain Models—Rarity and Exponentiality.
29. *de Veubeke:* A Course in Elasticity.
30. *Sniatycki:* Geometric Quantization and Quantum Mechanics.
31. *Reid:* Sturmian Theory for Ordinary Differential Equations.
32. *Meis/Markowitz:* Numerical Solution of Partial Differential Equations.
33. *Grenander:* Regular Structures: Lectures in Pattern Theory, Vol. III.
34. *Kevorkian/Cole:* Perturbation Methods in Applied Mathematics.
35. *Carr:* Applications of Centre Manifold Theory.
36. *Bengtsson/Ghil/Källén:* Dynamic Meteorology: Data Assimilation Methods.
37. *Saperstone:* Semidynamical Systems in Infinite Dimensional Spaces.
38. *Lichtenberg/Lieberman:* Regular and Chaotic Dynamics, 2nd ed.
39. *Piccini/Stampacchia/Vidossich:* Ordinary Differential Equations in $\mathbf{R}^n$.
40. *Naylor/Sell:* Linear Operator Theory in Engineering and Science.
41. *Sparrow:* The Lorenz Equations: Bifurcations, Chaos, and Strange Attractors.
42. *Guckenheimer/Holmes:* Nonlinear Oscillations, Dynamical Systems, and Bifurcations of Vector Fields.
43. *Ockendon/Taylor:* Inviscid Fluid Flows.
44. *Pazy:* Semigroups of Linear Operators and Applications to Partial Differential Equations.
45. *Glashoff/Gustafson:* Linear Operations and Approximation: An Introduction to the Theoretical Analysis and Numerical Treatment of Semi-Infinite Programs.
46. *Wilcox:* Scattering Theory for Diffraction Gratings.
47. *Hale/Magalhães/Oliva:* Dynamics in Infinite Dimensions, 2nd ed.
48. *Murray:* Asymptotic Analysis.
49. *Ladyzhenskaya:* The Boundary-Value Problems of Mathematical Physics.
50. *Wilcox:* Sound Propagation in Stratified Fluids.
51. *Golubitsky/Schaeffer:* Bifurcation and Groups in Bifurcation Theory, Vol. I.
52. *Chipot:* Variational Inequalities and Flow in Porous Media.
53. *Majda:* Compressible Fluid Flow and System of Conservation Laws in Several Space Variables.
54. *Wasow:* Linear Turning Point Theory.
55. *Yosida:* Operational Calculus: A Theory of Hyperfunctions.
56. *Chang/Howes:* Nonlinear Singular Perturbation Phenomena: Theory and Applications.
57. *Reinhardt:* Analysis of Approximation Methods for Differential and Integral Equations.
58. *Dwoyer/Hussaini/Voigt (eds):* Theoretical Approaches to Turbulence.
59. *Sanders/Verhulst:* Averaging Methods in Nonlinear Dynamical Systems.

(continued following index)

Bernard Chalmond

Modeling and Inverse Problems in Imaging Analysis

With 110 Figures

 Springer

Library
University of Texas
of San Antonio

Bernard Chalmond
Department of Physics
University of Cergy-Pontoise
Neuville sur Oise
95031 Cergy-Pontoise Cedex
France
bernard.chalmond@cmla.ens-cachan.fr

Editors

S.S. Antman
Department of Mathematics
and
Institute for Physical Science
and Technology
University of Maryland
College Park, MD 20742-4015
USA
ssa@math.umd.edu

J.E. Marsden
Control and Dynamical
Systems, 107-81
California Institute of
Technology
Pasadena, CA 91125
USA
marsden@cds.caltech.edu

L. Sirovich
Division of Applied
Mathematics
Brown University
Providence, RI 02912
USA
chico@camelot.mssm.edu

Mathematics Subject Classification (2000): 69U10, 6802, 6207, 68Uxx

Library of Congress Cataloging-in-Publication Data
Chalmond, Bernard, 1951–
    [Eléments de modélisation pour l'analyse d'images. English]
    Modeling and inverse problems in image analysis / Bernard Chalmond.
        p.   cm. — (Applied mathematical sciences ; 155)
    Includes bibliographical references and index.
    ISBN 0-387-95547-X (alk. paper)
        1. Image processing—Digital techniques—Mathematical models.   2. Image
analysis—Mathematical models.   3. Inverse problems (Differential equations)   I. Title   II.
Applied mathematical sciences (Springer-Verlag New York Inc.) ; v. 155
    QA1.A647
    [TA1637]
    006.4′2—dc21   2002026662

ISBN 0-387-95547-X   Printed on acid-free paper.

© 2003 Springer-Verlag New York, Inc.
All rights reserved. This work may not be translated or copied in whole or in part without the
written permission of the publisher (Springer-Verlag New York, Inc., 175 Fifth Avenue, New York,
NY 10010, USA), except for brief excerpts in connection with reviews or scholarly analysis. Use
in connection with any form of information storage and retrieval, electronic adaptation, computer
software, or by similar or dissimilar methodology now known or hereafter developed is forbidden.
The use in this publication of trade names, trademarks, service marks, and similar terms, even if
they are not identified as such, is not to be taken as an expression of opinion as to whether or not
they are subject to proprietary rights.

Printed in the United States of America.

9 8 7 6 5 4 3 2 1   SPIN 10887014

Typesetting: Pages created by the author using a Springer TEX macro package.

www.springer-ny.com

Springer-Verlag   New York Berlin Heidelberg
A member of BertelsmannSpringer Science+Business Media GmbH

Library
University of Texas
at San Antonio

Foreword

In the last decade of the past century we witnessed an exceptional participation of mathematicians in the development of digital image processing as a science. These contributions have found a natural place at the low level of processing, with the advent of mathematical morphology, differential equations, Markov random fields, and wavelet theory. They are also reflected in the increasingly important role modeling has played in solving complex problems.

Although modeling is often a hidden stage of the solution, the benefits of correctly modeling an image processing problem are huge. Modeling is the very place where "sensitivity" steals a lead over "geometry", to put it in Pascal's words. Correct modeling introduces information that cannot be expressed by data or deduced by equations, but reflects the subtle dependency or causality between the ingredients. Modeling has more to do with pots and pans than recipes and spices, to draw out the culinary metaphor. Bernard Chalmond's work is mainly dedicated to modeling issues. It does not fully cover this field, since it is mostly concerned with two types of model: Bayesian models issued from probability theory and energy-based models derived from physics and mechanics. Within the scope of these models, the book deeply explores the various consequences of the choice of a model; it compares their hypotheses, discusses their merits, explores their validity, and suggests possible fields of application.

Chalmond's work falls into three parts. The first deals with the processing of spline functions, interpolation, classification, and auto-associative models. Although splines are usually presented with their mechanical interpretation, Bernard Chalmond provides the Bayesian interpretation as well. The last chapter of this part, devoted to auto-associative models and pursuit problems, provides an original view of this field. The second part is concerned with inverse problems considered as Markovian energy optimization. Consistent attention is paid to the choice of potentials, and optimization is classically presented either as a deterministic or as a stochastic issue. The chapter on parameter estimation is greatly appreciated, since this aspect of Markovian modeling is too often neglected in textbooks. In the last part, Bernard Chalmond takes the time to subtly dissect some image processing problems in order to develop consistent modeling. These particular problems are concerned with denoising, deblurring, detecting noisy lines, estimating parameters, etc. They have been specifically chosen to cover a wide variety of situations encountered in robot vision or image processing. They can easily be adapted to the reader's own requirements.

This book fulfills a need in the field of computer science research and education. It is not intended for professional mathematicians, but it undoubtedly deals with applied mathematics. Most of the expectations of the topic are fulfilled: precision, exactness, completeness, and excellent references to the original historical works. However, for the sake of readability, many demonstrations are omitted. It is not a book on practical image processing, of which so many abound, although all that it teaches is directly concerned with image analysis and image restoration. It is the perfect resource for any advanced scientist concerned with a better understanding of the theoretical models underlying the methods that have efficiently solved numerous issues in robot vision and picture processing.

HENRI MAÎTRE

Paris, France

Acknowledgments

This book describes a methodology and a practical approach for the effective building of models in image analysis. The practical knowledge applied was obtained through projects carried out in an industrial context, and benefited from the collaboration of

R. Azencott, F. Coldefy, J.M. Dinten,
S.C. Girard, B. Lavayssière, and J.P. Wang,

to all of whom I extend my warmest thanks. Without their participation, this work could not have achieved full maturity.

I am obliged to several groups of students who have participated over the years in the course on modeling and inverse problem methods at the University of Paris (Orsay), the Ecole Normale Supérieure (Cachan), and the University of Cergy-Pontoise, in mathematics, physics, and engineering, and have contributed to the evolution of the present material. The English manuscript was prepared by Kari Foster for the publishers, and I am very grateful to her.

Contents

Foreword by Henri Maître   vii

Acknowledgments   ix

List of Figures   xiii

Notation and Symbols   xix

1   Introduction   1
   1.1   About Modeling 3
       1.1.1   Bayesian Approach 3
       1.1.2   Inverse Problem 8
       1.1.3   Energy-Based Formulation 10
       1.1.4   Models 11
   1.2   Structure of the Book 14

I   Spline Models   21

2   Nonparametric Spline Models   23
   2.1   Definition 23
   2.2   Optimization 26
       2.2.1   Bending Spline 26
       2.2.2   Spline Under Tension 28

	2.2.3	Robustness	31
2.3	Bayesian Interpretation		34
2.4	Choice of Regularization Parameter		36
2.5	Approximation Using a Surface		39
	2.5.1	L-Spline Surface	40
	2.5.2	Quadratic Energy	43
	2.5.3	Finite Element Optimization	46

3 Parametric Spline Models **51**

3.1	Representation on a Basis of B-Splines		51
	3.1.1	Approximation Spline	53
	3.1.2	Construction of B-Splines	54
3.2	Extensions		57
	3.2.1	Multidimensional Case	57
	3.2.2	Heteroscedasticity	62
3.3	High-Dimensional Splines		67
	3.3.1	Revealing Directions	68
	3.3.2	Projection Pursuit Regression	70

4 Auto-Associative Models **75**

4.1	Analysis of Multidimensional Data		75
	4.1.1	A Classical Approach	76
	4.1.2	Toward an Alternative Approach	80
4.2	Auto-Associative Composite Models		82
	4.2.1	Model and Algorithm	82
	4.2.2	Properties	84
4.3	Projection Pursuit and Spline Smoothing		86
	4.3.1	Projection Index	87
	4.3.2	Spline Smoothing	90
4.4	Illustration		93

II   Markov Models **97**

5 Fundamental Aspects **99**

5.1	Definitions		99
	5.1.1	Finite Markov Fields	100
	5.1.2	Gibbs Fields	101
5.2	Markov–Gibbs Equivalence		103
5.3	Examples		106
	5.3.1	Bending Energy	106
	5.3.2	Bernoulli Energy	107
	5.3.3	Gaussian Energy	108
5.4	Consistency Problem		109

6 Bayesian Estimation   **113**
   6.1   Principle   113
   6.2   Cost Functions   118
           6.2.1   Cost Function Examples   119
           6.2.2   Calculation Problems   121

7 Simulation and Optimization   **123**
   7.1   Simulation   124
           7.1.1   Homogeneous Markov Chain   124
           7.1.2   Metropolis Dynamic   125
           7.1.3   Simulated Gibbs Distribution   127
   7.2   Stochastic Optimization   130
   7.3   Probabilistic Aspects   134
   7.4   Deterministic Optimization   138
           7.4.1   ICM Algorithm   138
           7.4.2   Relaxation Algorithms   141

8 Parameter Estimation   **147**
   8.1   Complete Data   148
           8.1.1   Maximum Likelihood   149
           8.1.2   Maximum Pseudolikelihood   150
           8.1.3   Logistic Estimation   153
   8.2   Incomplete Data   156
           8.2.1   Maximum Likelihood   157
           8.2.2   Gibbsian EM Algorithm   161
           8.2.3   Bayesian Calibration   170

III   Modeling in Action   **175**

9 Model-Building   **177**
   9.1   Multiple Spline Approximation   177
           9.1.1   Choice of Data and Image Characteristics   179
           9.1.2   Definition of the Hidden Field   181
           9.1.3   Building an Energy   183
   9.2   Markov Modeling Methodology   185
           9.2.1   Details for Implementation   185

10 Degradation in Imaging   **189**
   10.1   Denoising   190
           10.1.1 Models with Explicit Discontinuities   190
           10.1.2 Models with Implicit Discontinuities   198
   10.2   Deblurring   201
           10.2.1 A Particularly Ill-Posed Problem   202
           10.2.2 Model with Implicit Discontinuities   204

10.3 Scatter 205
      10.3.1 Direct Problem 206
      10.3.2 Inverse Problem 211
10.4 Sensitivity Functions and Image Fusion 216
      10.4.1 A Restoration Problem 217
      10.4.2 Transfer Function Estimation 221
      10.4.3 Estimation of Stained Transfer Function 224

11 Detection of Filamentary Entities   **227**
11.1 Valley Detection Principle 228
      11.1.1 Definitions 228
      11.1.2 Bayes–Markov Formulation 230
11.2 Building the Prior Energy 231
      11.2.1 Detection Term 231
      11.2.2 Regularization Term 234
11.3 Optimization 236
11.4 Extension to the Case of an Image Pair 239

12 Reconstruction and Projections   **243**
12.1 Projection Model 243
      12.1.1 Transmission Tomography 243
      12.1.2 Emission Tomography 246
12.2 Regularized Reconstruction 247
      12.2.1 Regularization with Explicit Discontinuities . . . 248
      12.2.2 Three-Dimensional Reconstruction 252
12.3 Reconstruction with a Single View 256
      12.3.1 Generalized Cylinder 256
      12.3.2 Training the Deformations 259
      12.3.3 Reconstruction in the Presence of Occlusion . . . 261

13 Matching   **269**
13.1 Template and Hidden Outline 270
      13.1.1 Rigid Transformations 270
      13.1.2 Spline Model of a Template 272
13.2 Elastic Deformations 276
      13.2.1 Continuous Random Fields 276
      13.2.2 Probabilistic Aspects 282

References   **289**

Author Index   **301**

Subject Index   **305**

List of Figures

1.1   Non-uniform lighting.   2

1.2   Field of deformations.   2

1.3   Face modeling.   3

1.4   Filamentary entity detection.   4

1.5   Image sequence analysis.   4

1.6   Tomographic reconstruction.   5

1.7   Hidden field.   6

1.8   Image characteristics.   14

2.1   Smoothing function.   32

2.2   Robust smoothing.   33

2.3   Effect of regularization parameter.   36

2.4   Bias/variance dilemma.   37

2.5   Approximation of a field of deformations.   43

2.6   Image deformation.   43

3.1   B-spline function.   52

3.2   Effect of distance between knots on spline approximation.   55

3.3   Tensor product of B-splines.   58

3.4   Smoothing by two-dimensional regression spline.   59

3.5   Suppression of a luminous gradient.   60

4.1   PCA and AAC of translated curves.   79

4.2   Various approximation situations.   81

4.3   Diagram showing travel through regions.   89
4.4   Choice of number of knots.   92
4.5   Simulation with an incorrect number of knots.   93
4.6   PCA and AAC approximation of faces.   95

5.1   Bernoulli fields.   108

7.1   Distance L^1 between two probability distributions.   135
7.2   Surface sampled regularly and irregularly.   143
7.3   Approximation of an irregularly sampled surface.   145

8.1   Segmentation of a radiographic image.   169

9.1   Multiple spline approximation.   178
9.2   Spots distributed in a sequence of images.   180
9.3   Gaussian approximation of spots.   181
9.4   Configuration of pointers.   182
9.5   Pointers undergoing regularization.   184
9.6   Extreme local configurations.   187
9.7   Regularization of 3-D paths.   188

10.1 Mixed grid.   191
10.2 Mixed neighborhood system.   192
10.3 Edge configurations on fourth-order cliques.   193
10.4 Denoising an NMR image.   196
10.5 Examples of photon paths in a material.   206
10.6 Collision parameters of a probabilistic transport model. .   208
10.7 Simulated radiant image.   211
10.8 Collision parameters of the analytical model.   213
10.9 Restoration-deconvolution.   216
10.10 Simulation of an artificial degradation.   218
10.11 Restoration-fusion.   219
10.12 Stained transfer function.   223
10.13 X-ray imaging for nondestructive testing.   224

11.1 Estimation of the luminosity gradient.   228
11.2 A γ image after suppression of the luminosity gradient. .   229
11.3 Two examples of curved surfaces.   232
11.4 Estimation of valleys on a single image.   237
11.5 Estimation of valleys with fusion of an image pair.   240

12.1 Transmission tomography apparatus.   245
12.2 Emission tomography apparatus.   247
12.3 3-D transmission tomography apparatus.   253
12.4 Four radiographic views of two inclusions.   255

12.5 3-D tomographic reconstruction. 255
12.6 Generalized cylinder with rectangular section. 257
12.7 Radiographic projection of a generalized cylinder. 258
12.8 Electron microscope views of soldered tabs. 264
12.9 Estimation of the weld of a soldered tab from a single view. 265

13.1 Outline and visible contour. 271
13.2 Rigid realignment of three images using a template. . . . 272
13.3 Matching of a template with three radiographic projections. 275
13.4 Elastic deformations of a template. 280
13.5 Elastic matching. 282

Notation and Symbols

AAC	auto-associative composite model.
$B(t)$	B-spline function.
$\mathbf{B}$	matrix of B-splines.
$\mathbf{B}^\tau$	matrix of B-splines along the τ axis.
$\mathrm{Cov}(X, Y)$	covariance.
$\mathcal{C}^m$	functions that are m times continuously differentiable.
$D^m g$, $\frac{d^m g}{dt^m}$	mth derivative of $g(t)$.
$D^{i,j} g$, $\frac{\partial^{i+j} g}{\partial t^i \partial u^j}$	partial derivative of $g(t, u)$.
Δ	Laplacian.
$\Delta^m g$	mth-order difference operator.
$\mathrm{E}[X]$	mathematical expectation of the random variable X.
$\mathcal{E}_\ell$	space of finite elements e of diameter ℓ.
$g(t)$	curve.
$g(t, u)$	surface.
$g\,|e$	restriction of the function g to e.
$G(x) = 0$	equation of a manifold.
$\mathcal{G}$	two-dimensional grid.
$\mathcal{G}^{\mathrm{theo}}$, $\mathcal{G}^{\mathrm{emp}}$	generalization error.
H^m	mth-order Sobolev space.

$I(a, \mathcal{X})$	index of the AAC of a for the cloud $\mathcal{X}$.
$\dot{I}(a, \mathcal{X})$	index of the PCA of a for the cloud $\mathcal{X}$.
$I_K(a, \mathcal{X})$	Kullback index of a for the cloud $\mathcal{X}$.
Id	identity matrix or application.
$LG(\mu, V)$	Laplace–Gauss distribution of expectation μ and variance V.
L	linear differential operator.
L^2	square integrable function space.
$L(\theta)$	likelihood of θ.
M', ^tM	transpose matrix.
∇f	gradient of f.
$\mathcal{O}(X, W)$	operator for degradation of X by W.
$P(x)$	probability distribution.
P^a	projection operator with parameter a.
PCA	principal component analysis.
PPR	projection pursuit regression.
R	real numbers.
S^b	smoothing operator with parameter b.
$\mathcal{S}(\mathcal{T})$	space of cubic splines with knots $\mathcal{T}$.
$\mathcal{S}_N(\mathcal{T})$	space of natural cubic splines with knots $\mathcal{T}$.
σ^2	variance of a noise, of residuals, etc.
$T = \langle a, X \rangle$	coordinate of X on the a-axis.
$U(x)$	prior energy (or internal energy).
$U(y \mid x)$	fidelity energy y (or external energy).
$\mathrm{Var}(X)$	variance or variance–covariance matrix.
W	noise, residual, etc.
X, x	random variable X and its occurrence x.
$\mathcal{X}$	cloud of points in $I\!R^n$.
Ξ_k	symmetry.
Y, y	vector or field of observations.
Z	relative integers.
$\mathbf{1}_A$, $\mathbf{1}[A]$	indicator function.
$\mid A \mid$	cardinality of A if A is a set, or absolute value if A is a number.
$A \doteq B$	by definition B is written as A.
$< s, t >$	s and t are neighbors.
$\langle X, Y \rangle$	scalar product.
$\left(x_i^j\right)^2$	a raised index and an exponent can be present at the same time, and should be distinguishable according to the context.
$\vec{y}$, $\underline{y}$, $\mathbf{y}$	to indicate that y is a vector.
$Y \mid X = x$	probabilistic conditioning.
$\widehat{\theta}$	estimation of θ, or the argument of the minimum of a function.

Part II

c	clique.
C	set of cliques of a Markov random field.
E	set of states.
E_s	set of states at site s.
Γ	cost function.
H_i	local energy difference.
$\mathbf{H} = (H_1, \ldots, H_q)'$	vector of local energy differences.
ICM	iterated conditional mode algorithm.
$\mathcal{L}(\theta)$, $\mathcal{P}_\theta(x)$	pseudolikelihood of θ.
N_s	set of neighborhood sites to s.
$N_s(x)$	values in the neighborhood N_s of s.
p_{yx}	probability of transition of y to x of a Markov chain.
$\mathcal{P}$	transition matrix of a Markov chain.
$\mathcal{P}_\theta(x)$	pseudolikelihood of θ for the field x.
Π	discrete probability distribution.
s	site.
S	set of sites.
T_n	simulated annealing temperature.
$\mathbf{U}(y)$	energy $\mathbf{U}(x)$ limited to a subset y of x.
$V_c(x)$	potential of the clique c.
$\breve{x}_s$	a field x deprived of the value x_s.
y^O	a specific observation of Y.
$\{Z_n\}$	Markov chain.
$\mathcal{Z}$	normalization constant of a Gibbs distribution.
$\lceil a \rceil$	integer part of a.

Part III

C, C_i, $C_{\ell,i}$	set of cliques of a Markov random field.
$C(t)$	curve.
Δf	Laplacian of f.
$\mathcal{F}$	radiographic film.
ϕ	smoothing function.
$\varphi(.)$	scattered flux.
$\{\phi_i\}$	vector basis.
GC	generalized cylinder.
$G(s_1, s_2)$, $g(s_1, s_2)$	surface.

$G(s_1, s_2, \ell)$	stained transfer function at s of level ℓ.
$\mathcal{G}, \mathcal{G}_\theta$	projection of a GC.
$\vec{\mathcal{G}}$	continuous deformation field.
Γ_s	indicator of curvature line at s.
$H(\ell)$	transfer function for the gray level ℓ.
$H_i(.,.), H_{i,j}(.,.)$	local energy difference.
$\mathcal{H}, \mathbb{H}$	convolution kernel.
$I, I_0, \mathcal{I}$	an image.
$K^{(n)}$	kernel of $n-$step transition probability.
λ, Λ	energy of a photon.
Λ_s	parameters of a confidence region.
NDT	nondestructive testing.
$\mathcal{N}_s(\ell)$	set of neighboring sites to s with value ℓ.
$\aleph$	outline.
PSF	point spread function.
R_s	confidence region in s.
R_d^ℓ	ray between source ℓ and detector d.
$\mathcal{R}_x(.,.)$	Radon transform.
S, S^p, S^b	set of sites.
$\mathcal{S}$	spatial position.
SNR	signal-to-noise ratio.
STF	stained transfer function.
TF	transfer function.
T_s	degradation by a stain at site s.
$T(s, a)$	intersection between an object and a straight line (s, a).
$\mathrm{T}_\rho, \mathrm{T}_\tau, \mathrm{T}_\varsigma$	rigid transformations.
VBL	valley bottom line.
X^p, x^p	field in gray levels.
X^b, x^b	edge field or segmentation field.
X^ℓ, x^ℓ	VBL field.
$\{Z_n\}$	Markov chain.
$Z(t)$	Brownian process.

1
Introduction

Inverse problems in digital image analysis are the main theme of this book. Typically, we have image-type data (an image, a pair of images, an image sequence, a set of projection images, etc.) from which we wish to extract information that is of interest to the user. Here are a few examples:

1. Our perception of the image is disturbed by an *illumination fault*. This fault must be removed to provide a more readable image (Figure 1.1).

2. For a pair of similar images, we wish to estimate the *deformation field* that transforms one image into the other (Figure 1.2).

3. From a set of images of a family of objects, we wish to estimate a model representing the *variations* of these objects (Figure 1.3).

4. On a very noisy image, we wish to detect specific entities such as surface *rupture lines* (Figure 1.4).

5. On a sequence of images, we wish to estimate the *path* of objects that are moving and being deformed throughout the image sequence (Figure 1.5).

6. From several radiographic projections of an object, we wish to create a three-dimensional reconstruction of the *contents* of the object (Figure 1.6).

In these examples, the relevant information comes in the form of specific entities: fault illumination, deformation field, variations, rupture lines,

Figure 1.1. Non-uniform lighting.

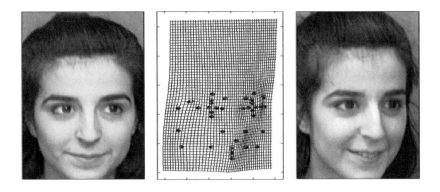

Figure 1.2. Deformation field for the matching of two images.

paths, contents of an object. Extracting one of these entities consists in taking characteristic properties of the entity, which we summarize in a model, and the model is then adjusted to fit the image data. Example 3 is different from the others in that it gives us a training basis; we have repeated observations of the same "scene". In the other examples, we have only one observation (an image, a pair, a sequence, etc.).

Image analysis and the associated modeling techniques play an important part in many areas of science and technology. Although our examples use certain types of image, the methods given can be adapted for use in other imaging contexts.

Models are developed on the basis of general and fundamental aspects, as well as concrete aspects arising from real applications. Many of the general aspects of model-building were actually developed outside the field of imaging, in some cases even before imaging emerged as a field in its own right. Spline functions, discussed in Part I of this book, are one example of

Figure 1.3. Database for the modeling of faces.

this. These general aspects provide tools to use on image analysis problems, but on their own they are not sufficient to build effective models. In Part III, we will see that modeling calls for a great deal of knowledge and skill to overcome the technical difficulties of certain applications. The solutions found may later be used to enhance the basic tools, but the problem-solving process may also reveal new fundamental problems. This occurred in the case of Markov random fields, which underwent considerable development when it was realized that they could be very useful in image analysis. These models are presented in Part II.

1.1 About Modeling

1.1.1 Bayesian Approach

HIDDEN ENTITY

Modeling is the key step in the following fundamental approach. Consider a random vector $Y = \{Y_s, \ s \in S\}$ representing image-type data. In the example in Figure 1.7(a), Y_s is simply the gray level at site s of the image sampled at the nodes of a rectangular grid S.

For the application under consideration, any observed value y of Y contains a hidden entity x that we want to find, where x is itself the occurrence of a random vector X. The search for hidden entities is the main theme

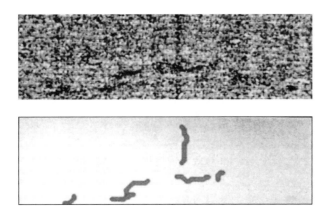

Figure 1.4. Search for filamentary entities in a noisy image.

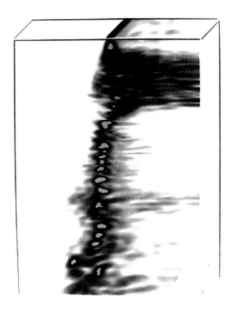

Figure 1.5. Monitoring the deformation of stains in a sequence of images.

of this book. In our example, x is a binary image $\{x_s, \ s \in S\}$ encoded on $\{-1, 1\}$ that represents a partition of S, since the aim of the underlying application is to seek alphanumeric characters in noisy images: $x_s = 1$ if s belongs to the alphanumeric character, and $x_s = -1$ otherwise. So, in Figure 1.7(a), we can just make out the presence of the letter **S**, or possibly

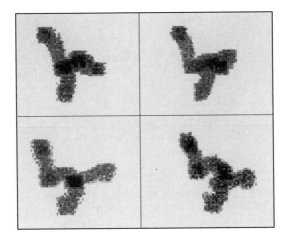

Figure 1.6. Four radiographic projections of a 3D object for its tomographic reconstruction.

the number **8** or the number **6**. The search for the hidden entity therefore involves looking for the partition x that best fits the observed image y.

To do this, we define a set of properties that the estimated hidden image $\hat{x}$ must possess. These properties are described by a probability distribution $P^{\text{prior}}(x)$. For any occurrence x, the probability $P^{\text{prior}}(x)$ quantifies how well x satisfies these properties. In our example, one of the properties would demand that the partition $\hat{x}$ consists of strongly connected regions, i.e. consisting of large black and white regions. The configuration x given in Figure 1.7(b) is therefore less probable that the one in Figure 1.7(c), which is in turn less probable than the one in Figure 1.7(d) with respect to the chosen property.

We also define a second set of properties to quantify how well any occurrence x fits y. To do this, we use a probability distribution $P(y \mid x)$. In our example, P would quantify the homogeneity of the gray levels within the regions of the partition. By applying the Bayes formula, we obtain an expression for the posterior distribution with respect to the prior distribution[1] P^{prior}:

$$P^{\text{post}}(x \mid y) \quad \propto \quad P^{\text{prior}}(x)\, P(y \mid x), \tag{1.1}$$

The search for the hidden entity then boils down to an optimization problem: we choose the occurrence $\hat{x}$ that optimizes a selection criterion

---

[1]In this book the terms *prior* and *posterior* stand for *a priori* and *a posteriori*, respectively.

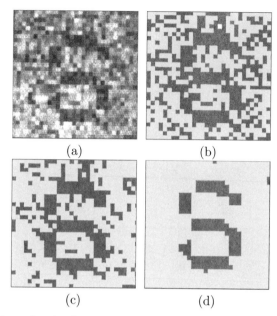

(a)   (b)

(c)   (d)

Figure 1.7. Example of a hidden entity. (a) Image in which an alphanumeric character appears to be present. (b)-(c) In view of the connectedness, the image *c* is more probable than the image *b*. (d) Bayesian estimation of the hidden entity.

based on $P^{\mathrm{post}}(x \mid y)$, and the most immediate of these is

$$\widehat{x} = \arg\max_x P^{\mathrm{post}}(x \mid y). \qquad (1.2)$$

We have then performed a Bayesian estimation of the hidden field relative to the distribution P^{prior}. In our example, after the optimization phase we obtained the partition $\widehat{x}$ shown in Figure 1.7(d). This result $\widehat{x}$ is what we expected, because in this experiment y was obtained by artificially degrading an image of the letter **S** with noise.

This approach is based on an important modeling phase that leads to choosing the probability distributions $P^{\mathrm{prior}}(x)$ and $P(y \mid x)$. In fact, the models of these distributions depend on parameters that we write as θ_1 and θ_2. We can therefore rewrite (1.1) as

$$P^{\mathrm{post}}_\theta(x \mid y) \propto P^{\mathrm{prior}}_{\theta_1}(x)\, P_{\theta_2}(y \mid x), \qquad (1.3)$$

where $\theta = (\theta_1, \theta_2)$. In our example, if we assume that the degradation of the alphanumeric characters is caused by an independent Gaussian noise with distributions $\mathcal{LG}(\mu_1, \sigma^2)$ and $\mathcal{LG}(\mu_{-1}, \sigma^2)$ in the white and black regions respectively, then $\theta_2 = (\mu_{-1}, \mu_1, \sigma^2)$ and

$$P_{\theta_2}(y \mid x) = \prod_{s \in S} \frac{1}{\sqrt{2\pi}\sigma} \exp{-\frac{(y_s - \mu_{x_s})^2}{\sigma^2}}. \qquad (1.4)$$

Furthermore, in Chapter 5 we will see how to model a distribution $P_{\theta_1}^{\text{prior}}(x)$ that favors partitions x consisting of large white and black regions. This is particularly true of the distribution

$$P_{\theta_1}^{\text{prior}}(x) \;=\; \frac{1}{\mathcal{Z}}\exp-\sum_{<s,t>}\theta_1 x_s x_t, \qquad (1.5)$$

when θ_1 is negative and sufficiently far from zero. In this model, $<s,t>$ denotes two sites that are immediate neighbors in S, and we sum over all the pairs of neighboring sites $<s,t>$. The estimation of these parameters is therefore a crucial point. Even a very good model can lead to a highly erroneous configuration $\widehat{x}$ when its parameters are badly estimated, as Figures 1.7(b) and 1.7(c) illustrate.

This Bayesian approach is not specific to image analysis, but it has proven to be particularly effective in this field. Before it came into use, the choice of P^{prior} was very limited. This book is chiefly concerned with this approach; only a few of the methods mentioned come from a different context.

MAXIMUM LIKELIHOOD AND BAYESIAN ESTIMATION

The Bayesian estimation of x is in fact the general case of another very popular method: the Bayesian estimation of the parameters of a random vector distribution. In general terms, let $P_{\xi^*}(y)$ be the probability distribution for a random vector Y whose expression depends on an unknown vector ξ^*. First, we consider the case where ξ^* is a vector of parameters. For clarity, we temporarily use ξ^* instead of ξ to denote the true value of the vector of parameters that we wish to estimate. The vector ξ^* belongs to a domain $\mathcal{D}$. Estimation is performed starting with an observed value y^O of Y.

Definition 1.1. *Let y^O be an observed value of a random vector Y, and let ξ^* be the unknown vector of its distribution parameters. To estimate ξ^*, we use the principle of maximum likelihood, which consists in choosing the vector $\widehat{\xi}$ that makes y^O the most probable observed value*

$$\widehat{\xi} \;=\; \arg\max_{\xi}\; P_{\xi}(y^O).$$

"Likelihood" is the function $L(\xi) = P_{\xi}(y^O)$ defined for all ξ in $\mathcal{D}$.

The Bayesian approach does not assume that there is only one unknown vector ξ^*, but that the ground state ξ obeys a prior probability distribution $P^{\text{prior}}(\xi)$ in $\mathcal{D}$ whose expression is known. We then write $P(y \mid \xi)$ instead of $P_{\xi}(y)$. The posterior distribution is then written $P^{\text{post}}(\xi \mid y) = \frac{1}{P(y)}P^{\text{prior}}(\xi)P(y \mid \xi)$. The posterior likelihood is strictly equivalent to the posterior distribution

$$L^{\text{post}}(\xi) \;=\; P^{\text{post}}(\xi \mid y^O).$$

To apply the principle of maximum posterior likelihood, we choose the ground state

$$\widehat{\xi} = \arg\max_{\xi} L^{\text{post}}(\xi).$$

We say that we have performed a Bayesian estimation with respect to the prior distribution $P^{\text{prior}}(\xi)$, where $P^{\text{post}}(\xi \mid y^O)$ was the criterion to be optimized. Note that $\widehat{\xi}$ is no longer related to a particular ξ^*, but to the choice of $P^{\text{prior}}(\xi)$.

PENALIZED LIKELIHOOD AND TRADEOFF

Most of the time, we use the logarithm of the likelihood in order to deal with simpler expressions

$$
\begin{aligned}
\widehat{\xi} &= \arg\max_{\xi} \left[\log P^{\text{post}}(\xi \mid y^O) \right] \\
&= \arg\max_{\xi} \left[\log P^{\text{prior}}(\xi) + \log P(y^O \mid \xi) \right] \quad (1.6) \\
&= \arg\max_{\xi} \left[\log P^{\text{prior}}(\xi) + \log L(\xi) \right]. \quad (1.7)
\end{aligned}
$$

When we have no prior information about ξ, L^{post} is reduced to L. Otherwise, L^{post} is seen as a penalized likelihood, where the penalty on L is defined by P^{prior} as shown in (1.7). If we set $U = -\log P$, (1.6) is written as

$$
\begin{aligned}
\widehat{\xi} &= \arg\min_{\xi} \left[U^{\text{prior}}(\xi) + U(y^O \mid \xi) \right] \quad (1.8) \\
&\doteq \arg\min_{\xi} \left[U^{\text{post}}(\xi \mid y^O) \right].
\end{aligned}
$$

These expressions will be used extensively later on. From this point of view, by focusing on (1.1) for which $\xi \equiv x$, we find the selection criterion (1.2), which is written as follows for our example:

$$\widehat{x} = \arg\min_{x} \left[\sum_s \frac{(y_s - \mu_{x_s})^2}{\sigma^2} + \theta_1 \sum_{<s,t>} x_s x_t \right], \quad (1.9)$$

in view of (1.4) and (1.5). This criterion selects the partition $\widehat{x}$ that provides a *tradeoff* between the fit and the connectedness; the first sum quantifies the difference between y and x, and the second sum quantifies the connectedness. The tradeoff is a weighted sum of these two terms, where θ_1 is the weighting parameter.

1.1.2   Inverse Problem

The Bayesian approach deals with inverse problems. The term "inverse problem" comes from the fact that an operator $\mathcal{O}$ linking X to Y is sometimes available. Historically, this term is related to the "well-posed" problem according to the definition given by Hadamard for continuous linear operators H in the deterministic case. A problem is said to be well-posed if $g = Hx$ has a unique and stable solution in the sense that $\forall x, \tilde{x} : g = Hx, \tilde{g} = H\tilde{x} \Rightarrow \tilde{x} \to x$ when $\tilde{g} \to g$. Otherwise, the problem is said to be ill-posed, and constraints on x must be defined in order to stabilize the solution. As an example, consider the model representing a blurred image on the continuous domain $\mathcal{S} = [0, 1]^2$:

$$g(s) = [\mathcal{O}(x)](s) = \int h(u)x(s - u)\, du.$$

The original image x is blurred because of the convolution by h, which is usually a characteristic of the image acquisition system used. When h is a Gaussian density, this model is the heat equation. The *direct* problem consists in calculating g from x, whereas the inverse problem is an attempt to calculate x from g, in other words, to invert the heat diffusion process. We can show that two very similar blurred images g and g' can correspond to two very different images x and x', respectively, which explains the ill-posed nature of the deblurring problem. A more realistic model of blur is as follows:

$$Y(s) = [\mathcal{O}(x, W)](s) = \int h(u)x(s - u)\, du + W(s), \qquad (1.10)$$

where the $W(s)$ are random variables representing a noise on g. This model also applies to the example in Figure 1.7(a). The expression (1.4) corresponds to the model

$$Y_s = [\mathcal{O}(x, W)]_s = x_s + W_s,$$

where the W_s are independent Gaussian random variables.

To alleviate the ambiguity of an ill-posed problem, we then constrain the solution space. The constraints on x are expressed as follows:

(i) Explicitly by reducing the space (for example, by modeling x with a polynomial surface), and the solution is obtained by projection on this space; or

(ii) Implicitly through a prior distribution $P^{\text{prior}}(x)$. Given this distribution as well as the distribution $P(y \mid x)$ obtained from the probability distribution $W(s)$, a Bayesian estimation of x can be calculated.

These two approaches will be introduced in Chapters 2 and 3 through their application to the problem of smoothing noisy data, and from then on they will be an essential part of our modeling equipment.

1.1.3 Energy-Based Formulation

The implicit formulation of constraints can be expressed from a deterministic, rather than a probabilistic, viewpoint, as Equation (1.9) indicates. In this case, we seek to construct a functional, referred to as an energy, of the form

$$U^{\text{post}}(x \mid y) = U^{\text{prior}}(x) + U(y \mid x), \tag{1.11}$$

which is not necessarily interpreted as a log-likelihood, as in (1.8). The external energy, or "fidelity energy", U quantifies the deviation between y and x. The prior energy U^{prior}, also known as internal energy, represents the constraints on the solution being sought. The solution $\hat{x}$ to the inverse problem is obtained by minimizing $U^{\text{post}}(x \mid y)$. The prior energy $U^{\text{prior}}(x)$ is called the regularizing energy, because its purpose is to stabilize the solution of the ill-posed inverse problem. We therefore say that the inverse problem is solved by regularization.

Following the example of noisy alphanumeric characters in Figure 1.7, let us consider another example, which was one of the first known energy-based formulations. We wish to approximate a set of points in the plane with ordinates $y = (y_1, \dots, y_n)$ and abscissas $(t_1, \dots, t_n)$, with a discretized curve with ordinates $x = (x_1, \dots, x_n)$. If we impose a regularity property on x, quantified by its variations in slope, then it is natural to consider the following energy (disregarding boundary effects):

$$U^{\text{post}}(x \mid y) = \theta \sum_i [(x_{i+1} - x_i) - (x_i - x_{i-1})]^2 + \sum_i (y_i - x_i)^2, \tag{1.12}$$

where the parameter θ allows a tradeoff between the regularity of x and the fidelity of x to y.

Outside of the probabilistic context, another way to formalize the inverse problem is to use the variational approach, expressed in continuous terms. The energy (1.12) is then rewritten as

$$U^{\text{post}}(x \mid y) = \theta \int x''(t)^2 dt + \sum_i (y_i - x(t_i))^2, \tag{1.13}$$

where $\int x''(t)^2 dt$ quantifies the *bending* of the curve x. Note that for the variational approach, we must demonstrate the existence of a minimum for the energy under consideration. This was not necessary for the discrete formulation because minimization was performed on a finite set. The following example is a case of edge detection in a noisy image y. If x denotes the nonnoisy image and C denotes the hidden edges, i.e. the set of discontinuities in the surface x, then the energy is written as

$$U^{\text{post}}(x, C \mid y) = \theta_1 \iint_{S \setminus C} \|\nabla x\|^2 ds + |C| + \theta_2 \iint_S (y - x)^2 ds,$$

where $\mathcal{S} = [0, 1]^2$. The first term in U^{post} is a smoothing term, and $|C|$ is the length of the edges. We search for $(\hat{x}, \hat{C})$ such that $\hat{x}$ best fits y, its gradient being low outside $\hat{C}$. This energy is the continuous version of the discrete energy

$$U^{\text{post}}(x, C \mid y) = \theta_1 \sum_{\substack{<i,j>: \\ C_{(ij)} = 0}} (x_i - x_j)^2 + \sum_{<k^*,l^*>} C_{<k^*,l^*>} + \theta_2 \sum_{i \in S} (y_i - x_i)^2,$$

(1.14)

where S is the grid with the observed values y_i at its nodes, and C is a binary field on the dual grid S^* of S. If an edge passes through a site of S^*, then C takes the value 1; otherwise, its value is 0. For each pair of neighboring sites $<i, j>$ in S, a site $k^* = (ij)$ is placed in S^* midway between i and j (see Figure 10.1, page 191).

Finally, we will see in Chapter 13 that the continuous formulation also has a probabilistic interpretation.

1.1.4 Models

DEFINITIONS

The term "model" is widely used in scientific contexts, but often without reference to a well-established definition. Typically, a model is an analytical expression that provides an approximate representation of either an observable phenomenon such as y or a virtual entity such as C. We distinguish two types of model.

The first category is that of *exogenous* models,

$$Y = f(x, \theta) + W. \tag{1.15}$$

These models seek a function f with parameter θ that gives an approximation of the relationship between x and y. They are used to make a prediction: We predict y based on the exogenous variable x. In terms of the quadratic error, the best prediction is $E[Y \mid X = x]$, since for a given x and for any $\tilde{f}$ we have

$$E\left[(Y - \tilde{f}(x))^2 \mid X = x\right] \geq E\left[(Y - E[Y \mid X = x])^2 \mid X = x\right],$$

where E is the mathematical expectation. Usually, X is observable, and in this case, the hidden entity is simply a parameter θ. This parameter is calculated by minimizing an external energy such as the empirical quadratic error

$$U(y \mid \theta) = \sum_i (y_i - f(x_i, \theta))^2,$$

or a log-likelihood $L(\theta)$ when available. This is what happens for parametric splines (Chapter 3). With inverse problems

$$Y = \mathcal{O}(x, W, \theta),$$

the operator $\mathcal{O}$ stands for f, but x and θ are both unknown. We will call this type of model a *degradation model* because it models the degradation that x has undergone, for example its degradation by blurring and noise (1.10). This model will give us an external energy $U(y \mid x, \theta)$, denoted by $U_\theta(y \mid x)$.

The second category is that of *endogenous* models. These models are self-predicting, meaning that they seek to predict X based on X itself:

$$X = f(X, \theta) + W. \tag{1.16}$$

Markov processes on $\mathbf{Z}$, like the following, are an example of this type of model:

$$X_t = \sum_{i>0} \theta_i X_{t-i} + W_t.$$

This is a discretized version of a stochastic differential equation. For a given x, θ is estimated by minimizing an internal energy $U(x \mid \theta)$ such as the quadratic error of prediction

$$U(x \mid \theta) = \sum_t \left[x_t - \sum_{i>0} \theta_i x_{t-i} \right]^2, \tag{1.17}$$

or, more generally, a log-likelihood. The dimensionality reduction of a random vector X in $\mathbf{R}^n$ is another case. For example, when $f(X, \theta)$ is the orthogonal projection of X on a subspace of $\mathbf{R}^n$ with basis $\theta = (\theta_1, \dots, \theta_d)$, $d < n$, we have

$$f(X, \theta) = \sum_j \theta_j \langle \theta_j, X \rangle,$$

which is equivalent to predicting X from its projection. This is called an auto-associative model and it is described in Chapter 4. It approximates the observed values x of X by the manifold whose equation is $x - f(x, \theta) = 0$.

In this book, every expression of energy, whether internal or external, will be called a model. So, for a given θ, (1.17) is an internal energy in x that indicates prior knowledge of regularity on x. Prior energies lead to an implicit representation of x. For example, let us consider the internal energy in (1.13), which is $U^{\mathrm{prior}}(x) = \int x''(t)^2 dt$. Given this energy, we represent x by a third-order piecewise polynomial curve called a cubic spline. We can show that

$$\forall \; \{\zeta_i \, , \; i = 1, \ldots, n\} \, , \quad \widehat{x} = \arg \min_{x \, : \, \{x(t_i) = \zeta_i\}} \int x''(t)^2 dt,$$

is one such cubic spline.

A posterior energy $U^{\text{post}}(x \mid y)$ is the combination of an endogenous model $U_{\theta_1}^{\text{prior}}(x)$ and an exogenous model $U_{\theta_2}(y \mid x)$, each with its own set of parameters θ. Here again, the representation is implicit. For example, the minimum of the posterior energy (1.13)

$$\widehat{x} = \arg \min_x \left\{ \theta \int x''(t)^2 dt + \sum_i (y_i - x(t_i))^2 \right\},$$

is still a cubic spline.

Every specific expression of posterior energy will be called a model. Similarly, to define $P_{\theta_1}^{\text{prior}}(x)$ and $P_{\theta_2}(y \mid x)$ requires building a model that is a specific analytic expression of the distributions: $P_{\theta_1}^{\text{prior}}(x)$ for the prior model and $P_{\theta_2}(y \mid x)$ for the degradation model. We must emphasize that this type of model is not intended to provide a faithful representation of reality, but merely to quantify the fidelity of x to the prior properties and to the observed values y. This means that a simulation of $P_{\theta_1}^{\text{prior}}(x)$ does not necessarily give images x that can be visually interpreted, whereas $P^{\text{post}}(x \mid y)$ tends to deliver images x that are close to y and satisfy the prior properties.

MODELING

Although the model U is simple in our example (1.12), in other applications it often comes out of a long and laborious modeling procedure, resulting in a complicated expression such as (1.14). Models are often presented in their final form, without any explanation of how the expressions were obtained. As a result, the knowledge and skills brought to bear on these difficult modeling problems tend to be hidden from the reader. We attempt to remedy this situation in Part III, where several examples of model construction are given in step-by-step detail (at a slight cost to the readability of the text).

We will see in particular that image data y are not necessarily composed of the original image to be analyzed, but rather of *features* extracted from the image, such as the local entropy or the local Fourier coefficients. These image characteristics are chosen to quantify various local attributes in the images, such as regularity for the entropy, or texture for the Fourier coefficients. Defining and selecting these features is a crucial stage in modeling. Figure 1.8 gives another example: edge orientation characteristics obtained by Gabor filtering.

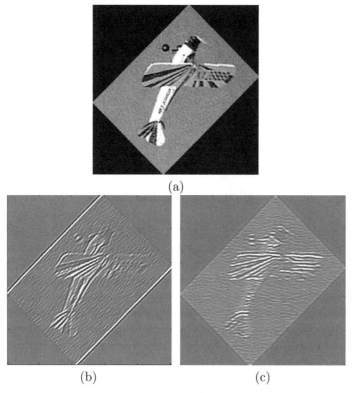

(a)

(b)   (c)

Figure 1.8. Edge orientation characteristic (by Gabor frequency filtering). (a) Original image in gray levels. (b) Orientation of $\pi/4$. (c) Orientation of π.

1.2 Structure of the Book

We advise the reader to approach this book in the following sequence. To become acquainted with modeling and inverse problems, read Chapters 1 and 2 first. These should be followed by Chapters 5 and 6, and finally Chapter 9 and Section 10.1, dealing with applications. Next, go back to Chapters 7 and 8 for an introduction to two tricky problems of the energy-based approach: optimization and parameter estimation. A reader who is concerned mainly with the imaging-related subjects discussed in the other chapters can, however, skip Chapters 7 and 8 at first reading. With these suggestions in mind, the rest of the book can be read sequentially or sampled according to the reader's interest.

Because there are so many types of image analysis problems, this book does not attempt to be exhaustive or to provide a catalog of methods. It aims on the one hand to explain an approach to modeling for real applications, and on the other hand to give a detailed description of the main tools and concepts used, references to which are currently scattered throughout the literature. We therefore give an in-depth presentation of two broad types of technique that are widely used in image analysis: spline representations and Markov models. Although the discretized expression of splines is a special case of Markov energy, splines will also be used in other forms: exogenous models (parametric splines) and auto-associative models, which lead to the subject of pattern recognition.

Part I

Nonparametric Splines (Chapter 2). This book discusses modeling mainly from a stochastic point of view using discrete energies. The energy of smoothing spline functions (1.12), however, is a special case of Markov energy. We shall start in Chapter 2 with the variational approach to continuous splines, which is necessary for the following three reasons: (i) to use a simple example to address, from the outset, most of the issues inherent in modeling (Bayesian estimation, robustness, estimation of the regularization parameter θ, optimization); (ii) to introduce the variational approach, an important method that we will come across again later (Section 13.1); and (iii) to give a full introduction to spline functions, which will be one of the basic ingredients of the many models used for both image analysis (Chapter 9, Section 10.4, Chapter 11, Section 13.1) and pattern recognition (Chapter 4, Section 12.3).

Parametric Splines (Chapter 3). As we have already pointed out (point (i), Section 1.1.2), regularity constraints on x can be expressed explicitly by space reduction. This is the subject of Chapter 3, where x is constrained to belong to a vector space of spline functions, and x is broken down on a basis B of splines: $x = B\theta$. This leads to the exogenous model

$$Y_t = f(t, \theta) + W_t = B_t\theta + W_t,$$

where W_t are centered random variables $\mathrm{E}[W_t] = 0$. The solution in θ is obtained via the least squares criterion, i.e., by minimizing the external energy, which is the quadratic error $U(y \mid \theta) = \|y - B\theta\|^2$. This model is parametric in that it provides a dimensionality reduction: $\dim(y) > \dim(\theta)$. This is an important point, because in many modeling tasks the spline representation is a powerful and easy-to-use tool. The two examples given involve nonstationarity factors due to non-uniform lighting during image acquisition (see Example 1, page 1). The degradation model given in Section 10.4 and Chapter 11 is written as

$$Y = \mathcal{O}\left(x, \widetilde{W}, \theta\right), \quad \text{with} \quad \mathrm{E}\left[\widetilde{W}\right] = B\theta.$$

Nonuniformity therefore takes the form of a luminosity gradient, represented by $B\theta$. This is simply the nonstationary mean of the noise $\widetilde{W}$. We therefore have θ as one of the unknown parameters of the posterior energy. Thanks to the simplicity of the spline representation, the posterior energy expression does not become unnecessarily complicated.

Auto-Associative Models (Chapter 4). This book mainly covers situations in which only one observation is available: a single image, one pair of images, one sequence of images, etc. The problem consists in searching for hidden entities in the given observation. Here, however, the situation is different: we have several observations $\{x^{(1)}, x^{(2)}, \dots\}$ of the same random phenomenon X. These might be several views of an object subjected to deformations or a database of faces (see Example 3, page 1). We must establish an auto-associative model of the type (1.16) $X - f(X) = W$ such that $x - f(x, \theta) = 0$ defines a small manifold representing the variability of the observations and therefore also of the phenomenon. This means that the manifold provides an approximation (a smoothing) of the cloud of points $\{x^{(j)}\}$ in $\mathbb{R}^n$. To perform this smoothing (using parametric spline functions), we come up against the problem of the unorganized points in the scatter of points. This arises in the context of pattern recognition: given a new observation x^*, can we decide to which $x^{(j)}$ it is nearest?

Part II

The Bayesian approach to inverse problems is mentioned in this introductory chapter, illustrated with smoothing splines (Chapter 2), and then discussed in detail in the context of Markov random fields (Chapter 6). At that stage, inverse problems will be formally defined in terms of Markov energies.

We will then see that solving an inverse problem to find $\hat{x}$ such that

$$\hat{x} = \arg \min_x [U_{\theta_1}(x) + U_{\theta_2}(y \mid x)], \tag{1.18}$$

requires a huge number of calculation steps because the set of configurations x is very large (Chapter 7). The unknown parameters θ must also be estimated (Chapter 8). These two problems, especially the latter, are a handicap in modeling and for energetic approaches in general. The calculation of $\hat{x}$ according to the criterion of posterior likelihood (1.18) or any other derived criterion can be approached via two types of technique:

(i) stochastic optimization techniques, or

(ii) deterministic optimization techniques.

Most of these techniques are numerical relaxation algorithms. Algorithms are iterative, starting off with an initial solution $x^{(0)}$. The running solution is modified only on some coordinates, and the position of the coordinates changes cyclically with the iterations. In the deterministic case, this is what happens with the coordinate-by-coordinate *gradient descent* algorithm. Note that relaxation is related to regularization; we usually regularize the running solution locally at each change. Because of this connection, regularization methods are often called *relaxation methods*.

Although both types of technique are also used in Parts I and III, a more detailed presentation of stochastic algorithms is given in Chapter 7 because the estimation methods for θ parameters lead to stochastic algorithms of the same family (Chapter 8) as those of optimization. Furthermore, the Markov chains that form the basis of these algorithms turn out to be useful tools for image simulation (Section 10.3). Simulation, optimization, and estimation of parameters are all advantages of the probabilistic approach.

Part III

In Part III we examine inverse problems in image analysis within a Bayesian framework. The main approach applied is discrete and is based on the Markov formalism of Part II. We will, however, also mention the continuous approach, both variational and probabilistic. In most cases, we will be trying to build an energy dedicated to the application being considered. This construction will be based on a general methodology (Chapter 9) that should give the reader an idea of how to proceed. Although bending energies (Chapter 2) are a special case of energy, dedicated to smoothing, parametric spline models (Chapter 3) have a special status of their own. They will be use to parametrize secondary entities involved in the composition of certain energies, for example, the nonuniformity of lighting.

Most of the applications presented were developed for radiographic nondestructive testing, but the resulting methods and models can be applied more generally. In Part III most of the methodological elements of the first two parts will be used, but in no particular order because the image analysis tasks are the main point. We do not want to give the impression that any systematic processing procedures exist; instead, we wish to show that a well-founded and sufficiently flexible working method will make it possible to develop dedicated models.

Model-Building (Chapter 9). This chapter gives the approach to building a discretized energy for a real application. Although the subsequent chapters can be read in any order, we advise the reader to start with this chapter. Here, the imaging problem is similar to the "spot tracking" of random spots in a temporal sequence of images (see Example 5, page 1). This is a generalization of the spline approximation problem (Chapter 2). The idea is to estimate a family of regular curves of unknown number and length that will link sparse variable-shaped spots distributed in a sequence of images.

Degradation in Imaging (Chapter 10). Four radiographic imaging problems are discussed: denoising, deblurring, scatter analysis, and sensitivity function estimation. The first two are classical and provide an ideal opportunity to apply Markov-type approaches. Denoising that allows for explicit discontinuities is a typical example. The suggested energy model allows a clear analysis of the interactions between the components of the prior energy, which is not always possible in such detail. The blurring problem provides an accurate illustration of the concept of an ill-posed problem, by analogy with the heat diffusion process. The other two problems, scatter and sensitivity functions, involve far more advanced image analysis tasks. Scatter is a probabilistic problem for which we use a complicated degradation model. Other degradation models tend to be simple, but in this case the physical aspects must be considered in greater detail. Estimation of the sensitivity function will involve the most highly developed version of the tricubic spline model (Chapter 3).

Detection of Filamentary Entities (Chapter 11). Here again, the aim is to estimate a family of regular curves of unknown number and length that will best fit a highly degraded image made up of filamentary entities (see Example 4, page 1). This is not related to a spline-type approximation, however, because there is no preferred order as there was before for the temporal sequence of images. In imaging terminology, this is a valley (or peak) detection problem. Of the models presented here, the discretized energy model is the most highly developed.

Reconstruction from Projections (Chapter 12). The degradation operator $\mathcal{O}$ arises from radiographic projections (see Example 6, page 1). Like the blurring problem, the problem of reconstruction from a small number of projections is very ill-posed. To obtain a stable solution, we must therefore use Markov regularization. We will also look at the special case where only one projection is available. To solve the inverse problem in this case, we must have a parametric model of the randomly shaped object to be reconstructed. As in Chapter 4, the parametric estimation will be based on a set of observations of this object. This new situation brings us into the context of pattern recognition, which is also relevant to the following chapter.

Matching (Chapter 13). We take the template of an object, i.e., a model of its apparent outline. We wish to deform this template to make it match an object that is hidden in the image to be analyzed. Two Bayesian approaches are examined. They are different from those seen in previous chapters, because they use continuous energies. The first, a variational approach, considers a spline representation of the template as in Chapter 2. The second approach introduces a new random field concept: continuous

Gaussian fields. This allows us to obtain deformations of the continuous domain $\widetilde{S}$ containing the template, and therefore also global deformations of the template.

REVIEW OF THE LITERATURE

The following brief overview of general works related to the subject should give the reader a few guidelines for choosing among the references listed in the bibliography.

The literature concerning spline approximation is vast, covering both the numerical analysis and statistical points of view. For analytical aspects, refer to Laurent [126] and Ahlberg, Nilson and Walsh [2]. DeBoor [65] is one of the very first works on numerical aspects. Cinquin [51] examines the interest of splines in imaging. Dierckx [69] is a comprehensive work covering the use of parametric splines. The statistical point of view is mainly found in the journals, but we should mention Eubank [75]. The high-dimensionality spline representation is relatively recent, and the reader will need to consult Friedman on this subject [81, 108].

The concept of Markov random fields dates back beyond their use in image analysis. A rigorous mathematical introduction is given in Griffeath [103]. For a broad introduction to the mathematical aspects of random fields, refer to Guyon [104], which gives the basics required for an understanding of parametric estimation. Hammersley and Handscomb [105] gives an introduction to Monte Carlo methods. Duflo [74] provides very broad coverage of stochastic algorithms, and Winkler [182] focuses on Monte Carlo dynamics in image analysis.

D. Geman [89] clearly explains the basics of the Markov approach to inverse problems in image analysis. An exhaustive review of Markov methods in imaging can be found in Li [129], a book that could well serve as an introduction to image analysis, as could the book by Figueiredo, Leitao, and Jain [77]. The variational approach, which is a secondary topic in the present work, is one of the main themes (alongside partial differential equations) of the book by Aubert and Kornprobst [5]. We recommend that book to any reader who wishes to gain a broader view of modeling in image analysis. Finally, the book by Grenander [102] is a groundbreaking work on pattern recognition.

Part I

Spline Models

2

Nonparametric Spline Models

Smoothing by spline functions is a tried and tested method that is used in many different fields. As mentioned in the introductory chapter, the variational approach of one-dimensional splines provides an immediate introduction to the major issues inherent in modeling (regularization, Bayesian estimation, robustness, estimation of the regularization parameter, optimization) thanks to the simplicity of one-dimensional splines. Approximation splines also provide a first formalization of the intuitive concepts of smoothing and adaptive surfaces and curves. These are two essential concepts in imaging.

2.1 Definition

We have observed values y_i at points with abscissas $t_i \in \mathsf{R}$, $i = 1, \ldots, n$, written as $\mathcal{T} = \{t_i\}$. These observed values have a natural order induced by $t_1 < t_2 < \cdots < t_n$. In other situations, where the natural order is unknown, we will see that the basic problem is to find this order (Chapter 4). Each observed value is assumed to be the result of a random variable Y_i that satisfies the endogenous model

$$Y_i = g(t_i) + W_i, \qquad (2.1)$$

where W_i are uncorrelated random variables with a zero mean. The variance σ^2 of the W_i values is taken to be constant, but this assumption will be relaxed later (Section 3.2.2).

The function $g(t)$ is considered to be a deterministic function on R. Its purpose is to approximate y_i with random deviations at each t_i, represented by W_i (also called errors or residuals). Taking an approximation model means imposing certain preselected properties on g. We can therefore impose the requirement that $g(t)$ must be a smooth curve. This property can be quantified by the sum of the local variations of g:

$$U_1(g) = \int \left(D^m g(t)\right)^2 dt, \tag{2.2}$$

where $D^m g$ is the mth derivative of g. Furthermore, for any approximation function g, an empirical approximation error is the quadratic error

$$U_2(y \mid g) = \sum_{i=1}^{n} \left(y_i - g(t_i)\right)^2.$$

Finally, the selection criterion for g is the following functional:

Model 2.1. *The bending energy of approximation of y by a smooth curve g of order m is*

$$\begin{aligned} U(g \mid y) &= \alpha U_1(g) + U_2(y \mid g) \\ &= \alpha \int \left(D^m g(t)\right)^2 dt + \sum_i \left(y_i - g(t_i)\right)^2, \end{aligned} \tag{2.3}$$

where α is a weighting parameter.[1]

The approximation is of order m, and $m + 1$ is the degree of the curve. Using the terminology presented in the Introduction, $U_1(g)$ is a prior (or internal) energy, $U_2(y \mid g)$ is a fidelity (or external) energy, and $U(g \mid y)$ is a posterior energy. The specific energy U_1 is here referred to as *bending energy*. The regularization parameter α allows a tradeoff between bending and approximation. The method for determining this parameter will be discussed later (Section 2.4). For a fixed value of α, the problem is solved as follows:

$$\hat{g} = \arg\min_g U(g \mid y), \tag{2.4}$$

where minimization is performed in a space of sufficiently regular functions g. In applications where the aim is essentially a spline smoothing, $m = 2$ is usually chosen. On the other hand, in the discrete formulation of Parts II and III of this book, where smoothing is associated with other objectives, a first-order smoothing is often chosen. In the rest of the present chapter we will use

$$m = 2.$$

---

[1] This parameter is here labeled α, instead of θ as it was in Chapter 1.

We therefore show [154] that the curve $\widehat{g}$ has the following properties.

Theorem 2.1.

(i) The curve $\widehat{g}$ is a polynomial of degree $m+1$ on each interval (t_i, t_{i+1}).

(ii) At each point t_i, the first m derivatives of $\widehat{g}$ are continuous, while the $(m+1)$th can be discontinuous.

(iii) In each of the intervals $(-\infty, t_1)$ and (t_n, ∞), $D^m\widehat{g}(t) = 0$.

Note that these properties are not directly imposed on $\widehat{g}$; they are an implicit result of $U_1(g)$. Because g is not defined by an explicit parametric model, these splines can be called "nonparametric".

Definition 2.1. *A curve that satisfies* (i) *and* (ii) *with* $m = 2$ *is called a cubic spline with knots* $\mathcal{T} = \{t_i\}$ *or bending spline. If it also satisfies* (iii), *it is called a natural cubic spline.*

The curve $\widehat{g}$ is an approximation spline or smoothing spline. When we impose the condition that the spline must pass through the n points $\{(t_i, y_i), i = 1, \ldots, n\}$, i.e., when we relax $U_2(g \mid y)$, $\widehat{g}$ is an interpolation spline. We write $\mathcal{S}(\mathcal{T})$ to denote the set of cubic splines on $\mathcal{T}$. We must emphasize the importance of reaching a tradeoff between bending and approximation, as mentioned in (2.3). It was Whittaker [179] who first proposed a functional of the type (2.3) but with discrete bending energy $U_1(g)$:

$$U_1(g) = \sum_i \left(\Delta^m g(t_i)\right)^2, \tag{2.5}$$

where Δ^m is the mth difference operator of g:

$$\Delta^1 g(t_i) = g(t_{i+1}) - g(t_i),$$
$$\Delta^k g(t_i) = \Delta^{k-1} g(t_{i+1}) - \Delta^{k-1} g(t_i).$$

Note that g is no longer a continuous function on $\mathbb{R}$, although we continue to write $g(t_i)$ instead of using the discrete notation g_i. For the special case of $m = 2$, this energy is therefore

$$U_1(g) = \sum_i \left(2g(t_i) - g(t_{i-1}) - g(t_{i+1})\right)^2. \tag{2.6}$$

We then obtain the discrete version of the energy (2.3).

Model 2.2. *The bending energy of approximation of* y *by a smooth discretized curve* g *of order* $m = 2$ *is*

$$U(g \mid y) = \alpha U_1(g) + U_2(g \mid y) \tag{2.7}$$
$$= \alpha \sum_i \left(2g(t_i) - g(t_{i-1}) - g(t_{i+1})\right)^2 + \sum_i \left(y_i - g(t_i)\right)^2.$$

This type of model is a special case of the class of models described in Chapter 5. Once again we stress that finding a tradeoff between several constraints will be a recurring theme throughout this book. Most of the models presented will apply this concept.

The model (2.7) with $m = 2$ is a special case of the more general model of *splines under tension*:

$$U(g \mid y) = \alpha_1 U_{1,1}(g) + \alpha_2 U_{1,2}(g) + U_2(g \mid y) \tag{2.8}$$
$$= \alpha_1 \int \left(D^1 g(t) \right)^2 dt + \alpha_2 \int \left(D^2 g(t) \right)^2 dt + \int \phi\big(y(t), g(t)\big) \ dt,$$

where $U_{1,1}(g)$ is a term that quantifies the elongation of g, and ϕ is a positive regular function. The parameters α_1 and α_2 control the elastic properties of g. If we made these parameters a function of t, we could impose a higher stiffness at some parts of the spline than at others. This model will be looked at again in Section 13.1.2, where we represent the outline of a shape for the purpose of pattern recognition.

2.2   Optimization

In Section 2.2.1 we will solve $\hat{g} = \arg\min_g U(g \mid y)$ for the continuous formulation (2.3) of bending splines. With this energy, the solution is analytical [2, 65] and follows the solution of the interpolation problem. In Section 2.2.2 we will deal with the model of splines under tension (2.8). This presentation uses the calculus of variations and is applicable to the general case, unlike the previous presentation of this model.

2.2.1   Bending Spline

• INTERPOLATION

Proposition 2.1. *Let $\{\zeta_i\}$ be n values set according to $\{t_i\}$. There is only one natural spline g interpolating these ζ_i values:*

$$g(t_i) = \zeta_i, \quad \forall i = 1, \ldots, n.$$

Proof:
The expression for g can be a function of the second derivatives of g (written as M_i) for which the condition $M_1 = M_n = 0$ is imposed:

$$M_i = D^2 g(t_i), \quad \forall i = 1, \ldots, n,$$
$$M_1 = M_n = 0.$$

Let us now determine this expression. For any interval (t_i, t_{i+1}), g is written as

$$g(t) = a_i(t - t_i)^3 + b_i(t - t_i)^2 + c_i(t - t_i) + d_i. \tag{2.9}$$

We set $h_{i+1} = t_{i+1} - t_i$. By differentiating g and imposing the M_i values on the knots $D^2 g(t_i) = M_i$, $D^2 g(t_{i+1}) = M_{i+1}$, etc., we obtain the following coefficients in turn:

$$b_i = M_i/2, \qquad a_i = (M_{i+1} - M_i)/(6 h_{i+1}),$$

$$d_i = \zeta_i, \qquad c_i = \frac{\zeta_{i+1} - \zeta_i}{h_{i+1}} - h_{i+1} \frac{2M_i + M_{i+1}}{6}.$$

The last coefficient is obtained by considering $g(t_{i+1})$ and entering the expressions for a_i, b_i, and d_i into it. Because $D^1 g$ is continuous, we can enter the coefficients for $i = 2, \ldots, (n-1)$ in terms of M_i into the continuity condition $D^1 g(t_i^-) = D^1 g(t_i^+)$, to obtain

$$M_{i-1} h_i + 2 M_i (h_i + h_{i+1}) + M_{i+1} h_{i+1} = 6 \left(\frac{\zeta_{i+1} - \zeta_i}{h_{i+1}} - \frac{\zeta_i - \zeta_{i-1}}{h_i} \right).$$

The continuity constraints have therefore given us $n - 2$ equations in $n - 2$ unknowns M_i, since we have $M_1 = M_n = 0$. These equations give rise to the linear system denoted by

$$RM = Q\zeta, \tag{2.10}$$

where M is an $(n-2)$-vector, ζ is an n-vector, R is a tridiagonal $(n-2) \times (n-2)$ matrix with generic row $(\ldots, h_i, 2(h_i + h_{i+1}), h_{i+1}, \ldots)$, and Q is a tridiagonal $(n-2) \times n$ matrix with generic row

$$6 \left(\ldots, \frac{1}{h_i}, \; -\left(\frac{1}{h_i} + \frac{1}{h_{i+1}} \right), \; \frac{1}{h_{i+1}}, \ldots \right).$$

This gives the following solution:

$$M = R^{-1} Q\zeta. \tag{2.11}$$

In conclusion, any series of ordinates ζ_i has an associated cubic spline defined by its second derivatives M_i given by (2.11). $\qquad \square$

• APPROXIMATION

We now come to the matter of smoothing (2.4). In view of the above development, this problem can be handled with respect to ζ instead of g. We simply need to determine the series $\zeta_1, \ldots, \zeta_n$ that gives the minimum of U via its interpolation spline. For a function f that is linear on $(0, h)$, we have $\int_0^h f(t)^2 dt = \frac{h}{3} \left[f(0)^2 + f(0)f(h) + f(h)^2 \right]$. Because $D^2 g$ is itself linear throughout (t_i, t_{i+1}), we can deduce a new expression for U_1 and therefore also U,

$$U(g \mid y) \equiv U(\zeta \mid y)$$

$$= \frac{\alpha}{3} \sum_{i=1}^{n-1} h_{i+1} \left[M_i^2 + M_i M_{i+1} + M_{i+1}^2 \right] + \sum_i (y_i - \zeta_i)^2,$$

or in matrix form,[2]

$$U(\zeta \mid y) = \alpha \frac{2}{3} M' R M + (y - \zeta)'(y - \zeta).$$

Using $M = R^{-1} Q \zeta$, we then obtain

$$U(\zeta \mid y) = \alpha \frac{2}{3} \zeta' \left(R^{-1} Q \right)' R \left(R^{-1} Q \right) \zeta + (y - \zeta)'(y - \zeta).$$

Because $\left(R^{-1} Q \right)' R \left(R^{-1} Q \right)$ gives a symmetric positive definite result, the minimum of $U(\zeta \mid y)$, written $\widehat{\zeta}$, is the vector that cancels the gradient

$$\frac{2}{3} \alpha (R^{-1} Q)' R (R^{-1} Q) \widehat{\zeta} - \left(y - \widehat{\zeta} \right) = 0, \qquad (2.12)$$

or, taking into account $\left(R^{-1} \right)' = R^{-1}$,

$$\frac{2}{3} \alpha Q' R^{-1} Q \widehat{\zeta} = y - \widehat{\zeta}, \qquad (2.13)$$

which is a symmetric pentadiagonal system. By solving it, we finally reach the expected solution $\widehat{\zeta}$. The solution $\widehat{\zeta}_1, \ldots, \widehat{\zeta}_n$ is the same as the desired solution $\widehat{g}(t_1), \ldots, \widehat{g}(t_n)$. The continuous expression for $\widehat{g}$ in the interval $[t_1, t_n]$ is obtained by using the values of $\widehat{M_i}$ associated with $\widehat{\zeta_i}$ in equation (2.11): $M = R^{-1} Q \widehat{\zeta}$. Finally, the expression for $\widehat{g}(t)$ is obtained by substituting this into (2.9).

2.2.2 Spline Under Tension

We need to minimize the energy (2.8), which is rewritten as[3]

$$U(g \mid y) = \int_0^1 F\big(t, g(t), g'(t), g''(t)\big) dt$$

$$\text{with } F\big(t, g, g', g''\big) = \phi(y, g) + \alpha_1 (g')^2 + \alpha_2 (g'')^2,$$

where we have set $g' = D^1 g = \frac{d}{dt} g$ and $g'' = D^2 g = \frac{d^2}{dt^2} g$. One necessary condition for the existence of a minimum of U is given by the following classic result [59]:

---

[2] From now on, M' and $^t M$ will both denote the transpose of the matrix M.

[3] In this section, derivatives are temporarily written as g', g'', $\ldots$, $g^{(n)}$ instead of $D^{(n)} g$.

Proposition 2.2. *Let be a functional $U(x) = \int_0^1 F\left(t, x, x', x'', \ldots, x^{(m)}\right) dt$.*
If the derivatives up to order $m - 1$ have fixed values at the limits 0
and 1, then one necessary condition for a minimum at x is given by the
Euler–Lagrange equation

$$\frac{\partial F}{\partial x} - \frac{d}{dt}\frac{\partial F}{\partial x'} + \frac{d^2}{dt^2}\frac{\partial F}{\partial x''} + \cdots + (-1)^m \frac{d^m}{dt^m}\frac{\partial F}{\partial x^{(m)}} = 0,$$

provided that the various derivatives exist.

Proof:
The proof involves a calculus of the variations of U. Let us perform this
calculation for the simple case of $m = 1$. We assume that $x(0) = x_0$,
$x(1) = x_1$, and x'' is continuous. Let $\hat{x}$ be a minimum of U, and let
$x(t) = \hat{x}(t) + \xi h(t)$, where h is continuously differentiable and satisfies
$h(0) = h(1) = 0$. The function ξh is seen as a variation of $\hat{x}$ with which
the following relationship is associated:

$$U_\xi \doteq U(\hat{x} + \xi h) = \int_0^1 F(t, \hat{x} + \xi h, \hat{x}' + \xi h')dt.$$

The necessary condition for the minimum is obtained by setting its deriva-
tive to zero at the point $\xi = 0$, i.e., $\frac{dU_\xi}{d\xi}|_{\xi=0} = 0$. This derivative can be
expressed as follows:

$$\frac{dU_\xi}{d\xi} = \int_0^1 \left[\frac{\partial F}{\partial(\hat{x} + \xi h)}h + \frac{\partial F}{\partial(\hat{x}' + \xi h')}h'\right] dt,$$

$$\frac{dU_\xi}{d\xi}\bigg|_{\xi=0} = \int_0^1 \left[\frac{\partial F}{\partial \hat{x}}h + \frac{\partial F}{\partial \hat{x}'}h'\right] dt.$$

Integration by parts then gives

$$\int_0^1 \frac{\partial F}{\partial \hat{x}'}h'dt = \frac{\partial F}{\partial \hat{x}'}h\,|_0^1 - \int_0^1 \frac{d}{dt}\frac{\partial F}{\partial \hat{x}'}h\,dt = -\int_0^1 \frac{d}{dt}\frac{\partial F}{\partial \hat{x}'}h\,dt,$$

since $h(0) = h(1) = 0$. Finally,

$$\frac{dU_\xi}{d\xi}\bigg|_{\xi=0} = \int_0^1 \left[\frac{\partial F}{\partial \hat{x}} - \frac{d}{dt}\frac{\partial F}{\partial \hat{x}'}\right] h\,dt. \tag{2.14}$$

We can complete this calculation by applying the following lemma: If f is
a continuous function over $[0, 1]$ and if for any continuous function h such
that $h(0) = h(1) = 0$, we have $\int_0^1 fh\,dt = 0$, then $f \equiv 0$.
  In our calculation h is arbitrary, and we therefore obtain

$$\frac{\partial F}{\partial \hat{x}} - \frac{d}{dt}\frac{\partial F}{\partial \hat{x}'} = 0,$$

or, in more detailed form,

$$\frac{\partial F}{\partial \hat{x}} - \frac{\partial^2 F}{\partial t\partial \hat{x}'} - \frac{\partial^2 F}{\partial \hat{x}\partial \hat{x}'}\hat{x}' - \frac{\partial^2 F}{\partial^2 \hat{x}'}\hat{x}'' = 0,$$

where the notation $\frac{\partial F}{\partial \hat{x}}$ means $\frac{\partial F}{\partial x}|_{x=\hat{x}}$. The differential expression is the variational derivative of U with respect to x, and is analogous to the gradient used to optimize functions.   □

Now let us look at how to minimize the energy $U(g \mid y)$ (2.8). A curve g is the local minimum of this energy if it satisfies the Euler–Lagrange equation with the boundary conditions $g(0) = g_0$, $g(1) = g_1$, $g'(0) = g'_0$, and $g'(1) = g'_1$:

$$\frac{d\phi}{dg}(y, g) - \frac{d}{dt}(2\alpha_1 g') + \frac{d^2}{dt^2}(2\alpha_2 g'') = 0.$$

Note that defining and selecting the boundary conditions is a crucial point for concrete applications. We can solve this equation using the finite difference method, which proceeds by discretization. If we write the discretized form of g in steps of $\delta = 1/n$ as $\{g_i = g(i\delta), \ i = 0, \ldots, n\}$, the Euler–Lagrange equation becomes

$$\frac{d\phi(y_i, g_i)}{dg} - 2\frac{\alpha_1}{\delta^2}(2g_i - g_{i-1} - g_{i+1})$$

$$+ 2\frac{\alpha_2}{\delta^4}(6g_i - 4g_{i-1} - 4g_{i+1} + g_{i-2} + g_{i+2}) = 0.$$

This is rewritten in matrix form $\underline{F} - \underline{A}\,\underline{g} = 0$, where $\underline{F}$ is dependent on y_i and $\underline{A}$ is a pentadiagonal matrix.

Finally, we study a more general case that is typical of certain contour detection problems in imaging (Section 13.1.2). The function y is no longer a curve, but an "image" defined on $[0, 1] \times \mathbb{R}$ (in practice, $[0, 1] \times [0, 1]$), so that ϕ is rewritten $\phi[y(g(t))]$. The expression $-\phi[y(g(t))]$ is normally used to designate an image feature at the site $(t, g(t))$, and we wish to find a spline under tension $g : [0, 1] \mapsto \mathbb{R}$, passing through points with high image features. One solution to this variational problem is obtained using an evolution equation as follows. We write

$$\frac{dU}{dg} = \frac{d\phi}{dg}(y, g) - \frac{d}{dt}(2\alpha_1 g') + \frac{d^2}{dt^2}(2\alpha_2 g''),$$

the variational derivative of U. As in the above proof (2.14), for h, an infinitesimal variation of g, we have

$$U(g + h) - U(g) \approx \int_0^1 \frac{dU}{dg} h \, dt.$$

If we want to satisfy $U(g + h) \leq U(g)$, we must therefore choose h proportional to $-\frac{dU}{dg}$, i.e., $h = -\xi\frac{dU}{dg}$ with $\xi > 0$. This leads to an iterative procedure to minimize U. Setting $(g + h) - g = -\xi\frac{dU}{dg}$, for any iteration $\tau = 0, 1, 2, \ldots$, the running solution $g^{\tau+1}$ is $g + h$, in which g corresponds to g^τ. After discretization, this gives

$$g^{\tau+1} = g^\tau - \xi\left(\underline{F}^\tau - \underline{A}^\tau g^\tau\right), \tag{2.15}$$

where g^0 is an initial solution. This procedure is analogous to the gradient descent algorithm used to minimize functions (Section 7.4). It is often introduced using an evolution equation with partial derivatives [59]:

$$\frac{\partial g}{\partial \tau} = \frac{\partial \phi}{\partial g} - \frac{d}{dt}\left(2\alpha_1 g'\right) + \frac{d^2}{dt^2}\left(2\alpha_2 g''\right),$$

where $g(t,\tau)$ is now a function of one evolution variable τ. After discretization with finite differences in t and τ, we get back to (2.15). In image analysis this is called the *active contour* technique, and it uses curves of the type $g : [0,1] \mapsto R^2$, not simply $g : [0,1] \mapsto R$. The procedures described above can be directly extended to apply to this case. In Section 2.5 this spline approximation problem will be extended to surfaces, for the case of observed y values given discretely at the top of an *irregular* grid. With this type of observation grid, the method of finite differences will no longer be suitable, because it requires uniform discretization. We will then introduce the finite element method, for which the study of convergence is more advanced.

2.2.3 Robustness

Spline smoothing is appropriate when the model (2.1) agrees with observations $\{y_i\}$. This can be better understood if we take a specific probability distribution for random variables W_i. The choice of $U_2(y \mid g)$ implies that these variables are independently distributed according to a Gaussian distribution, $LG(0, \sigma^2)$, where the variance σ^2 is independent of time of observation t_i. In practice, one often finds data with certain observed values that do not mesh with this model. This occurs, for example, in the presence of outlying observed values y_i for which the variance of W_i is greater than σ^2 by an abnormally large margin. In this case, the previous smoothing must be modified to prevent these outliers from perturbing $\widehat{g}$. This is illustrated by Figures 2.2 and 11.1 (page 228). In Figure 11.1(a), the curve is fitted without taking outliers into consideration, unlike the fitting shown in Figure 11.1(b). For bending splines, the energy (2.3) is rewritten as

$$U(g \mid y) = \alpha \int \left(D^2 g(t)\right)^2 dt + \sum_i \phi\big(y_i - g(t_i)\big), \tag{2.16}$$

where the function $\phi(u)$ increases less rapidly than the quadratic function $\phi_0(u) = u^2$. The function $\phi(u)$ is often chosen to be convex for optimization

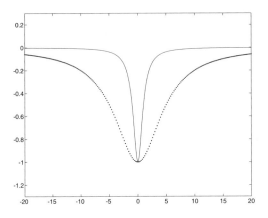

Figure 2.1. Smoothing function for $b = 1$ (——) and $b = 5$ ($\ldots$).

purposes. Huber [112] considers the function

$$\phi(u) = \begin{cases} u^2, & \text{if } |u| \leq b, \\ 2b|u| - b^2, & \text{otherwise} . \end{cases}$$

which consists in replacing parabolic sections $\{\phi_0(u), |u| > b\}$ by half-lines that satisfy 0-order and first-order continuity with the parabola at points $|u| = b$. Here, b acts as a threshold for outlying observed values. Many other functions have been suggested, such that depicted in Figure 2.1:

$$\phi(u) = \frac{-1}{1 + (u/b)^2}. \tag{2.17}$$

This function appears in Chapter 10, although in a slightly different context: regularization with implicit discontinuities.

The energy $U(g \mid y)$ defined in (2.16) is minimized by an iterative calculation. For $\phi_0(u) = u^2$, it should be noted, with reference to (2.13), that the smoothing spline defined by $(\widehat{g}(t_1), \ldots, \widehat{g}(t_n)) = \widehat{\zeta}$ is linear with respect to y: $\widehat{\zeta} = Ay$, and therefore

$$\widehat{g}(t_i) = \sum_j A_{ij} \, y_j. \tag{2.18}$$

When ϕ is not the quadratic function ϕ_0, robust smoothing is the result of an iterative algorithm that starts with nonrobust smoothing (2.18): $\widehat{g}^{(0)}(t_i) = \widehat{g}(t_i)$. After iteration k, y_i is replaced by a value $y_i^{(k+1)}$, so that the new error $\left|u_i^{(k+1)}\right| = \left|y_i^{(k+1)} - \widehat{g}^{(k)}(t_i)\right|$ is less than or equal to the error $\left|u_i^{(k)}\right| = \left|y_i^{(k)} - \widehat{g}^{(k)}(t_i)\right|$. For the Huber function, for example, we choose

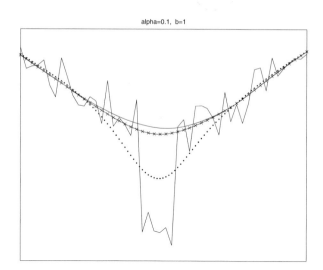

Figure 2.2. +++++ robust smoothing, ... nonrobust smoothing, ——— smoothing in the absence of outliers.

$y_i^{(k+1)}$ so that $\left| u_i^{(k+1)} \right| = b$ in the case of outliers:

$$
y_i^{(k+1)} = \begin{cases} y_i^{(k)}, & \text{if } \left| y_i^{(k)} - \widehat{g}^{(k)}(t_i) \right| \le b, \\ \widehat{g}^{(k)}(t_i) + b \times \operatorname{sgn}\left(y_i^{(k)} - \widehat{g}^{(k)}(t_i) \right), & \text{otherwise.} \end{cases}
$$

More generally, since $\phi(u)$ increases less rapidly than $\phi_0(u)$, we set $D^1\phi_0\left(u^{(k+1)}\right) = D^1\phi\left(u^{(k)}\right)$, i.e.,

$$
2\left(y_i^{(k+1)} - \widehat{g}^{(k)}(t_i) \right) = D^1\phi\left(y_i^{(k)} - \widehat{g}^{(k)}(t_i) \right).
$$

After applying this change, we calculate the smoothing spline of iteration $k+1$ as follows:

$$
\widehat{g}^{(k+1)}(t_i) = \sum_j A_{ij}\, y_j^{(k+1)}.
$$

This algorithm converges quickly. The accuracy of $\widehat{g}^{(k)}$ at convergence would then remain to be studied. Figure 2.2 illustrates this approach. The extreme values shown were created artificially by subtracting 3.5σ from the six middle observed values, which then become outliers. The continuous curve is the spline smoothing before the introduction of these values. The dotted curve is the nonrobust smoothing $\widehat{g}^{(0)}$ after the introduction of these values. The curve with crosses is the robust smoothing.

2.3   Bayesian Interpretation

The idea of looking at nonparametric smoothing curve estimation from a Bayesian standpoint goes back to Whittle [180]. This is only natural, since choosing to represent the data by a smooth curve with a certain degree of regularity constitutes the choice of a prior distribution.

Let $\mathcal{S}_N(\mathcal{T})$ be the set of all natural cubic splines (see definition, page 25). The set $\mathcal{S}_N(\mathcal{T})$ is an n-dimensional vector space. In each interval (t_i, t_{i+1}), there are four unknowns, a_i, b_i, c_i, d_i, except for the first and last intervals, where there are three unknowns because $M_1 = M_n = 0$. This makes a total of $4(n-1) - 2$ unknowns, from which $3(n-2)$ continuity constraints at $t_2, \ldots, t_{n-1}$ must be subtracted. We therefore have a total of n effective unknowns. Section 3.1 shows that a basis can be built for this space consisting of so-called B-splines, written as $B_j(t)$. For any spline $g \in \mathcal{S}_N(\mathcal{T})$, the linear representation on the basis under consideration can be expressed as follows:

$$g(t) = \sum_j \beta_j \; B_j(t).$$

The initial model (2.1) is then written as

$$Y_i = \sum_j \beta_j B_j(t_i) + W_i, \tag{2.19}$$

for which the unknowns are $\theta = (\sigma^2, \beta)$, where $\sigma^2 = \mathrm{Var}(W_i)$. Any spline g is therefore defined by a given n-vector β.

Note. The linear representation in (2.19) is not a parametric model of g because it has the same dimensionality as y, whereas a parametric model provides a dimensionality reduction, by definition.

This representation can be used to rewrite the energy terms (2.3) as a function of β instead of g. For the first term, this gives

$$U_1(g) \; = \; \int \left(D^2 g(t)\right)^2 dt \; = \; \beta' \Lambda \beta,$$

$$\text{with} \;\; \Lambda_{ij} \; = \; \int_{\mathsf{R}} \left(D^2 B_i(t)\right)\left(D^2 B_j(t)\right) dt.$$

It can be shown that Λ is a nonnegative definite matrix. The second term is as follows:

$$U_2(y \mid g) \; = \; \sum_i \left(y_i - g(t_i)\right)^2 = \sum_i \left(y_i - \sum_j \beta_j B_j(t_i)\right)^2 = \|y - \mathbf{B}\,\beta\|^2 ,$$

where the matrix of elements $\mathbf{B}_{ij} = B_j(t_i)$ is written $\mathbf{B} = (\mathbf{B}_{ij})$. The new expression for the energy $U(g \mid y)$ is therefore

$$U(g \mid y) \equiv U(\beta \mid y) = \alpha \, \beta'\Lambda\beta + \|y - \mathbf{B}\,\beta\|^2. \tag{2.20}$$

The solution that minimizes this energy is easily seen to be

$$\widehat{\beta} = (\mathbf{B}'\mathbf{B} + \alpha\Lambda)^{-1}\mathbf{B}'y.$$

To give a Bayesian interpretation of this, let us consider (2.19), which assumes that the vector β is random, and that variables W_i are random, independent, and Gaussian. With β and W assumed to be independent, we obtain the following for all σ^2:

$$P(w) = P(y \mid \beta) = \frac{1}{(2\pi\sigma^2)^{n/2}} \, \exp - \left[\frac{1}{2\sigma^2}\|y - \mathbf{B}\,\beta\|^2\right]. \tag{2.21}$$

Given an observed value y^O, the log-likelihood of β is, to within an additive constant,

$$\log L(\beta) \propto -\frac{1}{2\sigma^2}\left\|y^O - \mathbf{B}\,\beta\right\|^2.$$

Let us consider the following prior log-likelihood for β:

$$\log L^{\mathrm{prior}}(\beta) \propto -\frac{1}{2}\frac{\alpha}{\sigma^2}\beta'\Lambda\beta. \tag{2.22}$$

This corresponds to the logarithm of a Gaussian distribution $P^{\mathrm{prior}}(\beta)$. The Bayesian approach consists in writing $P^{\mathrm{post}}(\beta \mid y) = P(y \mid \beta) \, P^{\mathrm{prior}}(\beta)/P(y)$. In this case, the following posterior log-likelihood is obtained:

$$\log L^{\mathrm{post}}(\beta) \propto -\frac{1}{2}\frac{\alpha}{\sigma^2}\beta'\Lambda\beta - \frac{1}{2\sigma^2}\left\|y^O - \mathbf{B}\beta\right\|^2,$$

$$\propto -\alpha\beta'\Lambda\beta - \left\|y^O - \mathbf{B}\beta\right\|^2.$$

Except for its sign, this is the expression for $U\left(\beta \mid y^O\right)$, i.e., for $U\left(g \mid y^O\right)$. Note that the parameters that were initially $\theta = (\sigma^2, \beta)$ are now $\theta = (\alpha, \beta)$. The regularization parameter α must be adjusted for σ^2 according to the ratio α/σ^2 given in (2.22). The posterior log-likelihood can then be written as follows:

$$\log L^{\mathrm{post}}(\beta) \propto -\frac{1}{2}\beta'(\alpha\Lambda + \mathbf{B}'\mathbf{B})\beta + \left(y^O\right)'\mathbf{B}\beta.$$

This is the Gaussian log-likelihood over β, with mean μ and covariance matrix V^{-1}:

$$V = \alpha\Lambda + \mathbf{B}'\mathbf{B}, \quad \mu = V^{-1}\mathbf{B}'y^O.$$

Finally, note that a Bayesian interpretation of Model 2 in the discrete case would have been better. This will be discussed in Chapter 6.

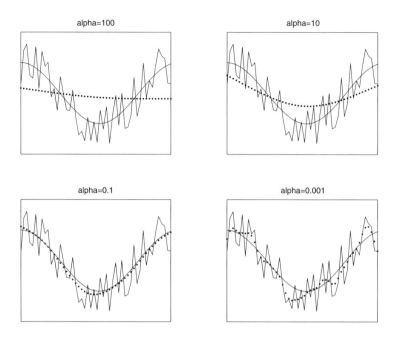

Figure 2.3. Effect of regularization parameter α. —— true function f, $\cdots$ spline $\widehat{g}_\alpha$ for different values α.

2.4   Choice of Regularization Parameter

For clarity, (2.1) can be rewritten as follows: $Y_i = f(t_i) + W_i$, where f now represents the true function, i.e. the function prior to degradation. The approximation spline $\widehat{g}$, which minimizes $U(g \mid y) = \alpha \int \left(D^2 g(t)\right)^2 dt + \sum_i \left(y_i - g(t_i)\right)^2$, is therefore a regularized version of f and will be written as $\widehat{g}_\alpha$ because it is dependent on the regularization parameter. This is a tradeoff between approximation and bending. How should α be chosen anywhere between the two extreme cases: no smoothing and significant bending on the one hand (i.e., $\alpha = 0$), and strong smoothing but minimal bending on the other hand (i.e., $\alpha = \infty$)? Figure 2.3 illustrates this situation for $f(t) = \cos(t)$, shown as a solid line. This is a fundamental problem of pattern recognition, known as the *bias/variance dilemma* [91, 175]. It is stated as follows:

Property 2.1. *The quadratic error of approximation allows the following breakdown:*

$$\mathrm{E}\left[\left(\widehat{G}_\alpha(t) - f(t)\right)^2\right] = \left(\mathrm{E}\left[\widehat{G}_\alpha(t)\right] - f(t)\right)^2 + \mathrm{E}\left[\left(\widehat{G}_\alpha(t) - \mathrm{E}\left[\widehat{G}_\alpha(t)\right]\right)^2\right]$$

$$\doteq \mathrm{Bias}\left(\widehat{G}_\alpha(t)\right) + \mathrm{Var}\left(\widehat{G}_\alpha(t)\right),$$

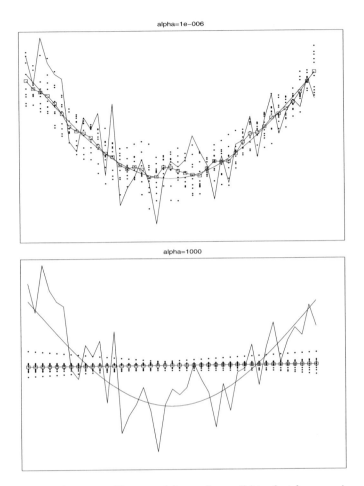

Figure 2.4. Bias/variance dilemma. (a) $\alpha \approx 0$: small bias but large variance. (b) $\alpha \approx \infty$: small variance but large bias.

——— true function f, broken solid line: one particular occurrence y^ℓ of Y,   ... $\{\widehat{g}_\alpha^{(l)}, \ell = 1, \ldots, N = 10\}$,   □-□-□ average of $\{\widehat{g}_\alpha^{(l)}, \ell = 1, \ldots, N\}$.

where $\widehat{G}_\alpha$ denotes the random function of which $\widehat{g}_\alpha$ is one occurrence.

A tradeoff between the bias and the variance must be reached. For a set error, decreasing the bias increases the variance and vice versa. Figure 2.4(a) suggests that a very small value of α results in a low bias and a large variance. This means that if we had a large number N of data sets $\{y_i^{(l)}\}$, $l = 1, \ldots, N$, the average of $\widehat{g}_\alpha^{(l)}(t)$ would be close to $f(t)$. These $\widehat{g}_\alpha^{(l)}(t)$ values would, however, fluctuate more, resulting in a large variance of the estimator. The consequences of this can be disastrous when it comes to making a prediction, because $\widehat{g}_\alpha$ very closely approximates the data on which its calculation was based, but it might be a very bad match for other sets of data. On the contrary, Figure 2.4(b) suggests that a very large value of α results in a large bias and a small variance. To make these intuitive observations more specific, we would have to express the bias and variance as functions of α, or use the integral of the bias and integral of the variance

$$\overline{\mathrm{Bias}}(\widehat{G}_\alpha) = \int \mathrm{Bias}(\widehat{G}_\alpha(t))\, dt \,, \quad \overline{\mathrm{Var}}(\widehat{G}_\alpha) = \int \mathrm{Var}(\widehat{G}_\alpha(t))\, dt.$$

This is difficult. Rice and Rosenblatt [157] applied this method for the energy $U(g \mid y) + [(y_1 + y_n) - (g(t_1) + g(t_n))]^2$. They obtain (approximately) $\overline{\mathrm{Bias}}(\widehat{G}_\alpha) \propto \alpha^{5/4}$ and $\overline{\mathrm{Var}}(\widehat{G}_\alpha) \propto \frac{\sigma^2}{n}\alpha^{-1/4}$. This result reinforces our observations.

One principle used to determine α automatically from just one set of data is to take a second approximation criterion, say $CV(\alpha)$. A classic method that is especially suited to Model 1 is known as "cross validation". For Model 2, the choice of α will be looked at later (Chapter 8, Chapter 9). Let $\widehat{g}_{\alpha,k}$ denote the approximation spline calculated from the initial set of data $\{(t_i, y_i)\}_{i=1}^n$ minus the point (t_k, y_k). The spline $\widehat{g}_{\alpha,k}$ minimizes the following expression:

$$\alpha \int \left(D^2 g(t)\right)^2 dt \; + \sum_{i:\, i\neq k} \left(y_i - g(t_i)\right)^2.$$

Cross validation [166] applies the idea that if α is a good choice, then $\widehat{g}_{\alpha,k}(t_k)$ should be a good approximation of y_k. Repeating this procedure for all t_k makes it possible to measure the quality of the approximation using the quadratic error

$$CV(\alpha) = \frac{1}{n} \sum_{k=1}^n \left(\widehat{g}_{\alpha,k}(t_k) - y_k\right)^2.$$

The chosen parameter is therefore $\widehat{\alpha} = \arg\min CV(\alpha)$. Note that this criterion does not only apply in the context of splines, but to the general case, unlike the refinement explained in the appendix below. In other contexts, however, the use of this criterion is very expensive in terms of computer time.

Appendix. This criterion was refined as follows [60]. Referring to (2.13), we see that $(\widehat{g}_\alpha(t_1), \ldots, \widehat{g}_\alpha(t_n)) = \widehat{\zeta}$ is linear with respect to y. Note that $\widehat{\zeta} = A(\alpha)y$. It is possible to show that

$$\widehat{g}_{\alpha,k}(t_k) - y_k = \frac{\widehat{g}_\alpha(t_k) - y_k}{1 - \frac{\partial}{\partial y_k}\widehat{g}_\alpha(t_k)}.$$

This result, combined with $\widehat{g}_\alpha(t_k) = \sum_j A_{kj}(\alpha)y_j$, gives

$$CV(\alpha) = \frac{1}{n} \sum_{k=1}^{n} \frac{\left(\widehat{g}_\alpha(t_k) - y_k\right)^2}{(1 - A_{kk}(\alpha))^2}.$$

For the ideal parameter α, when all $A_{kk}(\alpha)$ are equal, $CV(\alpha)$ is the "natural" estimator of σ^2, the variance of W_k. Because all values of $A_{kk}(\alpha)$ are not necessarily equal, $CV(\alpha)$ is not quite this estimator.

A change of coordinates can, however, reduce the problem to this case, giving the following new criterion:

$$GCV(\alpha) = \frac{1}{n} \sum_{k=1}^{n} \frac{\left(\widehat{g}_\alpha(t_k) - y_k\right)^2}{\left(1 - \frac{1}{n}\text{Trace } A(\alpha)\right)^2}.$$

This is known as the generalized cross validation criterion. According to [177], it produces a small gain for equally spaced values of t_i. The way to select the correct parameter is therefore to minimize $GCV(\alpha)$ for positive real numbers. See [173] for a practical calculation method.

2.5 Approximation Using a Surface

We now have observed values $\{y_k;\ k = 1, \ldots, n\}$ at points with coordinates $(t_k, u_k) \in \mathsf{R}^2$ that are not necessarily distributed at the nodes of a regular grid. To work on this type of grid is a major advantage for imaging applications, for example in stereo vision [168], deformation analysis [26], or velocity field analysis. We assume that these observed values are governed by the model (2.1):

$$Y_k = g(t_k, u_k) + W_k.$$

The function $g(t, u)$ represents a surface that approximates the data y_k. Let

$$D^{i,j} \doteq \frac{\partial^{i+j}}{\partial t^i \partial u^j},$$

such that $D^{2,0}g = \frac{\partial^2 g}{\partial t^2}$, $D^{1,1}g = \frac{\partial^2 g}{\partial t \partial u}$, and $D^{0,2}g = \frac{\partial^2 g}{\partial u^2}$. The regularity of the surface g can be represented through various energies U_1. We will examine two of these. A first expression is obtained by extending the energy (2.2) to the surfaces. The bending internal energy of a smooth surface g of order $m = 2$ is quadratic:

$$U_1(g) = \int_\Omega \left[\left(D^{2,0}g\right)^2 + 2\left(D^{1,1}g\right)^2 + \left(D^{0,2}g\right)^2 \right] dt\, du, \qquad (2.23)$$

where Ω is a closed region of R^2 containing (t_k, u_k). This energy is rotationally invariant. A second expression uses the linear differential operator

$\mathsf{L} = \sum_{i,j=0}^{m} a_{i,j} D^{i,j}$ with positive coefficients $a_{i,j}$:

$$U_1(g) = \int_\Omega \left(\mathsf{L}g\right)^2 dt\, du. \tag{2.24}$$

One special case occurs when L is the Laplacian $\Delta g = D^{2,0}g + D^{0,2}g$. When the variations of g are small, this Laplacian energy and the bending energy (2.23) are related to an integral of the surface curvature (see Appendix, Chapter 11). More specifically, the physical energy of a thin plate subjected to deformation is the integral of a quadratic form of the main curvatures of g ([59], page 250), which leads to the following equation (assuming small deformations):

$$U_1(g) = \int_\Omega \left[\left(\left(D^{2,0}g\right)^2 + 2\left(D^{1,1}g\right)^2 + \left(D^{0,2}g\right)^2 \right)\xi + (\Delta g)^2 (1 - \xi) \right] dt\, du,$$

where $\xi \in [0,1]$ is a constant of elasticity that is characteristic of the thin plate ([155], page 368). Whichever energy U_1 is used, (2.23) or (2.24), the energy of approximation of y by a smooth surface g is

$$U(g \mid y) = \alpha U_1(g) + \sum_k \left(y_k - g(t_k, u_k)\right)^2. \tag{2.25}$$

We will now look at how this approximation is applied, starting with the energy (2.24) that gives rise to surfaces called L-splines (Section 5.2.1). We will see that calculating the smoothing surface is advantageous when L has a Green's function. We will then continue with the energy (2.23). The mathematical framework of the associated variational problem is first described (Section 2.5.2). We then look at a specific computation (Section 2.5.3): the finite elements method. This also illustrates the relationship between this approach and those based on discrete energy. Although this computation is general in scope, it will look more complicated than the previous one.

2.5.1 L-Spline Surface

Let U_1 be the energy (2.24). In our experimental examples, we will consider the operator $\mathsf{L}g = \Delta g + a_0 g$ with $a_0 > 0$:

$$U_1(g) = \int_\Omega \left[\Delta g(s) + a_0 g(s)\right]^2 ds, \tag{2.26}$$

where we have set $s = (t, u)$. This energy is suitable for situations where the surface being sought has both a low amplitude and small local variations. We will use this energy to examine the interpolation and the approximation of y. Refer to the section entitled "Green's function and Covariance Function" (Section 13.2.2, page 284), and in particular to Lemmas 13.1 and 13.2. We define the Green's function Φ of L as the solution of $L\Phi(s, s') = \delta(s - s')$ in the sense of distributions, where δ is the Dirac function (Lemma 13.1). This function allows us to define a positive definite symmetric function $V(s, s') = \int \Phi(s, \tau)\Phi(s', \tau)d\tau$ (Lemma 13.2) which, in the case $L = \Delta + a_0 Id$, is expressed as $V(s, s') = \exp -(a_0\|s - s'\|)$. This function is interpreted as a spatial covariance function.

Proposition 2.3. *The surface $\widehat{g}$ that minimizes (2.26) among all the regular surfaces g that satisfy the interpolation constraint $\{g(s_k) = y_k \; ; \; k = 1, \ldots, n\}$ is written as*

$$\widehat{g}(s) = \sum_{k=1}^{n} \beta_k V(s, s_k), \tag{2.27}$$

where $\beta' = (\beta_1, \ldots, \beta_n)$ is the solution to the linear system $\mathbf{V}\beta = y$, in which the matrix $\mathbf{V}$ contains the element $\mathbf{V}_{kl} = V(s_k, s_l)$.

Proof:
We use the proof given in [121]. Let $h(s)$ be a variation around $\widehat{g}(s)$ such that $g(s) = \widehat{g}(s) + h(s)$ is an interpolation surface. We therefore have $g(s) = \sum \beta_k V(s, s_k) + h(s)$ with $h(s_k) = 0$ for all $k = 1, \ldots, n$. We can therefore write $U_1(g) = \int [Lg(s)]^2 ds$ as

$$U_1(g) = \int [L\widehat{g}(s)]^2 ds + \int [Lh(s)]^2 ds + 2\sum_k \beta_k \int LV(s, s_k) \, Lh(s) ds.$$

Using $L\Phi(s, s_k) = \delta(s - s_k)$, the integral in the last term is expanded as follows:

$$\int L\left[\int \Phi(s, \tau)\Phi(s_k, \tau)d\tau\right] Lh(s) ds = \int L\Phi(s, s_k)h(s)ds = h(s_k) = 0,$$

giving $U_1(g) \geq U_1(\widehat{g})$. This proposition is also valid for (2.24). $\square$

This technique is similar to "kriging", a well-known method in the field of geostatics [121]. We also note that in [118], this problem of interpolating a set of points is generalized to the interpolation of a set of curves or surfaces. The approximation is obtained in a similar way.

Proposition 2.4. *The surface $\widehat{g}$ that minimizes (2.25) among all regular surfaces g of energy (2.26) is written as*

$$\widehat{g}(s) = \sum_k \beta_k V(s, s_k),$$

where β is the solution of the linear system $(\mathbf{V} + \alpha \, Id)\beta = y$.

Proof:
As in Section 2.2.1, we use the result of the interpolation to obtain the approximation result. For any set of n values $\zeta_1, \ldots, \zeta_n$ chosen according to

s_k, (2.27) gives the interpolation surface g^ζ such that $\{g^\zeta(s_k) = \zeta_k,\ k = 1,\ldots,n\}$ written as $g^\zeta(s) = \sum_{k=1}^n \beta_k V(s, s_k)$, where β is a solution of $\mathbf{V}\beta = \zeta$. The internal energy of this surface is $U_1(g^\zeta) = \beta'\mathbf{V}\beta$. The approximation surface that minimizes the energy (2.25) is therefore obtained for the vector $\zeta = \mathbf{V}\beta$ that minimizes

$$U(g^\zeta \mid y) \equiv U(\beta \mid y) = \alpha\, \beta'\mathbf{V}\beta + \|y - \mathbf{V}\beta\|^2.$$

As for (2.20), the β that gives this minimum is the solution of $(\mathbf{V}+\alpha\,\mathrm{Id})\beta = y$. This proposition is also valid for (2.24).   □

EXAMPLE.
Let us illustrate this technique for the energy (2.26). The problem consists in finding the correspondence (see Section 13.2) between two sets of points in the plane (Figure 2.5(a)). The first set, marked with the symbol ^{o}, consists of feature points of the face[4] shown in Figure 2.6(a), image size 112×92. The second set, marked with the symbol *, is the same set of feature points except that the face has been turned (Figure 2.6(b)). The points ^{o} and * are therefore matched pairs: for each characteristic point k, there is a vector $(x(s_k), y(s_k))$ going from point k on the frontal view of the face to point k on the rotated face, where s_k is the position relative to the frontal view of the face. We consider the second set to be a deformation of the first, and we wish to estimate the field of spatial deformations $\{(x(s), y(s)),\ s \in \Omega\}$ approximating the observed values $\{(x(s_k), y(s_k)),\ k = 1, ..., n\}$. Here, Ω is the square $[0, 1]^2$. To perform this approximation, we apply Proposition 2.4 separately for x and y. We thus approximate y_k to obtain the field of vertical deformations $\{y(s),\ s \in \Omega\}$. Figure 2.5(b) gives the resulting field $\{(x(s), y(s)),\ s \in \Omega\}$ in which the ^{o} symbol is still used to designate the initial ^{o} points, although they have now been moved. If this field is applied to the grid of the image showing the frontal view of the face, we obtain the image shown in Figure 2.6(c). Processing was performed with a_0 and $1/\alpha$ equal to the sampling interval in $[0, 1]$. We interpret $1/\alpha$ as the variance of the adjustment deviations. We should also note that certain apparent inaccuracies on the face in Figure 2.6(c) on the right-hand side are caused by a display problem; the graphics display applies an image given on a nonsquare grid to a square grid.

---

[4]Olivetti and Oracle Research Laboratory
http://www.cam-orl.co.uk/face database.html.

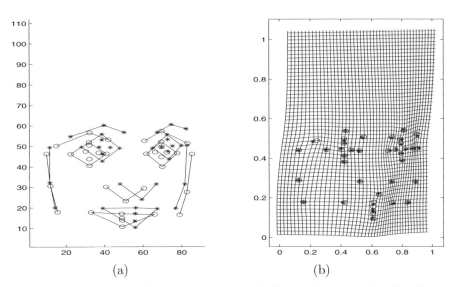

Figure 2.5. (a) Two sets of feature points in matched pairs (represented by * and ^{o}). (b) Field of deformations.

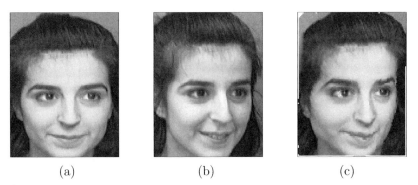

Figure 2.6. (a) Image associated with the first set of feature points *. (b) Image associated with the second set of feature points ^{o}. (c) Image resulting from the field of deformations applied to (a).

2.5.2   Quadratic Energy

[5] The energy (2.23) is now U_1. The minimization of $U(g \mid y)$ is given as the solution of the variational problem

$$\min_{g \in \mathbf{H}^2} U(g \mid y), \qquad (2.28)$$

---

[5]This subsection can be omitted at the first reading.

where H^2 is Sobolev space $H^2(\Omega)$ of degree $m = 2$ made up of functions g such that the $D^{i,j}g$ with $i + j \leq 2$ are in $L^2(\Omega)$, where $L^2(\Omega)$ is the space of square integrable functions. The space H^2 has the following norm

$$\|g\|_H = \left(\sum_{i,j \,:\, i+j\leq 2} \int \left(D^{i,j}g(t,u)\right)^2 dt\, du \right)^{1/2}.$$

An analytical expression for the optimum adjustment surface cannot be obtained. The approach used therefore consists in transforming the continuous problem into a discrete problem whose numerical solution approximates the exact solution. Because the energy under consideration is quadratic, the solution to the discrete problem will be the solution for a linear system. Ciarlet [35] describes this type of variational problem in great detail. We now provide a summary of this description.

Let us remove from $U(g \mid y)$ the constant $\sum y_k^2$. On the one hand, this constant is not involved in the optimization because it is independent of g, and on the other hand, this will allow us to define a linear form. The quadratic energy of our problem becomes

$$U(g) = \frac{1}{2}\alpha \int \left(\left(D^{2,0}g\right)^2 + 2\left(D^{1,1}g\right)^2 + \left(D^{0,2}g\right)^2 \right) dt\, du$$
$$+ \frac{1}{2}\sum_k \left(g(t_k, u_k)^2 - 2y_k g(t_k, u_k)\right),$$

where we have used $U(g) \doteq U(g \mid y)$. The factor of $\frac{1}{2}$ was introduced merely because it simplifies the writing of certain terms. This energy is also written as

$$U(g) = \frac{1}{2}A(g,g) - F(g), \tag{2.29}$$

$$\text{with } A(g,h) = \alpha \int \left(D^{2,0}g D^{2,0}h + 2D^{1,1}g D^{1,1}h + D^{0,2}g D^{0,2}h\right) dt\, du$$
$$+ \sum_k g(t_k, u_k)\, h(t_k, u_k),$$

$$\text{and } F(g) = \sum_k y_k\, g(t_k, u_k),$$

where A is a symmetric continuous bilinear form, and F is a continuous linear form. Assuming A to be elliptic, i.e.,

$$\exists\, a > 0 \,:\, A(g,g) \geq a\|g\|_H^2 \quad \forall g \in H^2,$$

we show that this energy has only one minimum. In practice, the data are such that this assumption is true.

We also establish a necessary and sufficient condition to obtain the minimum, as follows. If $\widehat{g}$ minimizes U, then $U(\widehat{g}) \leq U(\widehat{g}+\xi h)$ for any function $h \in \mathrm{H}^2$ and any scalar ξ. Let us calculate the variation

$$\delta U = U(\widehat{g} + \xi h) - U(\widehat{g}) = \xi[A(\widehat{g}, h) - F(h)] + \frac{1}{2}\xi^2 A(h, h).$$

The condition is obtained by setting the derivative at the point $\xi = 0$ to zero, i.e., $\lim_{\xi \to 0} \frac{\delta U}{\xi} = 0$. This leads to the variational equation $A(\widehat{g}, h) = F(h)$. More specifically, we have the following theorem.

Theorem 2.2. *If $\mathcal{E}$ is a closed subspace of H^2, then $\widetilde{g}$ is a solution of the restricted variational problem*

$$\min_{g \in \mathcal{E}} U(g) \tag{2.30}$$

if and only if it satisfies the variational equation

$$A(\widetilde{g}, h) = F(h), \quad \forall h \in \mathcal{E}. \tag{2.31}$$

When $\mathcal{E}$ is different from H^2, the purpose of $\mathcal{E}$ is to allow us to approximate the exact solution to the unrestricted problem (2.28) $\min_{g \in \mathrm{H}^2} U(g)$. This approximation is formalized as follows. The theorem applied to the case $\mathcal{E} = \mathrm{H}^2$ gives the condition $A(\widehat{g}, g) = F(g)$, $\forall g \in \mathrm{H}^2$, which makes it possible to rewrite U as

$$U(g) = \frac{1}{2}A(g, g) - F(g) = \frac{1}{2}A(g, g) - A(\widehat{g}, g)$$

$$= \frac{1}{2}A(g - \widehat{g}, g - \widehat{g}) - \frac{1}{2}A(\widehat{g}, \widehat{g}). \tag{2.32}$$

The last equation is a consequence of the symmetry of A. We then have

$$\widetilde{g} = \arg\min_{g \in \mathcal{E}} U(g) = \arg\min_{g \in \mathcal{E}} A(g - \widehat{g}, g - \widehat{g}). \tag{2.33}$$

We therefore interpret $\widetilde{g}$ as the projection of $\widehat{g}$ on $\mathcal{E}$ with respect to the scalar product $A(.,.)$. The approximation is characterized by the norm of the error $A(\widetilde{g} - \widehat{g}, \widetilde{g} - \widehat{g})^{1/2}$, which is comparable to an error in the energy. According to the Pythagorean theorem, we have $A(\widetilde{g} - \widehat{g}, \widetilde{g} - \widehat{g}) = A(\widehat{g}, \widehat{g}) - A(\widetilde{g}, \widetilde{g})$, and this expression is the same as $2(U(\widetilde{g}) - U(\widehat{g}))$ because $U(\widehat{g}) = -\frac{1}{2}A(\widehat{g}, \widehat{g})$ according to (2.32).

One way of ensuring that $\mathcal{E}$ is closed is to work with a space of finite size, as in the Ritz method:

$$\mathcal{E} = \left\{ h : h = \sum_{i=1}^{n} h_i \Phi_i \right\},$$

where Φ_i represents an n-dimensional basis in the space $\mathcal{E}$. By applying the above theorem, we then know that an element $g = (g_i)$ of $\mathcal{E}$ will be a solution if it satisfies the linear system

$$\sum_{j=1}^{n} A(\Phi_i, \Phi_j)g_j = F(\Phi_i), \quad 1 \le i \le n,$$

or, in matrix form,

$$\underline{A}\,\underline{g} = \underline{F}. \tag{2.34}$$

Equation (2.34) is known as the discrete variational equation, because g is here represented by a finite set of coefficients $g_1, \ldots, g_n$.

2.5.3 Finite Element Optimization

The choice of basis functions Φ_j with global support makes the Ritz method inappropriate for numerical processing. The finite element method is similar to the Ritz method, but it involves constructing a space $\mathcal{E}$ generated by basis functions with local support (Courant [58]). To do this, the region Ω, which we will assume in this case to be a polygon in $\mathbb{R}^2$, is partitioned according to a mesh $\mathcal{M}(\Omega)$ consisting of triangles or rectangles, called finite elements e, connected by their vertices,

$$\Omega = \bigcup_{e \in \mathcal{M}(\Omega)} e,$$

and such that for any pair of neighboring elements $<e_i, e_j>$, $e_i \cap e_j$ is either a vertex or an edge. The approximation space $\mathcal{E}$ is the space of the piecewise polynomial surfaces on the finite elements. For any $g \in \mathcal{E}$, the restriction $g|_e$ of g to e is a polynomial of degree greater than or equal to $m = 2$ at t and u, such that $g|_e \in \mathrm{H}^2(e)$. Let ℓ denote the maximum "diameter" of the elements on which $\mathcal{E}$ is built, and let $\mathcal{E}_\ell$ denote the space $\mathcal{E}$. Every space $\mathcal{E}_\ell$ therefore has an associated solution $\tilde{g}_\ell$ to the discrete variational equation obtained from (2.31):

$$A(\tilde{g}_\ell, h_\ell) = F(h_\ell), \quad \forall h_\ell \in \mathcal{E}_\ell.$$

One necessary condition for uniqueness and the convergence of $\tilde{g}_\ell$ to $\hat{g}$ when $\ell \to 0$ is that $\mathcal{E}_\ell$ be a subspace of H^2: $\mathcal{E}_\ell \subset \mathrm{H}^2$. For this, we impose the so-called conforming condition, which states that the $g_\ell \in \mathcal{E}_\ell$ are $m-1$ times continuously differentiable at the boundaries of the elements. This is a consequence of Cea's lemma [35].

Calculating $\tilde{g}_\ell$ raises problems due to the requirement to satisfy the conforming condition. This is why a proposal to relax the conforming condition has been made [167]. In this case, however, $\mathcal{E}_\ell$ is no longer a subspace of H^2, and A is no longer defined at the boundaries of the elements. We therefore approximate A with the bilinear form A_ℓ as follows:

$$A_\ell(.,.) = \sum_{e \in \mathcal{M}_\ell(\Omega)} A(.,.)|_e ,$$

where $\mathcal{M}_\ell(\Omega)$ is a mesh of diameter ℓ and $A(.,.)|_e$ is the restriction of the definition of A to e. The initial variational problem is replaced by the minimizing of the new energy

$$U^\ell(g_\ell) = \frac{1}{2} A_\ell(g_\ell, g_\ell) - F(g_\ell),$$

for which one necessary condition of a minimum at $\widetilde{g}_\ell$ is

$$A_\ell(\widetilde{g}_\ell, h_\ell) = F(h_\ell), \quad \forall h_\ell \in \mathcal{E}_\ell.$$

Sufficient conditions for the uniqueness and convergence of $\widetilde{g}_\ell$ are introduced in the appendix to this section. We shall now give some examples of nonconforming elements.

EXAMPLE.
The domain Ω is rectangular and has a square grid $\mathcal{M}_\ell(\Omega)$ whose squares are the finite elements of edge length ℓ. Any surface $g \in \mathcal{E}_\ell$ can be expressed as follows on the element e:

$$g|_e(t, u) = p(t, u),$$

$$\text{with } p(t,u) = \sum_{i+j \leq 3} \beta_{i,j} t^i u^j + \beta_{1,3} t u^3 + \beta_{3,1} t^3 u.$$

This includes twelve unknown parameters $\beta_{i,j}$. For any element e, the parameters $\beta_{i,j}$ are uniquely determined when g, $D^{1,0}g$, and $D^{0,1}g$ at the four vertices of e are given. These elements, named after Adini, are nonconforming, because only the continuity $\mathcal{C}^0$ is assured. Terzopoulos [168] considers the formalism of finite elements to handle a surface reconstruction problem in the context of stereo vision. The approximation surface calculation is performed as follows. The mesh $\mathcal{M}_\ell(\Omega)$ is the same as above. We consider the following relationship for any element e:

$$g|_e(t, u) = p(t, u) \quad \text{with} \quad p(t,u) = \sum_{0 \leq i+j \leq 2} \beta_{i,j} t^i u^j.$$

The parameters $\beta_{i,j}$ of $g|_e$ are expressed in terms of the nodal values of g. For example, for the element e whose vertices are the nodes $(0,0)$, $(\ell, 0)$, $(0, \ell)$, (ℓ, ℓ), the parameters are obtained by setting $p(0,0) = g(0,0)$, $p(\ell, 0) = g(\ell, 0)$, $p(0, \ell) = g(0, \ell)$, $p(\ell, \ell) = g(\ell, \ell)$, $p(-\ell, 0) = g(-\ell, 0)$,

$p(0, -\ell) = g(0, -\ell)$, and the same applies to all elements e. The solution is

$$\beta_{2,0}(e) = \frac{1}{2\ell^2}(g(\ell, 0) - 2g(0, 0) + g(-\ell, 0)), \tag{2.35}$$

$$\beta_{0,2}(e) = \frac{1}{2\ell^2}(g(0, \ell) - 2g(0, 0) + g(0, -\ell)),$$

$$\beta_{1,1}(e) = \frac{1}{\ell^2}(g(\ell, \ell) - g(0, \ell) - g(\ell, 0) + g(0, 0)),$$

$$\beta_{1,0}(e) = \frac{1}{2\ell}(g(\ell, 0) - g(-\ell, 0)),$$

$$\beta_{0,1}(e) = \frac{1}{2\ell}(g(0, \ell) - g(0, -\ell)),$$

$$\beta_{0,0}(e) = \frac{1}{2\ell}g(0, 0).$$

With these finite elements, the calculation code can be made simpler. Let us show how the solution $\tilde{g}_\ell$ is calculated. We use e_{ij} to denote the element whose lower left node is located at $(i\ell, j\ell)$. The energy approximated on the basis of the finite elements is

$$
\begin{aligned}
U^\ell(g) &= \frac{1}{2}\alpha\sum_e \int_e ((D^{2,0}g|e)^2 + 2(D^{1,1}g|e)^2 + (D^{0,2}g|e)^2) \\
&\quad + \frac{1}{2}\sum_k (g(t_k, u_k)^2 - 2y_k g(t_k, u_k)) \\
&= \frac{1}{2}\ell^2\alpha\sum_{ij} \left[(2\beta_{2,0}(e_{ij}))^2 + 2(\beta_{1,1}(e_{ij}))^2 + (2\beta_{0,2}(e_{ij}))^2\right] \\
&\quad + \frac{1}{2}\sum_k (g(t_k, u_k)^2 - 2y_k g(t_k, u_k)),
\end{aligned}
$$

or, if we again use the expressions (2.35) corresponding to e_{00},

$$
\begin{aligned}
U^\ell(g) = \frac{1}{2\ell^2}\alpha\sum_{ij} \Bigg(&\left[g((i+1)\ell, j\ell) - 2g(i\ell, j\ell) + g((i-1)\ell, j\ell)\right]^2 \\
&+ 2\Big[g((i+1)\ell, (j+1)\ell) - g(i\ell, (j+1)\ell) - \\
&\qquad g((i+1)\ell, j\ell) + g(i\ell, j\ell)\Big]^2 \\
&+ \left[g(i\ell, (j+1)\ell) - 2g(i\ell, j\ell) + g(i\ell, (j-1)\ell)\right]^2 \Bigg) \\
+ \frac{1}{2}\sum_k &(g(t_k, u_k)^2 - 2y_k\, g(t_k, u_k)). \tag{2.36}
\end{aligned}
$$

This discrete energy is written in vector form as follows:

$$U^\ell(g) = \frac{1}{2} \, \langle \underline{g} \, , \, \underline{A} \, \underline{g} \rangle - \langle \underline{F} \, , \, \underline{g} \rangle \, .$$

If N is the number of elements e, then $\underline{g}$ and $\underline{F}$ are N-vectors, where $\underline{F}$ contains the data $\{y_k\}$, and the rest of its components are zero. The ellipticity of A_ℓ means that $\underline{A}$ is a positive definite matrix. It is a block-diagonal sparse symmetric matrix. The minimum of the energy $\widetilde{g}$ is obtained by setting its gradient to zero:

$$\underline{A} \, \underline{\widetilde{g}} = \underline{F}.$$

This is the discrete variational equation (2.34). Its solution can be calculated by the Gauss-Seidel iterative method (Section 7.4.2). The use of this method is described on page 142, and Figures 7.2 and 7.3 illustrate this type of approximation.

Note. The expression (2.36) represents the finite difference approximation, which is a consequence of the choice of finite elements. It is also the generalization of the model (2.7). This example of nonconforming elements is therefore rather simplistic, because the above mathematical analysis goes beyond the requirements of the expression (2.36). This illustrates the connection between variational approaches and discrete Markov approaches. In fact, the discrete notation of the energy (2.36) corresponds exactly to a *potential energy* whose cliques would be the e_{ij} (Chapter 5).

• APPENDIX

If $\mathcal{E}_\ell$ has a norm $\|.\|_\ell$ and if, in this normed space, A_ℓ is elliptic, we then show [35] that the approximated variational problem allows a unique solution $\widetilde{g}_\ell$. Furthermore, if uniform ellipticity is imposed on the family $\{A_\ell\}$ in the following way,

$$\exists \, a > 0 \; : \;\; A_\ell(g_\ell, g_\ell) \geq a \, \|g_\ell\|_\ell^2 \, , \quad \forall g_\ell \in \mathcal{E}_\ell, \forall \ell,$$

with a independent of ℓ, then convergence to the solution $\widehat{g}$ of the unrestricted problem is characterized by the following lemma.

Lemma 2.1. (Strang). *If the $\{A_\ell\}$ are uniformly elliptic, then there exists a constant C that is independent of ℓ such that*

$$\|\widetilde{g}_\ell - \widehat{g}\|_\ell \; \leq \; C \left(\min_{h_\ell \in \mathcal{E}_\ell} \|h_\ell - \widehat{g}\|_\ell \; + \; \max_{h_\ell \in \mathcal{E}_\ell} \frac{|\, A_\ell(\widehat{g}, h_\ell) - F(h_\ell) \,|}{\|h_\ell\|_\ell} \right).$$

Proof:
Consider any surface $h_\ell \in \mathcal{E}_\ell$. As a consequence of the ellipticity of A_ℓ, we

have

$$
\begin{aligned}
a\|\widetilde{g}_\ell - h_\ell\|_\ell^2 &\leq A_\ell(\widetilde{g}_\ell - \widehat{g} + \widehat{g} - h_\ell, \widetilde{g}_\ell - h_\ell) \\
&= A_\ell(\widehat{g} - h_\ell, \widetilde{g}_\ell - h_\ell) + [A_\ell(\widehat{g}, \widetilde{g}_\ell - h_\ell) - F(\widetilde{g}_\ell - h_\ell)],
\end{aligned}
$$

and by the continuity of the A_ℓ, the existence of an M such that $|A_\ell(g, h)| \leq M\|g\|_\ell\|h\|_\ell$, lets us write

$$
\begin{aligned}
a\|\widetilde{g}_\ell - h_\ell\|_\ell &\leq M\|\widehat{g} - h_\ell\|_\ell + \frac{|\,A_\ell(\widehat{g}, \widetilde{g}_\ell - h_\ell) - F(\widetilde{g}_\ell - h_\ell)\,|}{\|\widetilde{g}_\ell - h_\ell\|_\ell} \\
&\leq M\|\widehat{g} - h_\ell\|_\ell + \max_{w_\ell \in \mathcal{E}_\ell} \frac{|\,A_\ell(\widehat{g}, w_\ell) - F(w_\ell)\,|}{\|w_\ell\|_\ell}.
\end{aligned}
$$

Finally, after adding $[a\|\widehat{g} - h_\ell\|_\ell]$ to both terms of this inequality, the triangular equality $\|\widetilde{g}_\ell - h_\ell\|_\ell + \|\widehat{g} - h_\ell\|_\ell \geq \|\widehat{g} - \widetilde{g}_\ell\|_\ell$ concludes the proof. $\square$

When $\mathcal{E}_\ell$ is a space of conforming elements, the differences $A_\ell(\widehat{g}, h_\ell) - F(h_\ell)$ are zero for all h_ℓ, which brings us back to Cea's lemma. For nonconforming elements, however, convergence is obtained under the consistency condition

$$
\lim_{\ell \to 0} \max_{h_\ell \in \mathcal{E}_\ell} \frac{|\,A_\ell(\widehat{g}, h_\ell) - F(h_\ell)\,|}{\|h_\ell\|_\ell} = 0. \tag{2.37}
$$

Before it was formalized in this way, this condition was recognized empirically by Irons [115] and gave rise to a convergence criterion known as the "patch test" [167].

3

Parametric Spline Models

We know from points (i) and (ii) of Section 1.1.2 that an inverse problem can be solved by constraining the solution either implicitly, via a prior energy $U_1(x)$ as in Chapter 2, or explicitly by dimensionality reduction of the space containing x. The latter approach is the subject of this chapter. It is completely different from regularization with a prior energy. Dimensionality reduction has two advantages: It produces a simple exogenous model, and it is easy to handle when used as a component of a more complicated model.

3.1   Representation on a Basis of B-Splines

We have n abscissas $\{t_i, \ i = 1, \ldots, n\} \dot{=} \mathcal{T}$, called knots. The set of cubic splines $\mathcal{S}(\mathcal{T})$ is obviously a vector space. Its dimension is

$$\dim(\mathcal{S}(\mathcal{T})) \ = \ n + 2,$$

because there are four polynomial coefficients per interval (t_i, t_{i+1}) and three continuity constraints $\mathcal{C}^2$ on each t_i, $i = 2, \ldots, n-1$, making a total of $4(n-1) - 3(n-2) = n+2$. Let us show a basis of this space. Let us assume that the knots are equidistant: $t_{i+1} - t_i = h$ for all $i = 1, \ldots, n-1$, and let us put down six additional knots spaced by h:

$$t_{-2} < t_{-1} < t_0 < t_1 \quad \text{and} \quad t_n < t_{n+1} < t_{n+2} < t_{n+3}.$$

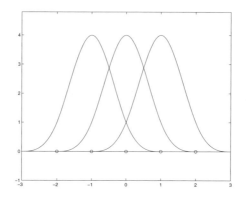

Figure 3.1. B-spline function for $h = 1$.

These new knots allow us to define $n+2$ cubic splines $\{B_i,\ i = 0, .., n+1\}$ that will constitute a basis of $\mathcal{S}(\mathcal{T})$.

Definition 3.1. *On equidistant knots $\{t_i\}$ such that $t_{i+1} - t_i = h$ for all i, the $4h$-support cubic splines are called B-spline functions. They are written as follows:*

$$B_i(t) =$$
$$\begin{cases}
(t - t_{i-2})^3, & \text{if } t \in [t_{i-2}, t_{i-1}], \\
h^3 + 3h^2(t - t_{i-1}) + 3h(t - t_{i-1})^2 - 3(t - t_{i-1})^3, & \text{if } t \in [t_{i-1}, t_i], \\
h^3 + 3h^2(t_{i+1} - t) + 3h(t_{i+1} - t)^2 - 3(t_{i+1} - t)^3, & \text{if } t \in [t_i, t_{i+1}], \\
(t_{i+2} - t)^3, & \text{if } t \in [t_{i+1}, t_{i+2}], \\
0, & \text{otherwise.}
\end{cases}$$

Later, we will explain how to construct these B-splines and how to extend them to the nonequidistant case. Clearly, $(B_i(t),\ t \in [t_1, t_n])$ belongs to $\mathcal{S}(\mathcal{T})$, and $B_i(t) = 0$ when $t \leq t_{i-2}$ or $t \geq t_{i+2}$. B-splines are the only nonzero cubic splines with the smallest compact support. They are linearly independent and therefore generate $\mathcal{S}(\mathcal{T})$. For any spline $g \in \mathcal{S}(\mathcal{T})$, we have

$$g(t) = \sum_{j=0}^{n+1} \beta_j B_j(t). \tag{3.1}$$

Interpolation was discussed in Section 2.2.1. Using equation (3.1), we can now give a very simple expression for the interpolation spline. To do this, instead of setting $D^2g(t)$ to zero outside the interval (t_1, t_n), we apply the constraint that imposes values for $Dg(t)$ at t_1 and t_n.

Proposition 3.1. *Let $\{\zeta_i\}$ be n values set according to the t_i values, and two values a_1 and a_n. There exists only one spline $g \in \mathcal{S}(\mathcal{T})$ that satisfies* the interpolation constraints $g(t_i) = \zeta_i$ for all $i = 1, \ldots, n$ and the boundary constraints $Dg(t_i) = a_i$ for $i = 1, n$.

Proof:
By entering g and Dg along the basis of the B-splines into these constraints, we can rewrite the constraints as follows:

$$\zeta_i = \sum_{j=0}^{n+1} \beta_j B_j(t_i), \quad i = 1, \ldots, n,$$

$$a_i = \sum_{j=0}^{n+1} \beta_j D^1 B_j(t_i), \quad i = 1, n.$$

This is an $(n+2)$-order linear system $(a_1, \zeta', a_n)' = \mathcal{B}\beta$. Note that these results do not depend on the knots being equidistant. For the equidistant case, we have the additional information $B_j(t_j) = 4h^3$, $B_j(t_{j-1}) = B_j(t_{j+1}) = h^3$, and $D^1 B_j(t_j) = 0$, $D^1 B_{j+1}(t_j) = -D^1 B_{j-1}(t_j) = 3h^2$, and therefore

$$\mathcal{B} = \begin{bmatrix} -3h^2 & 0 & 3h^2 & 0 & 0 & \cdots \\ h^3 & 4h^3 & h^3 & 0 & 0 & \cdots \\ 0 & h^3 & 4h^3 & h^3 & 0 & \cdots \\ & \cdot & & & & \\ & \cdot & & & & \\ & \cdot & & & & \\ 0 & \cdots & 0 & h^3 & 4h^3 & h^3 \\ 0 & \cdots & 0 & -3h^2 & 0 & 3h^2 \end{bmatrix}$$

$\square$

3.1.1 Approximation Spline

To approximate the data $\{(t_i, y_i), \; i = 1, \ldots, n\}$ with a cubic spline, we start by choosing a subset of $\mathcal{T}$, denoted by $\{\tau_1, \ldots, \tau_q\} \hat{=} \mathcal{T}_\tau$, such that $t_1 = \tau_1$, $t_n = \tau_q$. We then wish to approximate the data with a cubic spline that belongs to $\mathcal{S}(\mathcal{T}_\tau)$, where $\mathcal{T}_\tau$ is the set of approximation knots. Our approximation is based on the exogenous model in t:

$$Y_i = g(t_i) + W_i, \quad i = 1, \ldots, n, \tag{3.2}$$

$$g(t_i) = \sum_{j=0}^{q+1} \beta_j B_j(t_i),$$

where W_i are independent random variables. The B-splines B_j are built on $\mathcal{T}_\tau$. The model (3.2) is parametric because it provides a dimensionality reduction, since $q + 2$ is smaller than n. (It is *much* smaller in actual applications.) This linear model is written in vector form as follows:

$$Y = \mathbf{B}\beta + W, \tag{3.3}$$

where $\mathbf{B}$ is the matrix of elements $\mathbf{B}(i,j) = B_j(t_i)$. Assuming a Gaussian distribution $LG(0, \sigma^2)$ for the W_i, we apply the principle of maximum likelihood (Section 1.1) and estimate the parameters $\theta = (\sigma^2, \beta)$ by

$$\widehat{\theta} = \arg\min_{\theta} \left[\frac{1}{\sigma^2} \|y - \mathbf{B}\beta\|^2 + n \log \sigma^2 \right].$$

We therefore have, in particular,

$$\widehat{\beta} = \arg\min_{\beta} U(y \mid \beta),$$

$$\text{with } U(y \mid \beta) = \|y - \mathbf{B}\beta\|^2. \tag{3.4}$$

We usually call $\widehat{\beta}$ the "least squares estimator". The approximation spline is $\widehat{g} = \mathbf{B}\widehat{\beta}$. It is also called a "regression spline", because the model (3.2) is analogous with the regression models that are well known in the field of statistics [73]. The so-called normal equation has $\widehat{\beta}$ as its solution:

$$^t\mathbf{B}\mathbf{B}\widehat{\beta} = {}^t\mathbf{B}y.$$

This equation is obtained by multivariate differentiation of $\|y - \mathbf{B}\beta\|^2$.

In the case of equidistant $\{t_i\}$ and $\{\tau_j\}$, all intervals (τ_j, τ_{j+1}) contain the same number of t_i values. Let us call this number κ. We can then say that $\mathbf{B}$ is a block-diagonal matrix whose blocks are all equal and of size $\kappa \times 4$, because only four B-splines are nonzero in each interval.

Note. The number κ is interpreted as a smoothing parameter. When it has a small value ($\kappa = 2, 3, \dots$), $\widehat{g}(t)$ passes close to y_i, but when it is large, $\widehat{g}(t)$ goes very far from y_i. This means that κ is acting as the parameter α for nonparametric splines, as illustrated in Figure 3.2. From a practical standpoint, we can see that the empirical choice of κ is easier than choosing the regularization parameter in the variational approach.

3.1.2 Construction of B-Splines

[1] We begin by introducing the case of equidistant knots, which will then be extended to the general case [64]. Let us recall the definition of difference operators $\Delta^k F$ of a function F. For any series $\cdots < \tau_i < \tau_{i+1} < \cdots$, they are defined by

$$\begin{aligned} \left[\Delta^1 F\right](\tau_i) &= F(\tau_{i+1}) - F(\tau_i), \\ \left[\Delta^n F\right](\tau_i) &= [\Delta^{n-1} F](\tau_{i+1}) - [\Delta^{n-1} F](\tau_i). \end{aligned}$$

---

[1] This subsection can be omitted at the first reading.

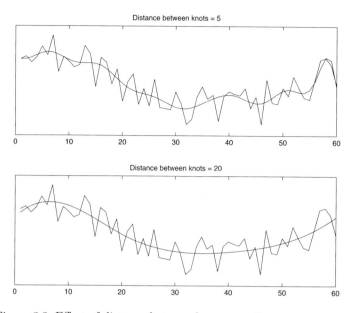

Figure 3.2. Effect of distance between knots on spline approximation.

For the fourth-order case and, more generally, for order n, their extended form is written as follows:

$$\left[\Delta^4 F\right](\tau_i) = F(\tau_{i+4}) - 4F(\tau_{i+3}) + 6F(\tau_{i+2}) - 4F(\tau_{i+1}) + F(\tau_i),$$

$$\left[\Delta^n F\right](\tau_i) = \sum_{j=0}^{n} (-1)^j C_n^j F(\tau_{i+j}).$$

These functions of τ_i have the following property.

Property 3.1. *If F is a polynomial of degree $(n-1)$ and if the τ_j are equidistant, then $[\Delta^n F](\tau_i) = 0$ for all i.*

The above property does not apply when the τ_i are not equidistant. The B-splines are built based on these differences. For every given τ, let us define the truncated cubic function

$$F_t(\tau) = (\tau - t)_+^3,$$

which is the same as

$$(\tau - t)_+^3 = \begin{cases} (\tau - t)^3, & \text{if } t \le \tau, \\ 0, & \text{otherwise.} \end{cases}$$

The functions $(\tau_i - t)_+^3$ are a basis of $\mathcal{S}(\mathcal{T})$. It is best to use a basis of functions with bounded support such as the basis of B-splines. To construct this basis, we set

$$K(t) \doteq [\Delta^4 F_t](\tau_i)$$
$$= (\tau_{i+4} - t)_+^3 - 4(\tau_{i+3} - t)_+^3 + 6(\tau_{i+2} - t)_+^3 - 4(\tau_{i+1} - t)_+^3 + (\tau_i - t)_+^3.$$

This brings us back to the definition of B-splines given earlier.

Proposition 3.2. *If the knots τ_i are equidistant such that $\tau_{i+1} - \tau_i = h$, we have $B_{i+2}(t) = K(t)$.*

Proof:
Let us take $K(t)$, a linear combination of the functions of t: $\{(\tau_{i+j} - t)_+^3, \ j = 0, \ldots, 4\}$. Each of these functions is a cubic spline with only one third-order discontinuity point at τ_{i+j}. Therefore, $K(t)$ is a cubic spline with five third-order discontinuity points at $\{\tau_{i+j}, \ j = 0, \ldots, 4\}$. Furthermore, $K(t) = 0$ when $t > \tau_{i+4}$. For any t set such that $t < \tau_i$, $F_t(\tau_i)$ is a third-degree polynomial, and therefore $[\Delta^4 F_t](\tau_i) = 0$ according to the above property. Next, we note that

$$K(t) = 0 \quad \text{if} \quad t \notin [t_i, t_{i+4}].$$

By expanding the truncated cubic functions on each of the intervals $[t_{i+j}, t_{i+j+1}]$ for $j = 0, \ldots, 3$, we obtain the B-spline expression introduced above. For example, on $[\tau_{i+2}, \tau_{i+3}]$, bearing in mind that $\tau_{i+4} = \tau_{i+3} + h$, we calculate

$$K(t) = (\tau_{i+4} - t)^3 - 4(\tau_{i+3} - t)^3,$$
$$= h^3 + 3h^2(\tau_{i+3} - t) + 3h(\tau_{i+3} - t)^2 - 3(\tau_{i+3} - t)^3,$$

which is the same as the expression for $B_{i+2}(t)$ given in the third line of the definition on page 52. In conclusion, we have

$$K(t) = [\Delta^4 F_t](\tau_i) = B_{i+2}(t).$$

$\square$

A useful expression for the B-spline is therefore

$$B_i(t) = [\Delta^4 F_t](\tau_{i-2}).$$

The advantage of using difference operators is that they make generalization possible. The above expression can immediately be generalized for splines that are not necessarily cubic. By setting $F_t(\tau) = (\tau - t)_+^{m+1}$ for $m = 0, 1, 2, \ldots$, the m-order B-splines can be deduced from

$$K(t) = [\Delta^{m+2} F_t](\tau) = \sum_{j=0}^{m+2} (-1)^j \, C_{m+2}^j \, (\tau_{i+j} - t)_+^{m+1}.$$

When the knots are not equidistant, we can show that

$$B_i(t) = (\tau_{i+m+2} - \tau_i) \sum_{j=0}^{m+2} \left(\frac{(\tau_{i+j} - t)_+^{m+1}}{\prod_{l=0:\ l \neq j}^{m+2} (\tau_{i+j} - \tau_{i+l})} \right).$$

Note that through homogeneity with this equation we could have defined the B-spline with equidistant knots using $B_i(t) = \frac{1}{h^3} \left[\Delta^4 F_t \right] (\tau_{i-2})$.

3.2 Extensions

3.2.1 Multidimensional Case

In contrast to Section 2.5, observed values are taken at the nodes of a regular grid $\{t_1, \ldots, t_n\} \times \{u_1, \ldots, u_n\} \equiv \mathcal{G}$. To keep the notation simple, we assume the same number of discretization points at t and u. We write these observed values as $y = \{y_{ij}, \ i, j = 1, \ldots, n\}$. In this case, we aim to approximate this scalar field with a continuous spline surface $g(t, u) \in \mathsf{R}$ with $t_1 \leq t \leq t_n$ and $u_1 \leq u \leq u_n$. We are concerned here with bicubic splines that are piecewise continuous polynomial surfaces as defined below.

Definition 3.2. *The function g is a bicubic spline on $\mathcal{G}$ if:*

(i) *The function g is a third-degree polynomial on R^2 at t and u respectively.*

(ii) *$g \in \mathcal{C}_2^4$, i.e., $D^{k,l}g$ is continuous for $0 \leq k, l \leq 2$, since the discontinuities of g are located on $\mathcal{G}$ only.*

The space of these splines is written as $\mathcal{S}(\mathcal{G})$.

We use $\mathcal{S}_N(\mathcal{G})$ to denote the space of bicubic splines subjected to the following boundary conditions:

$$\left. \begin{array}{ll} D^{2,0}g(t_i, u_j) = 0, & i = 1, \ldots, n, \quad j \in \{1, n\}, \\ D^{0,2}g(t_i, u_j) = 0, & j = 1, \ldots, n, \quad i \in \{1, n\}, \\ D^{2,2}g(t_i, u_j) = 0, & i, j \in \{1, n\}. \end{array} \right\} \quad (3.5)$$

To interpolate the y_{ij}, we have a theorem that is demonstrated in [2].

Theorem 3.1. *If $\widehat{g}$ is the solution of the equation*

$$\min_g \iint \left(D^{2,2}g(t, u) \right)^2 dt\, du,$$

where the minimum is taken from among the functions $g \in \mathcal{C}_2^4$ that satisfy (3.5), and

$$g(t_i, u_j) = y_{ij}, \quad 1 \leq i, j \leq n,$$

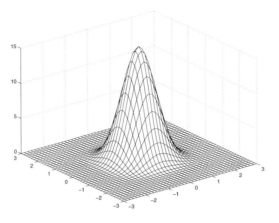

Figure 3.3. Tensor product of B-splines for $h = 1$.

then $\hat{g}$ belongs to $\mathcal{S}_N(\mathcal{G})$ and is unique.

We must emphasize that this energy, whose purpose is to quantify the regularity of the surface g, is not strictly a bending energy as was the energy (2.23) on page 39. In particular, it is not rotation invariant. To make the calculation more convenient, we use a representation based on B-splines as described below.

● Two-Dimensional Regression Spline

As in the one-dimensional case, we are looking for the bicubic approximation spline in a space of splines defined on a coarser grid than $\mathcal{G}$, which is, in practice, a subgrid of $\mathcal{G}$, denoted by

$$\{\tau_1, \ldots, \tau_q\} \times \{\lambda_1, \ldots, \lambda_q\} \doteq \mathcal{G}_{\tau, \lambda},$$

such that $\tau_1 = t_1$, $\tau_q = t_n$, $\lambda_1 = u_1$, $\lambda_q = u_n$. We have chosen the same number of knots in τ and λ to simplify the notation. We wish to approximate g with a spline belonging to $\mathcal{S}(\mathcal{G}_{\tau, \lambda})$. We show that the dimensionality of this space is $(q+2) \times (q+2)$ and that a basis $B_{kl}(t, u)$ is given by the "tensor" product of the B-splines $B_k^\tau(t)$ and $B_l^\lambda(u)$, constructed on $\{\tau_1, \ldots, \tau_q\}$ and $\{\lambda_1, \ldots, \lambda_q\}$ respectively:

$$B_{kl}(t, u) = B_k^\tau(t) \, B_l^\lambda(u), \quad k, l = 0, \ldots, q+1. \tag{3.6}$$

The approximation is based on the exogenous model in (t, u):

$$Y_{ij} = g(t_i, u_j) + W_{ij}, \quad i, j = 1, \ldots, n, \tag{3.7}$$

$$g(t_i, u_j) = \sum_{k=0}^{q+1} \sum_{l=0}^{q+1} \beta_{kl} \, B_k^\tau(t_i) \, B_l^\lambda(u_j),$$

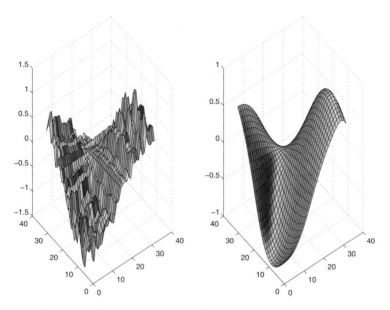

Figure 3.4. Smoothing by two-dimensional regression spline.

where W_{ij} are independent random variables. The advantage of this representation of g is its separability (3.6), which allows us to work in terms of matrices as follows:

$$Y = \mathbf{B}^\tau \beta \left({}^t\mathbf{B}^\lambda\right) + W. \tag{3.8}$$

The dimension of Y is $n \times n$, and the dimension of β is $(q+2) \times (q+2)$. The matrices $\mathbf{B}^\tau$ and $\mathbf{B}^\lambda$ are the $n \times (q+2)$ matrices of one-dimensional B-splines. When $\{\tau_1, \ldots, \tau_q\} \equiv \{\lambda_1, \ldots, \lambda_q\}$ we have $\mathbf{B}^\tau = \mathbf{B}^\lambda$. Without this feature of separability, the model would be expressed in vector form. This would result in very large dimensions, making a numerical calculation impossible. Figure 3.4 illustrates the fitting of this model to a noisy surface. We have y, a discrete sinusoidal surface with $n = 32$, degraded by a uniform noise on $[-0.5, 0.5]$. Smoothing was performed with $q = 3$. This fitting is obtained as follows. Assuming a Gaussian distribution for the W_{ij}, the principle of maximum likelihood leads us to a least-squares estimate

$$\widehat{\beta} = \arg \min_\beta U(y \mid \beta),$$

$$\text{with } U(y \mid \beta) = \|y - \mathbf{B}^\tau \beta \,{}^t(\mathbf{B}^\lambda)\|^2.$$

By differentiating, we obtain

$$\frac{d}{d\beta} U(y \mid \beta) = ({}^t\mathbf{B}^\tau) \, y \, \mathbf{B}^\lambda - ({}^t\mathbf{B}^\tau) \, \mathbf{B}^\tau \, \beta \, ({}^t\mathbf{B}^\lambda) \, \mathbf{B}^\lambda.$$

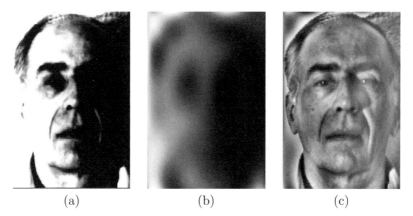

(a)   (b)   (c)

Figure 3.5. (a) Face in shadow. (b) Estimation of luminosity gradient. (c) Face after suppression of gradient.

We therefore know that $\widehat{\beta}$ is a solution of the normal equation

$$({}^t\mathbf{B}^\tau)\,\mathbf{B}^\tau\,\widehat{\beta}\,({}^t\mathbf{B}^\lambda)\,\mathbf{B}^\lambda \;=\; ({}^t\mathbf{B}^\tau)\,y\,\mathbf{B}^\lambda. \tag{3.9}$$

We can solve this system in two stages using the following breakdown:

$$({}^t\mathbf{B}^\tau)\,\mathbf{B}^\tau h = ({}^t\mathbf{B}^\tau)\,y\,\mathbf{B}^\lambda, \qquad \widehat{\beta}\,({}^t\mathbf{B}^\lambda)\,\mathbf{B}^\lambda = h.$$

We calculate h first, and then β, by applying Gauss's method. An example of a more exotic application would be the suppression of a shadow on a face. Figure 3.5[2] illustrates this application on an image of size 150×220 with a bicubic spline such that $q = 5$ on coordinate t and $q = 6$ on coordinate u.

• THREE-DIMENSIONAL REGRESSION SPLINE

This is a natural extension of the two-dimensional case. Our grid $\mathcal{G}$ is now a grid of coordinates (t, u, ν) whose dimensions have the general expression $(n_1 \times n_2 \times n_3)$. We wish to approximate y with a tricubic spline built on a $(q_1 \times q_2 \times q_3)$ subgrid. As in the two-dimensional case, these splines are expressed on a basis $B_{ijk}(t, u, \nu)$ obtained from a product of B-splines as follows:

$$B_{ijk}(t, u, \nu) = B_i^1(t)\,B_j^2(u)\,B_k^3(\nu),$$

where $\{B_i^1(t)\}$, $\{B_j^2(u)\}$, and $\{B_k^3(\nu)\}$ are the B-splines built on the nodes of the subgrid sides at t, u, and ν, respectively. The notation B_i^1 and B_j^2 refers to B_i^τ and B_j^λ (seen above) respectively. Their respective dimensions

---

[2]Yale face database
http://cvc.yale.edu/projects/yalefaces/yalefaces.html

are $(n_1 \times q_1)$, $(n_2 \times q_2)$, and $(n_3 \times q_3)$. The approximation is therefore based on the exogenous model

$$Y_{t,u,\nu} = g(t,u,\nu) + W_{t,u,\nu},$$
$$g(t,u,\nu) = \sum_{i,j,k} B_i^1(t)\, B_j^2(u)\, B_k^3(\nu)\, \beta_{i,j,k}. \tag{3.10}$$

The separability and the associated matrix representation are not as obvious as in the previous case. For these, we must define the products of matrix types $(.,.)_1$, $(.,.)_2$, and $(.,.)_3$ as

$$(\beta, \mathbf{B}^1)_1(i,j,k) = \sum_{\mu=1}^{q_1} \mathbf{B}^1(i,\mu)\beta(\mu,j,k), \quad (\beta, \mathbf{B}^1)_1 \in \mathcal{M}(n_1,q_2,q_3),$$

$$(\beta, \mathbf{B}^2)_2(i,j,k) = \sum_{\mu=1}^{q_2} \mathbf{B}^2(j,\mu)\beta(i,\mu,k), \quad (\beta, \mathbf{B}^2)_2 \in \mathcal{M}(q_1,n_2,q_3),$$

$$(\beta, \mathbf{B}^3)_3(i,j,k) = \sum_{\mu=1}^{q_3} \mathbf{B}^3(k,\mu)\beta(i,j,\mu), \quad (\beta, \mathbf{B}^3)_3 \in \mathcal{M}(q_1,q_2,n_3),$$

where $\mathcal{M}(p,q,r)$ refers to $p \times q \times r$ matrices. Using these definitions, the model (3.10) is written as

$$Y = \left(\left((\beta\,, \mathbf{B}^1)_1\,, \mathbf{B}^2 \right)_2 , \mathbf{B}^3 \right)_3 + W.$$

The maximum likelihood estimator is

$$\widehat{\beta} = \arg\min_{\beta} \left\| y - \left(\left((\beta\,, \mathbf{B}^1)_1\,, \mathbf{B}^2 \right)_2 , \mathbf{B}^3 \right)_3 \right\|^2,$$

and the solution to the normal equation is

$$\left(\left((\beta\,,^t \mathbf{B}^1\mathbf{B}^1)_1\,, ^t\mathbf{B}^2\mathbf{B}^2 \right)_2 , ^t\mathbf{B}^3\mathbf{B}^3 \right)_3 = \left(\left((y\,, ^t\mathbf{B}^1)_1\,, ^t\mathbf{B}^2 \right)_2 , ^t\mathbf{B}^3 \right)_3.$$

As in the two-dimensional case, the solution is calculated in three steps:

$$(f\,, ^t\mathbf{B}^3\mathbf{B}^3)_3 = \left(\left((y\,, ^t\mathbf{B}^1)_1\,, ^t\mathbf{B}^2 \right)_2 , ^t\mathbf{B}^3 \right)_3,$$
$$(h\,, ^t\mathbf{B}^2\mathbf{B}^2)_2 = f,$$
$$(\widehat{\beta}\,, ^t\mathbf{B}^1\mathbf{B}^1)_1 = h.$$

First we calculate f, then h, and finally $\widehat{\beta}$. At each stage we are solving a linear system whose matrix is block-diagonal.

3.2.2 *Heteroscedasticity*

[3] We now examine an unusual situation that occurs when the estimation variance is not constant, i.e., when σ^2 depends on the site (t, u, ν). This is known as heteroscedasticity. It occurs in radiography, for example, where the variance can depend on the intensity of the received signal (Section 10.4). Wang et al. [178, 11] describe this situation for the case of three-dimensional regression splines. Before we can deal with this type of problem, we need to introduce several classical concepts concerning the estimation of linear regression models.

● WEIGHTED LEAST SQUARES

Consider the classical linear regression model:

$$\vec{Y} = \mathcal{B}\vec{\beta} + \vec{W}.$$

Formally, this notation includes the regression models on one-, two-, and three-dimensional splines given in (3.2), (3.7), and (3.10), respectively. In the one-dimensional case, this notation is obvious. In Dimension 2, $\vec{Y}$ is a vector constructed by placing the rows of matrix Y end to end. By applying the same procedure to $\vec{W}$, we find a regression matrix $\mathcal{B}$ obtained from $B_{k,l}(t_i, u_j)$. This vector model grows to very large dimensions; it is of interest for its formal aspect only. We vectorize the three-dimensional model in a similar way. For this model, we set

$$\mathrm{E}(\vec{W}) = 0, \quad \text{and} \quad \mathrm{Var}(\vec{W}) = \Sigma.$$

If the random vector $\vec{W}$ follows a Gaussian distribution, the density of $\vec{Y}$ is

$$P_\theta(\vec{y}) = (2\pi)^{-n/2} |\Sigma|^{-1/2} \exp - \left[\frac{1}{2} \left\| \vec{y} - \mathcal{B}\vec{\beta} \right\|_{\Sigma^{-1}}^2 \right],$$

where $n = n_1 n_2 n_3$ and $\|A\|_{\Sigma^{-1}}^2 = {}^t A \Sigma^{-1} A$. The unknown parameters of the distribution are designated by θ. For example, $\theta = (\alpha, \beta)$ if Σ depends on parameters α, or $\theta = (\sigma^2, \beta)$ if $\Sigma = \sigma^2 \mathrm{Id}$. When we have an observed value of the random vector $\vec{Y}$, which we will call $\vec{y}^O$; we then need to fit the model to the data $\vec{y}^O$. This is done by estimating the parameters of the model.

We again consider the principle of maximum likelihood (Section 1.1). If θ designates the parameters of the model, the likelihood is given by the function $L(\theta) = P_\theta(\vec{y}^O)$. Estimation based on this principle involves choosing the parameter $\hat{\theta}$ that makes the observed value $\vec{y}^O$ the most

---

[3]This subsection can be omitted at the first reading.

probable:

$$\widehat{\theta} = \arg\max_{\theta} \ L(\theta).$$

The simplest case is $\Sigma = \sigma^2 \mathrm{Id}$, i.e., $\theta = (\sigma^2, \beta)$. The principle of maximum likelihood is written as

$$\widehat{\theta} = \arg\min_{\theta} \left[\frac{1}{\sigma^2} \left\| \vec{y}^O - \mathcal{B} \, \vec{\beta} \right\|^2 + n \log \sigma^2 \right],$$

which leads to the classic least-squares estimation

$$\widehat{\beta} = \left({}^t\mathcal{B}\mathcal{B} \right)^{-1} {}^t\mathcal{B}\vec{y}^O.$$

This is the situation described in the paragraphs above. A more general case would be $\Sigma = \sigma^2 V$, where V is known and nonsingular:

$$\widehat{\beta} \;\; = \;\; \arg\min_{\beta} \ \left\| \vec{y}^O - \mathcal{B}\vec{\beta} \right\|^2_{\Sigma^{-1}},$$

which gives the weighted least-squares estimation

$$\widehat{\beta} = \left({}^t\mathcal{B}V^{-1}\mathcal{B} \right)^{-1} {}^t\mathcal{B}V^{-1}\vec{y}^O. \tag{3.11}$$

This estimation is justified by the fundamental Gauss–Markov theorem:

Theorem 3.2. *The weighted least-squares estimator $({}^t\mathcal{B}V^{-1}\mathcal{B})^{-1} \, {}^t\mathcal{B}V^{-1}\vec{Y}$ is the estimator with the minimum variance among the unbiased estimators of β that are linear functions of Y_i's.*

Note that $\widehat{\beta}$ is unbiased if $\mathbf{E}(\widehat{\beta}) = \beta$. Here, minimum variance means that if $\widetilde{\beta}$ is another unbiased estimator, the matrix $\mathrm{Var}(\widetilde{\beta}) - \mathrm{Var}(\widehat{\beta})$ is positive definite, [23].

Finally, we consider a case that will be of interest later. Let Σ be a diagonal matrix with diagonal elements $\Sigma_{ii} = \sigma_i^2$ that depend on unknown parameters α. We then write $\Sigma(\alpha)$. The presence of α makes it difficult to calculate the maximum likelihood estimator. The simplest estimator is the ordinary least-squares estimator $({}^t\mathcal{B}\mathcal{B})^{-1} \, {}^t\mathcal{B}\vec{Y}$. It is unbiased, and under certain conditions it is consistent, which means that its variance tends to zero when the vector $\vec{Y}$ has infinite dimensions. These properties make it a good estimator, but we could improve it by trying to estimate α and β simultaneously. Since for any given α an estimate of β can be calculated using (3.11), the idea is to use a two-stage algorithm [32] that estimates α and β in turn as $(\alpha(l), \beta(l))$.

Algorithm 3.1. (Carroll)

- *Step 0*: Calculate the ordinary least-squares estimation $\beta(0)$.

- *Step l*:
  1. Calculate $\alpha(l)$ by minimizing a criterion $C(\alpha, \beta(l-1))$.
  2. Calculate $\Sigma(\alpha(l))$.
  3. According to (3.11), calculate
  $$\beta(l+1) = \left[{}^t\mathcal{B}\Sigma^{-1}(\alpha(l))\mathcal{B}\right]^{-1} {}^t\mathcal{B}\Sigma^{-1}(\alpha(l)) \, \vec{y}^{\,O}.$$

The criterion $C(\alpha, \beta(l-1))$ is based on the approximation errors. We will look at an example of this at the end of the section. Note that this algorithm does not necessarily minimize a joint criterion of (α, β). Carroll et al. [32] have examined the asymptotic behavior of $\beta(l)$ estimates according to the dimension n of the vector $\vec{Y}$, and have established the following properties:

Property 3.2. *The limit* $\lim_{n\to\infty} \mathrm{Var}[\sqrt{n}(\beta(1) - \beta)]$ *is the covariance matrix of the weighted least-squares estimator:* $\mathrm{Var}_{\mathrm{opt}} = ({}^t\mathcal{B}\Sigma^{-1}\mathcal{B})^{-1}$. *The estimator* $\beta(l)$ *becomes independent of l at order 3 in $n^{-1/2}$, for steps $l \geq 3$.*

From the Taylor expansion given by Carroll et al., for moderate n, we note that the variance of the estimator $\beta(l)$ stabilizes after three iterations. This is what we find in practice. For large n, one iteration is enough to obtain an estimator of optimum variance.

- VARIANCE WITH KNOWN ANISOTROPY

The noise in the model given by the independent random variables $W_{tu\nu}$ is said to be separable if there exist positive functions p_t, q_u, and r_ν such that for all (t, u, ν), $\mathrm{Var}(W_{tu\nu}) = p_t q_u r_\nu \sigma^2$. The functions p_t, q_u, and r_ν are weight functions, and we assume that they are known in this case. The approximation model is then

$$Y = \left(\left((\beta, \mathbf{B}^1)_1, \mathbf{B}^2 \right)_2, \mathbf{B}^3 \right)_3 + W,$$
$$\text{with } \mathrm{Var}(W_{tu\nu}) = p_t \, q_u \, r_\nu \, \sigma^2.$$

We write $P = \mathrm{diag}(p_1, \ldots, p_{n_1})$, $Q = \mathrm{diag}(q_1, \ldots, q_{n_2})$ and $R = \mathrm{diag}(r_1, \ldots, r_{n_3})$. Still assuming that the $W_{tu\nu}$ follow an independent Gaussian distribution, the principle of maximum likelihood leads to the weighted least-squares estimation

$$\widehat{\beta} = \arg\min_\beta \sum_{tu\nu} \frac{1}{p_t q_u r_\nu} \left(y_{tu\nu} - \sum_{i,j,k} B_i^1(t) \, B_j^2(u) \, B_k^3(\nu) \, \beta_{i,j,k} \right)^2,$$

$$\tag{3.12}$$

which can also be written as $\widehat{\beta} = \arg\min_\beta \|y - g\|_{PQR}^2$. Its solution is characterized as follows.

Proposition 3.3. *If the noise $W_{tu\nu}$ is separable, i.e., if $\mathrm{Var}(W_{tu\nu}) = p_t q_u r_\nu \sigma^2$, then the weighted least-squares solution is given by*

$$\left(\left((\beta, \,^t\mathbf{B}^1 P^{-1} \mathbf{B}^1)_1, \,^t\mathbf{B}^2 Q^{-1} \mathbf{B}^2 \right)_2, \,^t\mathbf{B}^3 R^{-1} \mathbf{B}^3 \right)_3$$

$$= \left(\left((y, \,^t\mathbf{B}^1 P^{-1})_1, \,^t\mathbf{B}^2 Q^{-1} \right)_2, \,^t\mathbf{B}^3 R^{-1} \right)_3. \quad (3.13)$$

Proof:
As we have seen, when the variance is constant, the least-squares estimate is the solution to the normal equation

$$\left(\left((\beta, \,^t\mathbf{B}^1 \mathbf{B}^1)_1, \,^t\mathbf{B}^2 \mathbf{B}^2 \right)_2, \,^t\mathbf{B}^3 \mathbf{B}^3 \right)_3 = \left(\left((y, \,^t\mathbf{B}^1)_1, \,^t\mathbf{B}^2 \right)_2, \,^t\mathbf{B}^3 \right)_3.$$

The weighted normal equation (3.13) can be obtained from this equation. By "normalizing" the initial model

$$Y = \left(\left((\beta, \mathbf{B}^1)_1, \mathbf{B}^2 \right)_2, \mathbf{B}^3 \right)_3 + W,$$

so that it has constant variance, we obtain

$$\left(\left((Y, P^{-1/2})_1, Q^{-1/2} \right)_2, R^{-1/2} \right)_3$$

$$= \left(\left((\beta, P^{-1/2}\mathbf{B}^1)_1, Q^{-1/2}\mathbf{B}^2 \right)_2, R^{-1/2}\mathbf{B}^3 \right)_3$$

$$+ \left(\left((W, P^{-1/2})_1, Q^{-1/2} \right)_2, R^{-1/2} \right)_3,$$

where $(((W, P^{-1/2})_1, Q^{-1/2})_2, R^{-1/2})_3$ now has constant variance. Consequently, its normal equation is (3.13). □

Note that an estimate of σ^2 can be calculated directly when the weight functions are known. The principle of maximum likelihood gives

$$\widehat{\sigma}^2 = \frac{1}{n_1 n_2 n_3} \|y - \widehat{g}\|_{PQR}^2 \,,$$

where $\widehat{g}$ comes from the solution of (3.13).

• VARIANCE WITH UNKNOWN ANISOTROPY

The situation with unknown weight functions is more complicated. We will illustrate this for the special case where only r_ν is unknown. The variance of the noise is

$$\text{Var}(W_{tu\nu}) = r_\nu \sigma^2 \doteq \sigma_\nu^2.$$

This situation is examined in Section 10.4. Previously, the unknown parameters of the model were (σ^2, β); now they are $(\{\sigma_\nu^2\}, \beta)$. To reduce the number of parameters to be estimated, we assume that the variations of σ_ν^2 are sufficiently regular to be represented by a spline model

$$\log \sigma_\nu^2 = \sum_k \alpha_k B_k(\nu).$$

The basis $\{B_k(\nu)\}$ is a basis of B-splines that are characteristic of this model, but not necessarily identical to the basis $\{B_k^3(\nu)\}$. The new parameters of the model are therefore (α, β). When the r_ν are known, the principle of maximum likelihood estimates β using (3.12), and σ_ν^2 using

$$\widehat{\sigma}_\nu^2 \doteq v_\nu^2 = \frac{1}{n_1 n_2} \sum_{t,u} \left(y_{tu\nu} - \widehat{g}(t, u, \nu)\right)^2,$$

where $\widehat{g}$ comes from $\widehat{\beta}$. Furthermore, for a Gaussian variable $Z \sim LG(\mu, \sigma^2)$, it is well known that the empirical variance of a sample of size m given by $V = \frac{1}{m}\sum_i (Z_i - \widehat{\mu})^2$ is related to the χ^2 distributions as follows: $(mV^2/\sigma^2) \sim \chi^2(m-1)$. This distribution can be approximated by the Gaussian distribution $LG(m, 2m)$ for sufficiently large m. If we assume that this result holds true for V_ν^2, we obtain

$$\frac{m\, V_\nu^2}{\sigma_\nu^2} \approx m + \sqrt{2m}\, \epsilon_\nu,$$

where ϵ follows the distribution $LG(0, 1)$ and $m = n_1 n_2$. We therefore obtain

$$\log V_\nu^2 = \log \sigma_\nu^2 + \log\left(1 + \sqrt{2/m}\,\epsilon\right) \approx \log \sigma_\nu^2 + \sqrt{2/m}\,\epsilon_\nu.$$

When the values of r_ν are unknown, we estimate (α, β) jointly according to Carroll's algorithm. At step l, we calculate

$$v_\nu^2(l) = \frac{1}{m} \sum_{t,u} \left(y_{tu\nu} - \widehat{g}^{(l-1)}(t, u, \nu)\right)^2,$$

where $\widehat{g}^{(l-1)}$ is the estimate of g calculated with $\beta(l-1)$. We then find $\alpha(l)$, the least-squares solution of

$$\log V_\nu^2(l) = \sum_k \alpha_k B_k(\nu) + \sqrt{2/m}\,\epsilon_\nu.$$

The criterion C that appears in Carroll's algorithm is therefore

$$C(\alpha, \beta(l-1)) = \sum_\nu \left[\log v_\nu^2(l) - \sum_k \alpha_k\, B_k(\nu)\right]^2.$$

To summarize, we have dealt with the following two models at the same time:

$$\begin{cases} Y_{t,u,\nu} = g(t,u,\nu) + W_{t,u,\nu}, \\ g(t,u,\nu) = \sum_{i,j,k} B_i^1(t)\, B_j^2(u)\, B_k^3(\nu)\, \beta_{i,j,k}, \end{cases}$$

and

$$\begin{cases} \mathrm{Var}(W_{t,u,\nu}) = \sigma_\nu^2, \\ \log \sigma_\nu^2 = \sum_k \alpha_k\, B_k(\nu). \end{cases}$$

3.3 High-Dimensional Splines

Thus far, the problem has consisted in approximating a vector of observed values y with a spline g parametrized along the orthogonal axes of R^n, and we limited ourselves to $n = 3$. [4] So, in the three-dimensional case, using $x = (x_1, x_2, x_3)'$ to denote the vector of coordinates in R^3 (the vector that was previously written as $(t, u, \nu)'$), the expression for the tricubic spline is

$$g(x) = \sum_{i,j,k} \beta_{ijk} B_{ijk}(x), \tag{3.14}$$

$$\text{with} \quad B_{ijk}(x) = B_i(x_1) B_j(x_2) B_k(x_3).$$

Note that the approximation consists in estimating the β_{ijk} given N pairs $(y^1, x^1), \ldots, (y^N, x^N)$, [5] using the exogenous model

$$Y = \sum_{i,j,k} \beta_{ijk} B_{ijk}(x) + W. \tag{3.15}$$

The points $x^1, \ldots, x^N$ are the nodes of a rectangular grid in Z^3, and the indices $\{1, \ldots, N\}$ correspond to alphabetical order on this grid.

We now look at a situation in which n is no longer limited to $n \leq 3$, but can take on large values. The set of vectors $\{x^1, \ldots, x^N\} = \mathcal{X}$ is considered to be a *sparse* cloud of points in R^n, where the points are not necessarily distributed at the nodes of a rectangular grid. The pairs (y^j, x^j) are seen as the result of a pair of random variables (Y, X) with values in R^{n+1}. We wish to find an approximate relationship between Y and X, or, in other words, to estimate an exogenous model, which in this case is simply a regression of Y on X. As in the model (3.15) where $\mathrm{E}(Y) = \sum_{i,j,k} \beta_{ijk} B_{ijk}(x)$, we

---

[4] Do not confuse the n used in this section with the notation n used in previous sections, which referred to the dimension of the vector y of observed values.

[5] In this section the indices for series of observed values and series of parameters are shown as exponents.

now proceed in a similar way to estimate

$$\mathrm{E}(Y \mid X = x) \doteq f(x).$$

To do this, we need to have a model $g(x)$ representing $f(x)$. This is difficult because of the large dimension n.

3.3.1   Revealing Directions

• THE CURSE OF DIMENSIONALITY

On a rectangular grid, we can in theory immediately extend the model (3.14) for a dimension greater than $n = 3$, but in practice this is difficult to do. The difficulty is caused by the *empty space* phenomenon known as the "curse of dimensionality". This means that as n increases, the number N of points required to fill the space R^n grows exponentially. We can illustrate this with a trivial example: Take the components $X_1, \ldots, X_n$ of the vector X, which are independent and uniformly distributed on $[0, 1]$. In this case, if A is an interval included in $[0, 1]$ and of width 0.1, we have $P(X_i \in A) = 0.1$. However, $P(X \in A^n) = 0.1^n$, which means that to satisfy $P(X \in A^n) = 0.1$ the width of A must be $0.1^{1/n}$, leading to a hypercube with a large volume. A similar example is the spline model (3.14) built with q knots on each of the axes. In this case, to estimate g requires the estimation of $(q + 2)^n$ parameters. Even for small dimensions n, this requires a very large amount of data: $N \gg (q + 2)^n$.

This problem of the scarcity of points x^j in R^n is a sampling problem of the random vector X. If X is distributed according to the above uniform distribution, then N must indeed be very large for the cloud $\mathcal{X}$ to fill R^n sufficiently, and so the situation is hopeless. However, if the theoretical distribution of X is concentrated in a limited region of R^n, then a cloud of moderate size could be enough to reveal the approximate structure of the distribution.

• REVEALING DIRECTIONS AND EXPLORATION

It is impossible to examine the structure of $\mathcal{X}$ visually if the dimension exceeds $n = 3$. Friedman and Tuckey [79] have therefore developed graphical display methods to explore $\mathcal{X}$ and to detect certain special structures, such as nonlinearity or the dispersion of $\mathcal{X}$ into several clouds. Exploration of this kind is based on the search for *revealing directions* in R^n, which are denoted by a and calculated by maximizing an arbitrary criterion $I(a, \mathcal{X})$, called the "index". To use the terminology presented in the introduction (Chapter 1), a can be interpreted as a hidden entity, and any index I can be interpreted as the opposite of an external energy

$$I(a, \mathcal{X}) = -U(\mathcal{X} \mid a).$$

Various indices have been envisaged [111]. We shall look at those built to reveal the possible non-Gaussian nature of the distribution of X, i.e., of the dispersed clouds. An example of this is the index built on Kullback information [124]. If a denotes the direction characteristic vector, then the index is theoretically defined by

$$I_K(a, X) = -\int \log\left(\frac{\varphi(t)}{p(t)}\right) p(t)\, dt, \tag{3.16}$$

where t is the abscissa on the axis given by a, and p is the density of X *projected* on this axis. We define φ as the Gaussian density with the same mean and variance as p. This choice is justified by the fact that $I_K(a, X) \geq 0$, where equality is achieved if and only if p is Gaussian. Using Jensen's inequality [153], equation (3.16) is written

$$I_K(a, X) = \mathrm{E}[-\log(\varphi/p)] \geq -\log \mathrm{E}[\varphi/p] = 0.$$

In practice, p is replaced by an estimate calculated on $\mathcal{X}$. Refer to Silverman [164] for density estimation methods. We now denote by $I_K(a, \mathcal{X})$ the empirical index. The first direction, which we may call $a^{(1)}$, is obtained by maximizing the empirical index

$$a^{(1)} = \arg\max_a I_K(a, \mathcal{X}).$$

For the index I_K, this is equivalent to looking for a such that the distribution of X projected on the axis a is as far as possible from being Gaussian. We then need to know how to pursue the exploration, in other words, how to determine the second revealing direction and then each of the successive ones. The principle is to "remove" from the cloud $\mathcal{X}$ the structure of interest that withheld $a^{(1)}$, and then again to maximize $I_K(a, \mathcal{X})$ on the transformed cloud. Friedman proceeds as follows. Because the index is zero if p is Gaussian, the idea is to transform the cloud so that the projected density is Gaussian on the axis $a^{(1)}$, while preserving the orthogonal directions. Let us write a instead of $a^{(1)}$ to simplify the notation, and let us consider $T = \langle a, X \rangle$, the random variable resulting from the projection of the random vector X on the axis a. Let $F_a(t)$ be its distribution function and Φ be the Gaussian function $LG(0,1)$. The operation

$$\widetilde{T} = \Phi^{-1}(F_a(T))$$

transforms T into $\widetilde{T}$, which follows a Gaussian distribution $LG(0,1)$. Let U be an orthonormal $n \times n$ matrix, whose first row is a. The transformation $R = UX$ applies a rotation such that $R_1 = \langle a, X \rangle = T$. Let $O = \mathrm{diag}(O_1, \ldots, O_n)$ be the transformation of R such that

$$O_1(R_1) = \Phi^{-1}(F_a(R_1)),$$
$$O_i(R_i) = R_i, \quad 2 \leq i \leq n.$$

Consequently, the operation

$$\widetilde{X} = U'OUX \tag{3.17}$$

transforms the projection T into a Gaussian $LG(0,1)$ without affecting the orthogonal directions. To determine the new revealing direction, called $a^{(2)}$, we transform the cloud according to (3.17) and then maximize the index on the transformed cloud:

$$a^{(2)} = \max_a I_K\left(a, \widetilde{X}\right).$$

This exploratory projection pursuit is quite similar to a recent method called "independent components analysis" (ICA). This method looks for a sequence of *orthogonal* projections that are as far from Gaussian as possible [56, 114].

3.3.2   Projection Pursuit Regression

We now return to the problem of approximating Y with X. Instead of performing a regression on the coordinates $x_1, \ldots, x_n$ as in (3.14), we wish to perform it on only a few revealing directions $a^{(1)}, \ldots, a^{(d)}$. This leads to the "projection pursuit regression" model (PPR). For simplicity, we write a^k instead of $a^{(k)}$.

Model 3.1. *The PPR model is the following exogenous model:*

$$Y = g(x) + W,$$

$$\text{with} \quad g(x) = \sum_{k=1}^{d} S^{b^k} P^{a^k}(x), \tag{3.18}$$

$$\text{and} \quad P^{a^k}(x) = \langle a^k, x \rangle,$$

where $P^a : R^n \mapsto R$ *is a projection operator, and* $S^b : R \mapsto R$ *is a one-dimensional spline smoothing operator parametrized in* $t = P^a(x)$. [6]

Because projection pursuit (PP) operates sequentially, so does the algorithm that fits the PPR model (3.18) to the data $(y^1, x^1), \ldots, (y^N, x^N)$. First, we fit the model with $d = 1$, and then, with the first component of $S^{b^1} P^{a^1}(x)$ estimated, we fit the model for $d = 2$, and so forth until an adequate approximation is obtained. Let us use $w^j(k)$ to denote the approximation errors at step k of the algorithm. This type of error is known as a residual. We have

$$w^j(k) = y^j - \sum_{l=1}^{k} S^{b^l} P^{a^l}(x^j),$$

---

[6]Note that $S^b P^a$ is simplified notation for $S^b \circ P^a$.

which leads to the following PP procedure:

Algorithm 3.2. (PPR).

1. *Initialize:*
   $k \leftarrow 0$, $w^j(0) \leftarrow y^j$, $j = 1, \ldots, N$.

2. *Calculate pursuit parameter and spline parameter:*

$$\left(a^{k+1}, b^{k+1}\right) = \arg\min_{a,b} \sum_{j=1}^{N} \left\| w^j(k) - S^b P^a(x^j) \right\|^2. \quad (3.19)$$

3. *Update residuals:* $w^j(k+1) \leftarrow w^j(k) - S^{b^{k+1}} P^{a^{k+1}}(x^j)$,
   $j = 1, \ldots, N$.

4. *Return to 2 if residuals are too large, with $k = k+1$;*
   Otherwise, $d = k$ and end.

The result is the expression of the form (3.18). By looping back up through the algorithm, we obtain

$$\begin{aligned}
w^j(d) &= w^j(d-1) - S^{b^{d-1}} P^{a^{d-1}}(x^j), \\
&= w^j(d-2) - S^{b^{d-2}} P^{a^{d-2}}(x^j) - S^{b^{d-1}} P^{a^{d-1}}(x^j),
\end{aligned}$$

$$\ldots$$

$$= y^j - \sum_{k=1}^{d} S^{b^k} P^{a^k}(x^j).$$

At each step $k+1$, spline smoothing S^b is performed on the values $w^j(k)$, which are ordered according to the projection values $P^a(x^j)$. It can be cubic spline-type smoothing, and if so, we face the problem of choosing the number and position of knots on the projection axis. This problem is presented in detail in Chapter 4. Note that in this algorithm the PP procedure does not use the exploration procedure described above, because in this case the residuals are guiding the pursuit. Formally, at each stage, the PPR algorithm operates with the following index:

$$I_r((a,b), w) = -U(w \mid a, b) = -\sum_{j=1}^{N} \left\| w^j - S^b P^a(x^j) \right\|^2. \quad (3.20)$$

When a is fixed, U is similar to the quadratic error of the least-squares criterion (3.4).

- APPENDIX

Models like (3.18) represent the function $f(x) = E(Y \mid X = x)$ by a sum of one-dimensional functions $g(x) = \sum_{k=1}^{d} f_k(\langle a^k, x \rangle)$ in which the functions f_k are regular nonlinear functions. This type of representation

has been studied by Diaconis and Shashahani [68]. To begin with, we have
the classic result as given in [4].

Theorem 3.3. *Functions on* $\mathbb{R}^n$ *of the form*

$$g(x) = \sum_{k=1}^{\infty} b^{1,k} \cos\left(2\pi\langle a^{1,k}, x\rangle\right) + b^{2,k} \sin\left(2\pi\langle a^{2,k}, x\rangle\right)$$

are dense in $L^2[0,1]^n$.

This means that any function $f(x)$ of $L^2[0,1]^n$ can be approximated by
its Fourier expansion. We also note that the function $f(x_1, x_2) = x_1 x_2$ has
an exact representation but that it is not unique:

$$\begin{aligned}
x_1 x_2 &= \frac{1}{4}(x_1 + x_2)^2 - \frac{1}{4}(x_1 - x_2)^2, \\
&= b(a_1 x_1 + a_2 x_2)^2 - b(a_1 x_1 - a_2 x_2)^2,
\end{aligned}$$

with $a_1 a_2 \neq 0$, $\|a\| = 1$, and $b = 1/(4a_1 a_2)$. If we set $a_1 = \cos\theta$ and
$a_2 = \sin\theta$, we see that all projection directions are possible except for the
orthogonal coordinate axes. It is therefore very difficult to interpret this
type of representation. This applies to polynomial functions only. More
specifically, we say that a function $f(x_1, x_2)$ is nonuniquely representable
if there exist two sets of parameters $\{a^k\}_{k=1}^{d}$ and $\{\tilde{a}^k\}_{k=1}^{\tilde{d}}$ of norm 1 that
are all distinct such that

$$f(x_1, x_2) = \sum_{k=1}^{d} f_k\left(\langle a^k, x\rangle\right) = \sum_{k=1}^{\tilde{d}} \tilde{f}_k\left(\langle \tilde{a}^k, x\rangle\right).$$

This brings us to the following theorem [68]:

Theorem 3.4. *A function* $f(x_1, x_2)$ *is nonuniquely representable if and
only if it is a polynomial.*

This result extends directly to $\mathbb{R}^n$. We might now wonder how this
nonuniqueness affects the PPR algorithm. To examine this question, we
consider the following experiment with an approximately polynomial surface [80]. We have $N = 200$ pairs $x^j = (x_1^j, x_2^j)$ which are simulated from a
pair of independent random variables with Gaussian distribution $LG(0,1)$
and written $X = (X_1, X_2)$. The discrete surface is then $y^j = x_1^j x_2^j + \epsilon^j$,
where ϵ^j is the occurrence of one $LG(0, 0.1)$. The PPR algorithm uses as
index (3.20). When the observations $\{(w^j, t^j = P^a(x^j)), \ j = 1, \ldots, N\}$
are sequenced in ascending order of t, $S^b(t_j)$ denotes the spline smoothing of w^j or other ways of smoothing such as local averaging. The PPR
algorithm essentially selects the two directions $(1,1)$ and $(1,-1)$ to two
decimal places. When this experiment is repeated many times it always
gives the same directions to two decimal places.

In the absence of the noise ϵ^j, this would be an unexpected result in view of the nonuniqueness theorem. An analytical study helps to explain it, as follows. We proceed in nonempirical fashion. The approximation of Y by a function of $\langle a, X \rangle$ is given for any direction $a = (a_1, a_2)$ by

$$\arg \min_c E[(Y - c)^2 \mid \langle a, X \rangle] = E[Y \mid \langle a, X \rangle] = E[X_1 X_2 \mid \langle a, X \rangle].$$

The first direction is a solution of

$$\min_a E\left[(Y - E[Y \mid \langle a, X \rangle])^2\right] = \min_a E\left[(X_1 X_2 - E[X_1 X_2 \mid \langle a, X \rangle])^2\right],$$

where minimization takes place with the constraint $\|a\| = 1$. The solutions are $a_1 = \pm a_2 = \pm 1/\sqrt{2}$, and they are obtained as follows. By setting $U = a_1 X_1 + a_2 X_2$ and $V = a_1 X_1 - a_2 X_2$, which are independent variables with a distribution $LG(0, 1)$, we obtain

$$X_1 X_2 = \frac{1}{4 a_1 a_2} \left(U^2 - V^2\right),$$

$$E[X_1 X_2 \mid \langle a, X \rangle] = \frac{1}{4 a_1 a_2} \left(U^2 - 1\right),$$

and therefore

$$E\left[(X_1 X_2 - E[X_1 X_2 \mid \langle a, X \rangle])^2\right] = \frac{1}{(4 a_1 a_2)^2} E\left[(V^2 - 1)^2\right].$$

The minimum of this expression is obtained for $a_1 = \pm a_2 = \pm 1/\sqrt{2}$.   □

This study has been extended to the case $(X_1 X_2)^2$, with X_1 and X_2 remaining Gaussian. Donoho has shown that the PPR algorithm chose four directions $(1, 1), (1, -1), (1, 0)$, and $(0, 1)$, but that the algorithm did not stop after four iterations even though $(x_1 x_2)^2$ satisfies the expression

$$(x_1 x_2)^2 = \sum_{k=1}^{4} b^k \left(x_k + a^k x_2\right)^4,$$

where the subsequent iterations use the four directions.

This situation is similar to "principal component analysis", which is described in the next chapter: From all possible bases, the algorithm selects the one that gives the best explanation in terms of variance.

4
Auto-Associative Models

The two previous chapters involved modeling a random phenomenon X for which only one observed value was available.[1] The information this value provided was not enough for us to deduce the random behavior of X, and this is why a prior information had to be introduced. This resulted in the use of a model for a smooth curve (image) degraded by random errors. When several observed values of the phenomenon are available, the situation is completely different, because we can then hope to deduce the variations in X without having to consider strong prior information.

4.1   Analysis of Multidimensional Data

In the fields of pattern recognition and artificial vision there are many situations where it is necessary to approximate multidimensional data with a model, as will be shown in Part III (Chapter 13). The problem can be set out in the following terms. Let $\mathcal{X} = \{x^j, \ 1 \leq j \leq N\}$ be a set of N points in R^n: $x^j = (x_i^j)_{1 \leq i \leq n}$.[2] Each x^j is considered to be the result of a random vector X. In practice, an x^j of this kind may designate a discrete curve or even an image. From a geometric point of view, we wish to approximate $\mathcal{X}$ with a manifold of dimension d, $d < n$, built using cubic splines. This

---

[1] We previously wrote X as Y.

[2] With the exceptions of $\|.\|^2$ and R^n, all the exponents in this chapter will in fact designate indices, not powers.

manifold will be given by an implicit equation [137]:

$$x - F(x, \theta) = 0,$$
$$\text{or} \quad G(x, \theta) = 0, \quad \forall x \in \mathsf{R}^n, \tag{4.1}$$

where θ is a parametric vector and $G(\theta, .)$ is a continuously differentiable function of R^n in R^n. The approximation is described by the endogenous model (Section 1.1.4)

$$X - F(X, \theta) = W,$$

where the random vector W represents the approximation error (or prediction error) of X by $F(X, \theta)$. This vector W is commonly known as the residual. The name "auto-associative model" is used because X is predicted by itself. Note that this term also crops up in other contexts. The most classical approach, principal component analysis (PCA) (Section 4.1.1), considers a linear approximation, the equation $G(x, \theta) = 0$, which then defines a subspace of R^n. This model works especially well when X is Gaussian. The purpose of Section 4.2 is to define and construct a model for situations in which a linear approximation is not sufficient.

4.1.1 A Classical Approach

The approximation of x^j seen as discrete curves was first suggested by Rice and Silverman [158] based on a PCA-type approach (see also [152]). This approach was later used with success in the field of imaging [57].

● PCA AS AUTO-ASSOCIATIVE MODEL

We now outline this method [133, 153]. To simplify the notation, we assume $\mathsf{E}(X) = 0$, (the cloud $\mathcal{X}$ is centered on the origin, i.e., $\sum_j x_i^j = 0$ for all i). The model is written as

$$X - F(X, \theta) = W,$$
$$\text{with} \quad F(X, \theta) = \sum_{k=1}^{d} \theta^k \left\langle \theta^k, X \right\rangle, \tag{4.2}$$

where the $\theta^k \in \mathsf{R}^n$ are orthogonal, and such that $\|\theta^k\| = 1$. The function $G(x, \theta) = x - F(x, \theta) = 0$ defines a d-dimensional vector subspace of R^n. The model provides a dimensionality reduction because the vector X in R^n has an associated vector $(\langle \theta^k, X \rangle)_{1 \leq k \leq d}$ in R^d. We therefore need to solve the problem of estimating $\theta = (\theta^1, \ldots, \theta^d)$ to optimize the prediction with respect to an error criterion $U(\mathcal{X} \mid \theta)$, $\hat{\theta}$ that gives the minimum of U.

We begin by defining U in terms of the random vector X. If $\sigma_i^2 = \mathrm{Var}(W_i)$ denotes the variance of the prediction error for X_i, then the quality of the prediction of X is given by the quadratic approximation (or prediction) error

$$U(X \mid \theta) \doteq \sum_{i=1}^{n} \sigma_i^2 = \text{trace} \text{Var}(X - F(X, \theta)),$$

which is the trace of the prediction error covariance matrix. Using this definition, the solution is as follows.

Proposition 4.1. *Let* $V = \mathrm{E}[X^t X]$ *be the covariance matrix of the centered vector* X *and* $U(X \mid \theta) = \text{trace} \text{Var}(X - F(X, \theta))$, *where* $\theta = (\theta_1, \dots, \theta_d)$ *is an orthonormal set. The solution*

$$\widehat{\theta} = \arg \min_{\theta} U(X \mid \theta)$$

consists of eigenvectors of V *associated with its* d *largest eigenvalues.*

The proof for the empirical case, i.e., for $U(\mathcal{X} \mid \theta)$, will be given later. We have $\text{Var}\left({}^t\widehat{\theta}^i X\right) = {}^t \widehat{\theta}^i V \widehat{\theta}^i = \lambda_i$ and $\text{Cov}\left({}^t\widehat{\theta}^i X, {}^t \widehat{\theta}^j X\right) = {}^t \widehat{\theta}^i V \widehat{\theta}^j = 0$ if $i \neq j$. When the X_i are correlated, the decorrelation of the $\langle \widehat{\theta}^j, X \rangle$ shows that the dimensionality reduction was well founded.

We now come to the estimation on $\mathcal{X}$. Consider the estimation of σ_i^2:

$$\widehat{\sigma}_i^2 = \frac{1}{N} \sum_{j=1}^{N} \left(w_i^j\right)^2,$$

which gives the empirical quadratic error

$$U(\mathcal{X} \mid \theta) = \sum_{i=1}^{n} \widehat{\sigma}_i^2 = \frac{1}{N} \sum_{j=1}^{N} \left\| w^j \right\|^2, \tag{4.3}$$

with respect to the observed prediction errors

$$w^j = x^j - F\left(x^j, \theta\right) = x^j - \sum_{k=1}^{d} \theta^k \langle \theta^k, x^j \rangle. \tag{4.4}$$

- **THE PCA TECHNIQUE**

This technique was introduced by Pearson [145]. According to the Pythagorean theorem, minimizing $U(\mathcal{X} \mid \theta)$ is equivalent to maximizing the following in θ:

$$\dot{I}(\theta, \mathcal{X}) = \frac{1}{N} \sum_{j=1}^{N} \left\| F(x^j, \theta) \right\|^2$$

where $\dot{I}(\theta, \mathcal{X})$ is called the *index* (see Section 3.3.1). This is the inertia of the cloud $\mathcal{X}$ projected onto the plane $x - F(x, \theta) = 0$. The solution follows from the theorem below.

Theorem 4.1. *Let H_d be the maximum inertia subspace of dimension d. The maximum inertia subspace H_{d+1} is then the direct sum of H_d and the 1-dimensional subspace that gives the maximum inertia among all the subspaces of dimension 1 that are orthogonal to H_d.*

Proof:
If H_d is the subspace generated by $(\theta^1, \ldots, \theta^d)$, we write $\dot{I}_{H_d} = \dot{I}(\theta, \mathcal{X})$. This theorem uses the property of inertia: $I_{H \oplus G} = \dot{I}_H + \dot{I}_G$.

Let F_{d+1} be any subspace of dimension $d+1$. To examine its inertia, we take a vector $b \in F_{d+1} \cap H_d^\perp$ (note that $\dim(F_{d+1} \cap H_d^\perp) \geq 1$; for example, when $d = 1$, F_2 can be included in $H_1^\perp$). From the breakdown $F_{d+1} = b \oplus G$ we obtain

$$\dot{I}_{F_{d+1}} = \dot{I}_b + \dot{I}_G.$$

But because H_d is the maximum inertia subspace among all d-dimensional subspaces, we have $\dot{I}_G \leq \dot{I}_{H_d}$ and therefore

$$\dot{I}_{F_{d+1}} \leq \dot{I}_b + \dot{I}_{H_d}.$$

The maximum inertia F_{d+1} is therefore obtained for $H_{d+1} = \dot{I}_{\tilde{b}} + \dot{I}_{H_d}$ where $\tilde{b}$ has maximum inertia.   □

To obtain H_d, we can thus proceed from one subspace to the next. We start by finding the maximum inertia 1-dimensional subspace, followed by the 1-dimensional subspace that is orthogonal to the previous one, and so on. The solutions H_d are therefore "nested", and the problem is reduced to finding the maximum inertia axes. If we choose a θ^k of norm 1 for the uniqueness of the representation, the inertia of any axis is written as

$$\dot{I}(\theta^k, \mathcal{X}) = \frac{1}{N} \sum_{j=1}^{N} \|\theta^k \langle \theta^k, x^j \rangle\|^2 \tag{4.5}$$

$$= {}^t\theta^k \left(\frac{1}{N} \sum_{j=1}^{N} x^j \, {}^t x^j \right) \theta^k$$

$$\doteq {}^t\theta^k \, \mathcal{V} \, \theta^k,$$

where $\mathcal{V}$ is the empirical covariance matrix of X calculated on $\mathcal{X}$. The following proposition is identical to Proposition 4.1.

Proposition 4.2. *The subspace H_d is generated by the d eigenvectors of $\mathcal{V}$ associated with the d largest eigenvalues.*

Proof:
Consider the first axis. To solve the inertia maximization problem with the constraint $\|\theta^1\|^2 = 1$, we use the Lagrangian $({}^t\theta^1 \mathcal{V} \theta^1) + \lambda \, ({}^t\theta^1 \theta^1 - 1)$,

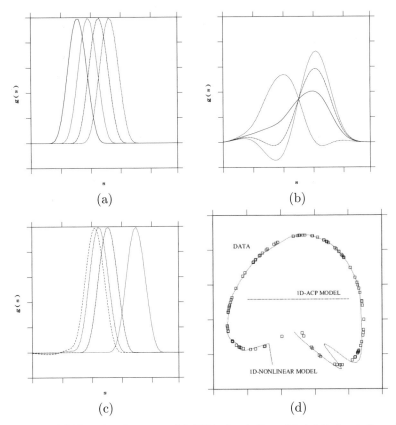

Figure 4.1. (a) Translated curves. (b) PCA simulation. (c) AAC simulation. (d) Projection of χ onto the first principal plane (θ^1, θ^2).

where λ is the Lagrange multiplier. When the gradient of the Lagrangian vanishes, the first axis is the solution of

$$\frac{\mathrm{d}}{\mathrm{d}\theta^1} \left({}^t\theta^1 \mathcal{V}\theta^1 \; + \; \lambda \, {}^t\theta^1\theta^1 \right) = 0,$$

i.e., $\mathcal{V}\theta^1 = \lambda\theta^1$. By substituting this into the inertia, we obtain ${}^t\theta^1\mathcal{V}\theta^1 = \lambda$. This means that maximizing the inertia of the first axis is equivalent to choosing the eigenvector of $\mathcal{V}$ associated with the largest eigenvalue. Furthermore, because the matrix $\mathcal{V}$ is symmetric, it has two-by-two orthogonal eigenvectors. This allows us to draw conclusions for the other axes.   □

To illustrate the limits of PCA, we consider the set of artificial data (Figure 4.1(a)) representing discrete curves in translation. Each curve has an associated vector x^j of its ordinates, seen as the result of a random vector X. For $d = 5$, the subspace H_d given by the PCA of the cloud of $\{x^j\}$ in R^n represents 95% of the inertia of the cloud. The resulting linear model allows us to simulate X by random selection of its components on each of the axes θ_k, $k = 1, \ldots, d$. Random selection on an axis is based on the density of the cloud projected on that axis. Simulation of X generates curves as shown in Figure 4.1(b). Note that this simulation does not reproduce curve deformations due to translations. The linear model given by PCA is therefore not suitable for such simple deformations as those resulting from a translation, because of the nonlinear dimensional cloud. See Figure 4.1(d).

Generalizations of PCA that would take nonlinearities into account have long been sought and suggested. A neural network approach to PCA-type auto-associative models used perceptron networks [15, 27, 119, 139]. The *principal curves* approach [107] is truly nonlinear, but it has the disadvantage of providing only representations of dimension $d \leq 2$ that are nonparametric.

4.1.2   Toward an Alternative Approach

At first glance, it looks like a good idea to generalize the model (4.2) according to an *additive* auto-associative model

$$F(x, \theta) \doteq \sum_{k=1}^{d} \mathsf{S}^{b^k} \mathsf{P}^{a^k}(x) , \qquad (4.6)$$

$$\text{with} \quad \mathsf{P}^{a^k}(x) = \langle a^k, x \rangle,$$

where we now have $\theta^k = (a^k, b^k)$. The projection function P^{a^k} on the axis a^k is also called the compression function. The smoothing function, also known as the restoration function, S^{b^k} is a function of R in R^n. It is continuously differentiable. For PCA, its application is limited to the line whose orientation is a^k: $\mathsf{S}^{b^k}\left(\langle a^k, x \rangle\right) = a^k \langle a^k, x \rangle$. Note that $\mathsf{S}^{b^k}\left(\langle a^k, x \rangle\right)$ is a curve in R^n parametrized in $t = \langle a^k, x \rangle$. We therefore distinguish between the parameter a^k and $t = \mathsf{P}^{a^k}(x)$, whose purpose is to order the x^j. We call a^k the *parametrizing axis*. In this chapter the restoration functions S^{b^k} will be cubic splines. Our problem, then, is to approximate a cloud of unorganized points in R^n with a manifold built using cubic splines. Figure 4.2 illustrates this context for $d = 1$. In Figure 4.2(a) the cloud is Gaussian, so the principal axis of the PCA is suitably positioned. In Figure 4.2(b) the principal axis orders the points in the cloud to make spline smoothing possible, unlike in Figure 4.2(c). Figure 4.2(d) depicts the same cloud as in Figure 4.2(c) but the approximation is better because the chosen axis retrieves the ordering of the points in the cloud.

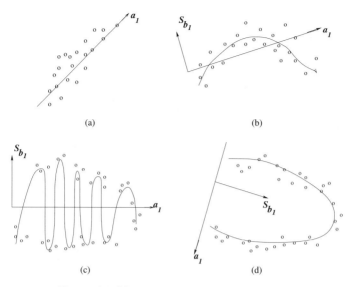

Figure 4.2. Various approximation situations.

• PROJECTION PURSUIT

This additive model can be considered as the endogenous version of the exogenous model (3.18):

$$y^j = \sum_{k=1}^{d} \mathrm{S}^{b^k} \mathrm{P}^{a^k} \left(x^j \right) + w^j(d), \qquad (4.7)$$

which approximates a random variable Y in R with a random vector X in R^n based on a sample $\left\{ (y^j, x^j), \ j = 1, \ldots, N \right\}$ of the pair (Y, X). The idea behind this model is to look for an approximation only in certain directions a^k that reveal structures of interest with respect to an index $I(a, \mathcal{X})$. The selection and optimization of this index constitute one of the main phases in the building of this type of model. In the context of endogenous models (Section 3.3.1), one example is the Kullback index I_K (3.16), which detects non-Gaussian structures. The PCA is another example. Its index $\dot{I}(a, \mathcal{X})$ is the variance of the cloud $\mathcal{X}$ projected on a. In the preceding section, this index was called inertia (4.5).

We recall the algorithm used in the PPR model (4.7). In Step 1, we limit ourselves to the model

$$y^j = \mathrm{S}^{b^1} \mathrm{P}^{a^1} \left(x^j \right) + w^j(1),$$

for which we calculate the estimate $\left(\widehat{a}^1, \widehat{b}^1\right)$ by minimizing the index (3.20) $I_r((a^1, b^1), w)$. In Step 2, we set

$$\widehat{w}^j(1) = y^j - S^{\widehat{b}^1} P^{\widehat{a}^1}(x^j),$$

to consider the exogenous model

$$\widehat{w}^j(1) = S^{b^2} P^{a^2}(x^j) + w^j(2),$$

and then we repeat Step 1 to obtain $\left(\widehat{a}^2, \widehat{b}^2\right)$ and so forth. This algorithm clearly leads to an estimation of the additive model (4.7).

One drawback of this model is that it is difficult to control the value of d and, more specifically, the complexity of the resulting approximation function. In (4.7), because the axes a^k are not necessarily orthogonal, d can be arbitrarily large if we require a very accurate approximation. A PPR model can therefore lead to approximations with extremely complicated expressions, even for a very simple relationship of Y to X (Section 3.3.1). This means that the PPR might not be immune to the curse of dimensionality after all. This is why a constraint on the angles formed by the axes a^k is introduced in [49].

4.2   Auto-Associative Composite Models

We now wish to build a nonlinear auto-associative model with mutually orthogonal axes. This property will help to control d and will allow us to propose an algorithm inspired by PPR that will be a generalization of PCA to cover the nonlinear case. The model (4.7) does not achieve this, but the auto-associative composite model (AAC) does. We now define the AAC with reference to the work of Girard et al. [45, 94, 96].

4.2.1   Model and Algorithm

Model 4.1. *We use the term auto-associative composite model to denote the following endogenous model:*

$$G(X, \theta) = W,$$

$$\text{where} \quad G(x, \theta) \doteq \left(\coprod_{k=d}^{1} G^k(\theta^k, .)\right)(x), \tag{4.8}$$

$$G^k(\theta^k, .) = \mathrm{Id}_{\mathbb{R}^n} - S^{b^k} P^{a^k}.$$

The symbol $\coprod$ denotes the composition product $\coprod_{k=d}^{1} G^k = G^d \circ \cdots \circ G^1$. Applying certain assumptions, we will show that the equation $G(x, \theta) = 0$

defines a d-dimensional manifold. The interpretation of (4.8) in the form of an algorithm is inspired by PPR.

Algorithm 4.1. (AAC algorithm).

1. *Initialize:*
   $k \leftarrow 0,\ w^j(0) \leftarrow x^j,\ j = 1, \dots, N.$

2. *Calculate parameter* $\theta^{k+1} = \left(a^{k+1}, b^{k+1}\right)$:

$$a^{k+1} = \arg\max_a I\left(a, \{w^j(k)\}_{j=1}^N\right), \qquad (4.9)$$

$$b^{k+1} = \arg\min_b \sum_{j=1}^N \left\|w^j(k) - S^b P^{a^{k+1}}\left(w^j(k)\right)\right\|^2. \, (4.10)$$

3. *Update residuals:*
   $w^j(k+1) \leftarrow w^j(k) - S^{b^{k+1}} P^{a^{k+1}}\left(w^j(k)\right),\ j = 1, \dots, N.$

4. *If residuals are too large, return to 2, with $k \leftarrow k + 1$; otherwise, $d \leftarrow k$ and end.*

Note that in dimension $d = 1$ the composite model (4.8) and the additive model (4.6) are equivalent. The algorithm is deduced from the following reasoning: If we note that at each step the PPR algorithm estimates an exogenous model $(r^j(k)$ as a function of $x^j)$, then in auto-associative mode it is appropriate at each step for the new algorithm to estimate an endogenous model $(r^j(k)$ as a function of $r^j(k))$. For any observed value x^j of X, we again designate the quantity $w^j(d) = G(x, \theta^j)$ by prediction error, or "residual". For example, the residual for $d = 2$ is of the form

$$w^j(2) = \left(\mathrm{Id}_{\mathbb{R}^n} - S^{b^2} P^{a^2}\right)\left(\mathrm{Id}_{\mathbb{R}^n} - S^{b^1} P^{a^1}\right)(x^j).$$

This residual is composed of a first residual $w^j(1) = \left(\mathrm{Id}_{\mathbb{R}^n} - S^{b^1} P^{a^1}\right)(x^j)$ obtained in auto-associative mode on x^j, which is then in turn processed in auto-associative mode according to $\left(\mathrm{Id}_{\mathbb{R}^n} - S^{b^2} P^{a^2}\right)(w^j(1))$ to give $w^j(2)$. More generally, we have

$$w^j(1) = \left(\mathrm{Id}_{\mathbb{R}^n} - S^{b^1} P^{a^1}\right)(x^j),$$
$$w^j(2) = \left(\mathrm{Id}_{\mathbb{R}^n} - S^{b^2} P^{a^2}\right)(w^j(1)),$$
$$\dots$$
$$w^j(d) = \left(\mathrm{Id}_{\mathbb{R}^n} - S^{b^d} P^{a^d}\right)(w^j(d-1)).$$

This step-by-step composition of $w^j(d)$ leads to the AAC algorithm where we process the residuals $w^j(k)$, $k = 1, \dots, d$ from one to the next. The definitions of the model and the algorithm will now be refined by adding certain constraints that were left out initially for the sake of simplicity.

• ORTHOGONALITY CONDITIONS

Lemma 4.1. *If we impose the natural condition that the restoration/compression operation must be the identity,*

$$\mathrm{P}^{a^k}\mathrm{S}^{b^k} = \mathrm{Id}_\mathsf{R},$$

then the residuals $w^j(k)$ are orthogonal to the axis a^k:

$$\mathrm{P}^{a^k}(w^j(k)) = 0, \quad \forall j. \tag{4.11}$$

Proof:
This is simple to prove

$$
\begin{aligned}
\mathrm{P}^{a^k}(w^j(k)) &= \mathrm{P}^{a^k}\left[x^j - \mathrm{S}^{b^k}\mathrm{P}^{a^k}(x^j)\right] \\
&= \mathrm{P}^{a^k}(x^j) - \mathrm{P}^{a^k}\mathrm{S}^{b^k}\mathrm{P}^{a^k}(x^j) \\
&= \mathrm{P}^{a^k}(x^j) - \mathrm{Id}_\mathsf{R}\mathrm{P}^{a^k}(x^j) = 0.
\end{aligned}
$$

□

Because of this, the algorithm is very simple to apply. After approximating according to an axis a^k, the residuals $w^j(k)$ are located in the subspace that is orthogonal to that axis. The new axis and the new component of the model will therefore be built in the subspace orthogonal to $a^k, a^{k-1}, \ldots, a^1$, which means that

$$\langle a^k, a^\ell \rangle = 0, \quad 1 \le \ell < k \le d, \tag{4.12}$$

$$\mathrm{P}^{a^\ell}\mathrm{S}^{b^k} = 0, \quad 1 \le \ell < k \le d. \tag{4.13}$$

These conditions must be added to the algorithm, with (4.12) affecting (4.9), and (4.13) affecting (4.10). Finally, we note that this algorithm is implemented incrementally, leading to a nested model. It has this feature in common with the PCA model.

4.2.2   Properties

We look again at the property (4.11) which justifies the definition used for residuals. This definition comes from writing $\mathrm{G}(x, \theta) = 0$ with $w^j(d) = \mathrm{G}(x^j, \theta)$. This way of defining approximation errors is usually ill-conditioned. For example, when $\mathrm{G}(x, \theta) = 0$ defines a conic section, w^j does not correspond to an orthogonal projection on the conic section (except for a circle), and in the extreme case w^j can be infinite for points near the conic section [25]. In our situation, however, with a fixed axis, we avoid this problem thanks to orthogonality (4.11). It is easy to show that this property gives

$$\mathrm{P}^{a^\ell}(w^j(k)) = 0 \quad \text{when} \quad 1 \le \ell \le k \le d. \tag{4.14}$$

This property contributes to the following proposition, which generalizes the properties of PCA:

Proposition 4.3. *AAC models satisfy the following properties:*

- ○ *With d axes, $G(x, \theta) = 0$ defines a d-dimensional manifold.*
- ○ *The approximation errors decrease.*
- ○ *For $d = n$, the model is exact.*

Proof [98]:

To prove the first statement, we place ourselves in the orthonormal basis formed by the vectors a^i, $i = 1, \ldots, n$, in which $x = \sum_i x_i a^i$, where $a^i = (0, \ldots, 0, 1, 0, \ldots, 0)'$, where 1 is in the ith place. In this basis, the model with d axes, which we will call $G^{[d]}$, is written as a function of the model with $d - 1$ axes:

$$G^{[d]}(x) = \left(\mathrm{Id}_{\mathbb{R}^n} - S^{b^d} P^{a^d}\right) G^{[d-1]}(x),$$
$$= G^{[d-1]}(x) - S^{b^d} G_d^{[d-1]}(x).$$

where $G_d^{[d-1]}(x) = P^{a^d}\left(G^{[d-1]}(x)\right)$ is the dth coordinate of the residual $G^{[d-1]}(x)$. Because the residuals $G^{[d]}(x)$ are orthogonal to a^i for $i = 1, \ldots, d$ (see (4.14)), the set of points represented by the model $G^{[d]}$ is the set of zeros common to the following $n - d$ functions:

$$G_i^{[d]}(x) = G_i^{[d-1]}(x) - S_i^{b^d}\left(G_d^{[d-1]}(x)\right), \quad i = d+1, \ldots, n, \quad (4.15)$$

One classic result [137] is that these functions define a submanifold of dimension d if their gradients are linearly independent. We will use a recurrence argument to show that the gradient of $G_i^{[d]}$ can be broken down as follows:

$$\nabla G_i^{[d]} = a^i + U^{[d,i]},$$
$$\text{with } U^{[d,i]} \in H\left(a^1, \ldots, a^d\right), \quad i = d+1, \ldots, n \qquad (\mathcal{H}_d)$$

where $H\left(a^1, \ldots, a^d\right)$ is the subspace generated by $a^1, \ldots, a^d$.

- ⋆ We know that $\mathcal{H}_1$ is true, since for $d = 1$, we have

$$\nabla G_i^{[1]} = a^i - \frac{\mathrm{d}S_i^{b^1}(x_1)}{\mathrm{d}t} a^1, \quad i = 2, \ldots, n,$$

where $\mathrm{d}t$ is actually the same as ∂x_1.

- ⋆ Let us assume that $\mathcal{H}_{d-1}$ is true, and prove $\mathcal{H}_d$. By differentiating equation (4.15), we obtain the following expression for $i = d+1, \ldots, n$:

$$\nabla G_i^{[d]}(x) = \nabla G_i^{[d-1]}(x) - \nabla G_d^{[d-1]}(x) \frac{\mathrm{d}S_i^{b^d}}{\mathrm{d}t}\left(G_d^{[d-1]}(x)\right).$$

By applying $\mathcal{H}_{d-1}$, we obtain $\nabla G_i^{[d-1]} = a^i + U^{[d-1,i]}$ and $\nabla G_d^{[d-1]} = a^d + U^{[d-1,d]}$ with $i = d+1, \ldots, n$. By setting $U^{[d,i]} = U^{[d-1,i]} - (a^d + U^{[d-1,d]}) \mathrm{d}S_i^{b^d}/\mathrm{d}t$, we obtain

$$\nabla G_i^{[d]} = a^i + U^{[d,i]}, \quad i = d+1, \ldots, n,$$

which establishes $\mathcal{H}_d$, since $U^{[d,i]} \in H(a^1, \ldots, a^d)$.

We will now form a linear combination of these vectors and show that it is zero if and only if each of its coefficients is zero. Let there be a family $\{\lambda_i\}_{d+1 \leq i \leq n}$ such that $\sum_{i=d+1}^{n} \lambda_i \nabla G_i^{[d]}(x) = 0$. Using the breakdown given above, we can write

$$\sum_{i=d+1}^{n} \lambda_i a^i + \sum_{i=d+1}^{n} \lambda_i U^{[d,i]} = 0.$$

Because $U^{[d,i]} \in H(a^1, \ldots, a^d)$, we have the following in particular:

$$\sum_{i=d+1}^{n} \lambda_i a^i = 0.$$

The only possible solution is the null vector, because the components a^i are linearly independent. We therefore have a d-dimensional manifold. $\square$

The AAC algorithm makes increasingly accurate approximations: For $d = 1$ the model is a curve, for $d = 2$ the model is a surface, and for higher values of d it is a manifold of dimension d. AAC models therefore have similar properties to the PCA model. We therefore define the fraction of information provided by the model of dimension d as follows:

$$K_d = 1 - \sum_j \|w^j(d)\|^2 / \sum_j \|x^j\|^2. \tag{4.16}$$

From this proposition we deduce that K_d is an increasing series and that $K_n = 1$. The series K_d allows us to choose the dimension of the model according to the fraction of information we wish to represent. Finally, the connection with additive models can be established. In [97] we show that an additive model built with the AAC algorithm is linear: $F(a, x) = \sum_{k=1}^{d} a^k P^{a^k}(x)$.

4.3 Projection Pursuit and Spline Smoothing

To be able to apply the AAC algorithm, we need to define the projection index $I(a^k, X)$, the smoothing functions S^{b^k}, and the model adjustment criteria other than the index. The essential points are:

(i) to define the projection index;

(ii) to control the amount of spline smoothing.

4.3.1 Projection Index

The aim is to find an axis selection criterion whose optimum value will give good ordering of the cloud of points x^j when such ordering exists. An example of good ordering is shown in Figure 4.2(d), whereas in Figure 4.2(c) the PCA axis turns out to be inadequate. For simplicity, we start at the first iteration of the algorithm and use a to denote a^1. A family of interesting indices has been suggested by Shepard and Carroll [161]. The projection function P^a is nonlinear and seeks approximately to preserve the distances between points: $\left\| x^j - x^i \right\| \simeq \left| P^a(x^j) - P^a(x^i) \right|$. This is what the PCA does in linear projection. To obtain a more suitable criterion, let us remember that conservation of distances is not important in and of itself; the important thing is to conserve the neighborhood structure. Points x^j that are neighbors in $\mathbb{R}^n$ must have neighboring projections in $\mathbb{R}$ and vice versa. This holds true in Figure 4.2(d) but not in Figure 4.2(c), where two neighboring points in $\mathbb{R}$ are not necessarily neighbors in $\mathbb{R}^n$. To simplify the criterion, we can abandon the requirement to preserve the entire neighborhood structure and simply keep the concept of *nearest neighbor*. In practice, it is not normally possible to satisfy conservation of neighborhoods for all points x^i. The chosen index is therefore naturally the number of points for which the constraints are satisfied:

$$I(a, \mathcal{X}) = \sum_{i=1}^{N} \sum_{j \neq i} \mathbf{1} \left[x^j \text{ nearest neighbor of } x^i \implies \right.$$

$$\left. P^a(x^j) \text{ nearest neighbor of } P^a(x^i) \right]. \quad (4.17)$$

If we define ϕ, the matching function of $\{1, \ldots, N\}$ in $\{1, \ldots, N\}$ by

$$\phi(i) = \arg \min_{j \neq i} \left\| x^i - x^j \right\|,$$

then the nearest neighbor of x^i is $x^{\phi(i)}$. We therefore wish to impose the condition that $P^a \left(x^{\phi(i)} \right)$ must be the nearest neighbor of $P^a(x^i)$ in $\mathbb{R}$. Consequently, a must satisfy $N - 1$ inequalities:

$$I(a, \mathcal{X}) = \sum_{i=1}^{N} \prod_{j \neq i} \mathbf{1} \left[\left| P^a(x^i) - P^a(x^{\phi(i)}) \right| \leq \left| P^a(x^i) - P^a(x^j) \right| \right]$$

$$= \sum_{i=1}^{N} \prod_{j \neq i} \mathbf{1} \left[\left| P^a(\bar{x}^{i\phi(i)}) \right| \leq \left| P^a(\bar{x}^{ij}) \right| \right],$$

where $\bar{x}^{ij} = x^i - x^j$. Using the fact that a and $-a$ define the same axis, we can write the index as follows:

Definition 4.1. *The index characterizing the parametrizing axes in terms of conservation of neighborhood structure with respect to the nearest neighbors is*

$$I(a, \mathcal{X}) = \sum_{i=1}^{N} \prod_{j \neq i} \mathbf{1}\left[\mathrm{P}^a\left(\bar{x}^{i\phi(i)} - \bar{x}^{ij} \right) \geq 0 \right] \times \mathbf{1}\left[\mathrm{P}^a\left(\bar{x}^{i\phi(i)} + \bar{x}^{ij} \right) \leq 0 \right].$$

Property 4.1. *The index satisfies the following invariance properties:*

- $I(a, \mathcal{X}) = I(a, \mathcal{X} + u)$, $u \in \mathsf{R}^n$;

- $I(a, s\mathcal{X}) = \begin{vmatrix} I(a, \mathcal{X}) & \text{if} & s > 0, \\ I(-a, \mathcal{X}) & \text{if} & s < 0; \end{vmatrix}$

- $I(Aa, A\mathcal{X}) = I(a, \mathcal{X})$ *with* $A'A = \mathrm{Id}$.

Here we write $\mathcal{O}\mathcal{X}$ to denote $\{\mathcal{O}(x^j)\}$. The first two properties, invariance of the axis in translation and in homothetic transformation, make this a Class III index as defined by Hubert [111], a type particularly suited to PP algorithms. The last property shows that the axis does not depend on the orientation of the cloud (rotation and reflection invariance).

- ● OPTIMIZING THE INDEX

Our aim is to maximize $I(a, \mathcal{X})$ in a. Note that $\mathrm{P}^a(x) = 0$, also written as $\langle a, x \rangle = 0$, defines a hyperplane in a orthogonal to x. With $I(a, \mathcal{X})$ we therefore have $N(N-1)$ hyperplanes, denoted by H_k:

$$H_k = \{a \in \mathsf{R}^n, \langle a, n_k \rangle = 0\}, \quad 1 \leq k \leq N(N-1),$$

$$n_k \in \left\{ \left(\bar{x}^{i\phi(i)} - \bar{x}^{ij} \right), \left(\bar{x}^{i\phi(i)} + \bar{x}^{ij} \right); i = 1, \ldots, N; j \neq i \right\}.$$

These $N(N-1)$ hyperplanes create a partition of R^n into a finite number of regions on which the index is constant. For a fixed region, however, we do not know how to build an axis inside the region. The basic principle of maximizing $I(a, \mathcal{X})$ is to perform a series of random visits of the regions to try to increase the value of the index at each visit. Each of these visits is defined by a hyperplane H_k obtained by random selection from the indices k. If $a(p-1)$ is the value of the axis at the penultimate visit, then the potential value of the new axis $a(p)$ is obtained by the symmetry Ξ_k that is orthogonal with respect to H_k:

$$\Xi_k(a) = a - 2\langle a, n_k \rangle n_k. \tag{4.18}$$

The algorithm for this procedure is as follows.

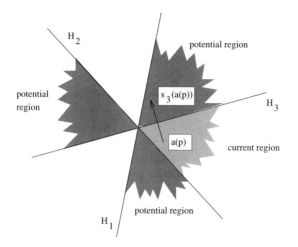

Figure 4.3. Diagram showing travel through regions.

Algorithm 4.2. (Search for Parametrizing Axis).

  1. *Initialize: Choose $a(0)$ randomly with norm 1.*

  2. *With $a(p)$ given, calculate $a(p+1)$ as follows:*
      - *Select k at random between 1 and $N(N-1)$.*
      - *Compare the regions separated by H_k:*
          $$a' \leftarrow \Xi_k(a(p)).$$
          If $I(a', \mathcal{X}) > I(a(p), \mathcal{X})$,
          then
              $$a(p+1) \leftarrow a'.$$
          Otherwise ,
              $$a(p+1) \leftarrow a(p).$$

  3. *Return to 2 until convergence.*

The cost of an iteration is the cost of calculating the index $I(a', \mathcal{X})$, i.e., $O\left(nN^2\right)$ (the differences $\bar{x}^{ij}$ are precalculated during initialization). Because the indices obtained form an increasing series, the algorithm converges to a local maximum. At that stage, all regions that are accessible via the symmetry operation Ξ_k (regions shown in dark gray in Figure 4.3) have an index value that is lower than the maximum (region shown in light gray in Figure 4.3). It is not guaranteed, however, that the value of the indices in the other regions (shown in white in Figure 4.3) is not higher. Because it is impossible to reach them directly by symmetry Ξ_k, one way to get them is to go through a region with a lower index (in our illustration a dark gray region) and to do so with a certain probability. This is the basic idea behind the *simulated annealing* algorithm, which is described in Section 7.3.1 and which we now consider.

Unlike the deterministic algorithm, which imposes $\Delta I \doteq [I(a(p+1), \mathcal{X}) - I(a(p), \mathcal{X})] > 0$ on the current iteration p, the stochastic algorithm allows a negative ΔI: $\Delta I > T_p \log \xi$, where $\xi \in]0, 1[$ and T_p is a series decreasing toward 0:

$$T_p = T_0 \lambda^p,$$

with $\lambda < 1$. As p increases, $T_p \log \xi \to 0$, and it becomes more and more unlikely that a region with a lower index than the previous region will be accepted.

Algorithm 4.3. (Stochastic Search for Parametrizing Axis).

1. *Initialize: Axis $a(0)$ is chosen randomly with norm 1.*

2. *With $a(p)$ given, calculate $a(p+1)$ as follows:*
   - *Select k at random between 1 and $N(N-1)$.*
   - *Compare the regions separated by H_k:*
     $a' \leftarrow \Xi_k(a(p))$.
     $\Delta I \leftarrow I(a', \mathcal{X}) - I(a(p), \mathcal{X})$.
     Take ξ according to a uniform density on $[0, 1]$.
     If $\Delta I > T_p \log \xi$,
     then
     $$a(p+1) \leftarrow a'.$$
     Otherwise,
     $$a(p+1) \leftarrow a(p).$$

3. $T_{p+1} \leftarrow \lambda T_p$.

4. *Return to 2 until convergence.*

4.3.2 Spline Smoothing

We now present the first iteration of the algorithm, for which (a^1, b^1) is written as (a, b). Note that here the axis a is known via its estimate $\hat{a}$, which will be written as a for simplicity. This is a classic situation: a smoothing problem in the form of a spline regression model, for which the number of knots needs to be determined.

We use $t^j = \mathrm{P}^a(x^j)$ to denote the projections of x^j. We must choose a type of smoothing function S^b and then estimate it while minimizing the quadratic error

$$U(\mathcal{X} \mid b) = \frac{1}{N} \sum_{j=1}^{N} \left\| w^j \right\|^2 = \frac{1}{N} \sum_{j=1}^{N} \left\| x^j - S^b(t^j) \right\|^2,$$

$$= \frac{1}{N} \sum_{i=1}^{n} \left[\sum_{j=1}^{N} \left(x_i^j - S_i^{b_i}(t^j) \right)^2 \right] = \sum_{i=1}^{n} U_i(\mathcal{X} \mid b_i). \quad (4.19)$$

Here $b = (b_1, \ldots, b_n)$, where b_i is the vector of parameters of the ith coordinate $S_i^{b_i}$ of S^b. The quadratic error $U(\mathcal{X} \mid b)$ is broken down into n independent quadratic errors $U_i(\mathcal{X} \mid b_i)$. We approximate the ith coordinate $\{x_i^1, \ldots, x_i^N\}$ of the cloud $\mathcal{X}$ according to the parametrization $\{t^1, \ldots, t^N\}$ with a cubic regression spline $S_i^{b_i}(t)$, $t \in [t_{\min}, t_{\max}]$. More specifically, this means that the parameters t^j must be placed in ascending order. The ordered series is written as $t^{\sigma(j)}$, and there is an associated series $x^{\sigma(j)}$. Smoothing of the ith coordinate is performed on the pairs $\left\{ \left(t^{\sigma(j)}, x_i^{\sigma(j)} \right), j = 1, \ldots, N \right\}$. For this type of representation we must choose discontinuity knots among the $t^{\sigma(j)}$ (Chapter 3). The number of knots defines the extent of smoothing. This choice is the same for the n coordinates. Let $\mathbf{B}$ be the matrix of B-splines for the chosen knots. Each of the n quadratic errors is then written as

$$\sum_{j=1}^{N} \left(x_i^{\sigma(j)} - S_i^{b_i}(t^{\sigma(j)}) \right)^2 = \| x_i - \mathbf{B} b_i \|_N^2, \quad 1 \leq i \leq n, \quad (4.20)$$

where $\|.\|_N$ is the Euclidean norm in $\mathbf{R}^N$. We write $\widehat{b}_i$ for the least-squares estimator obtained from (4.20). Let ν be the number of knots in the spline. We should emphasize that $\widehat{b}_i$ is calculated as soon as ν is fixed. This estimate is therefore a function of ν, written $\widehat{b}_i(\nu)$. We will apply a second criterion to determine an optimum number of knots.

• CHOOSING THE NUMBER OF KNOTS

The number of knots ν acts as the regularization parameter α of the energy (2.3) (see also the Note in Section 3.1.1). The bias/variance dilemma (Section 2.4) therefore comes into play. When ν is too large, the bias is low and the model closely approximates the data on which it was estimated. The variance, however, is high, and the estimated model is likely to give a very poor fit for other sets of data. In this case, we say that the model does not "generalize". With the quadratic prediction error criterion, we control the quality of the approximation at each point t^j and only at these points. At other locations, the simulation of X by this model can give very unrealistic results. The main reason for this is the curse of dimensionality.

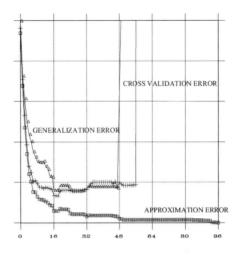

CROSS VALIDATION ERROR

GENERALIZATION ERROR

APPROXIMATION ERROR

Figure 4.4. Choice of number of knots.

We therefore need a criterion that takes the behavior of the restoration between the points t^j into consideration.

Definition 4.2. *Let $T = \langle a, X \rangle$ be the variable X projected onto a and let $f(t)$ be its probability density on $\mathbb{R}$. The theoretical generalization error is*

$$\mathcal{G}^{\text{theo}}(\nu) = \mathbf{E}\left[Q^2\left(X, \mathbf{S}^{\widehat{b}(\nu)}(T)\right)\right],$$

where Q^2 is the square of a distance and $\mathbf{E}$ is the mathematical expectation with respect to $f(t)$.

For practical purposes, we define an empirical version of $\mathcal{G}^{\text{theo}}$, written as $\mathcal{G}^{\text{emp}}$. Let $\widehat{f}(t)$ be an estimate of the density $f(t)$ (refer to [164]). To define the empirical generalization error, we simulate occurrences of T with density $\widehat{f}$. Let $\breve{t}$ be one of these occurrences, and $\phi(\breve{t})$ the t^j located nearest to $\breve{t}$ on the axis a: $\phi(\breve{t}) = \arg\min_j\left(\breve{t} - t^j\right)^2$. The empirical generalization error is then defined based on M occurrences of T, denoted by $\left(\breve{t}^1, \ldots, \breve{t}^M\right)$ as follows:

$$\mathcal{G}^{\text{emp}}(\nu) = \frac{1}{M}\sum_{k=1}^{M}\left\|x^{\phi\left(\breve{t}^k\right)} - \mathbf{S}^{\widehat{b}(\nu)}\left(\breve{t}^k\right)\right\|^2.$$

The resulting $\mathcal{G}^{\text{emp}}$ is also a quadratic error, but it represents the difference between the simulations $\mathbf{S}^{\widehat{b}(\nu)}\left(\breve{t}^k\right)$ of X and the data $\mathcal{X}$. We can also see how X is simulated via the random variable T. Finally, we choose the number ν of knots that minimizes this criterion. Note that for fixed ν, the

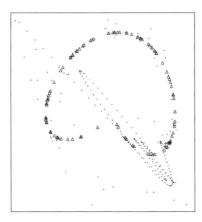

Figure 4.5. Simulation with an incorrect number of knots.

knots are placed such that all intervals defined by two consecutive knots have the same probability as given by $\widehat{f}(t)$.

4.4 Illustration

EXAMPLE 1.
We look again at the example shown on page 79. Figure 4.1(a) shows part of a set of curves obtained from the analytical expression for a curve in translation. We write this set as $\{g^j(s), j = 1, \ldots, N\}$, where $N = 100$ and j indicates a translation. These curves are all sampled on a *common* interval. This gives a vector $x^j = \left(g^j(s_1), \ldots, g^j(s_n)\right)$ for each curve, with $n = 50$. These points constitute a cloud $\mathcal{X}$ in R^n. They are distributed over a manifold of dimension $d = 1$. We can picture how this manifold would look by projecting the cloud on the principal plane of the PCA (Figure 4.1(d)). By principal component analysis, the representation of dimension $d = 1$ is given by a straight line in R^n. This representation is inadequate. For a good approximation, the first five axes that define a subspace representing 95% of the variance of the cloud must be selected. In this case, however, the linear representation has dimension $d = 5$, whereas X is in a manifold of dimension $d = 1$. The simulations therefore supply points of R^n that are not very representative of the set of examples (Figure 4.1(b)). This demonstrates that the PCA has very poor generalization capability.

We now consider AAC nonlinear modeling. The parametrizing axis was obtained by the simulated annealing version of the index optimization algorithm. This axis conserves the neighborhood around 93% of the points in the cloud. The generalization and cross validation errors have similar behaviors (Figure 4.4). With the optimum value $\nu = 18$, the representation of the cloud $\mathcal{X}$ is satisfactory, as shown in Figure 4.1(d). This figure shows the first factorial plane on which $M = 8000$ points $S^b(\check{t}^k)$ of R^n located on the manifold representing $\mathcal{X}$ were projected. These M projections constitute the "nonlinear model" curve, which looks continuous because of the large number of points. The corresponding simulated curves (Figure 4.1(c)) are consistent with the $\mathcal{X}$ curves (Figure 4.1(a)). Note that a poor choice of number of knots ($\nu \gg 18$) results in an incorrect representation because of the poor generalization behavior of the model (Figure 4.5).

EXAMPLE 2.
The set $\mathcal{X}$ comprises $N = 400$ images of the faces of 40 persons (10 images per person).[3] Figure 4.6(a) shows three different images of one of these individuals. In this case, the quality of the approximation is measured by the approximation error. To approximate $\mathcal{X}$ with a 20% error, PCA requires $d = 210$, and the AAC model requires $d = 89$ for $\nu = 2$ and $d = 65$ for $\nu = 3$. The PCA axes and the AAC axes are very close in this experiment. We are therefore in the situation depicted in Figure 4.2(b). Figures 4.6(b) and (c) show the approximation by PCA and AAC for $d = 89$. We have a 35% error for the PCA and a 20% error for the AAC. We can see that the AAC model preserves much more detail than the PCA model, which gives blurred faces.

● COMMENTS

Let us summarize the most important points about this method. Firstly, the projection pursuit axes are found by optimizing an index that favors the neighborhood structure of the cloud. When the optimized index is large, the resulting smoothing is valid because the axis gives a natural ordering of the points. Secondly, we have the ability to control the dimension of the manifold via the information fraction. Finally, we can determine the smoothing factor using a generalization criterion. These are automatic operations that require no manual adjustment of the parameters. When the index and the information fraction are close to their maximum values, the model is considered to be valid.

Note that modeling involves estimating many parameters: the axis parameters and the spline parameters. This does not mean, however, that excessive parametrizing is occurring. In fact, a dimensionality reduction is achieved. This becomes obvious when the method is used in compression. Once the model has been estimated on $\mathcal{X}$, the spline parameters are conserved once and for all. The compression of *any* new observed value x then

---

[3]Olivetti and Oracle Research Laboratory
http://www.cam-orl.co.uk/face database.html.

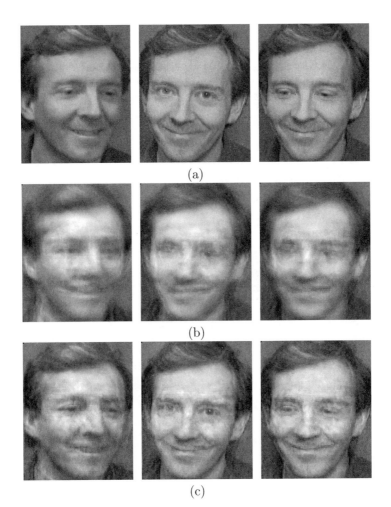

Figure 4.6. PCA and AAC approximation of faces. (a) Faces taken from the database. (b) PCA approximation. (c) AAC approximation.

requires only the transmission of d scalars, i.e., the projections of x on the d axes. Restoration is then given by the point of the manifold relative to the d scalars. This occurs thanks to the generalization properties of the model. The resulting compression ratio is very high.

Part II

Markov Models

5
Fundamental Aspects

Markov models are used to quantify the spatial interactions of observed values at the nodes of a grid S and give a probability for any configuration x on S. In this chapter we review the classic presentation of Markov random fields with the assumption that the observed values at any site $s \in S$ are discrete. This is true in most actual situations where sensors deliver digital images that are usually 8-bit encoded. This assumption simplifies the mathematics without limiting the applicability of these models.

5.1 Definitions

A stochastic interpretation of the models in Chapter 2 is found via the approximation error of the spline g with respect to the data. The spline is deterministic by nature, although a Bayesian interpretation of it was given (Section 2.3). With our new objective in mind, the important point from the previous chapter is the discrete energy model (2.7):

$$U(g \mid y) = \alpha U_1(g) + U_2(y \mid g) \tag{5.1}$$
$$= \alpha \sum_i \big(2g(t_i) - g(t_{i-1}) - g(t_{i+1})\big)^2 + \sum_i \big(y_i - g(t_i)\big)^2,$$

where U_1 quantifies the local variations of the discretized curve g. Each local variation $\big(2g(t_i) - g(t_{i-1}) - g(t_{i+1})\big)^2$ involves only a small number of *neighboring sites* $\{t_{i-1}, t_i, t_{i+1}\}$. As we will see, U_1 can be considered as the result of a specific distribution that is formalized as the Markov approach.

This approach focuses on the families of random variables $\{X_s\}_{s \in S}$ indexed by a *finite* set of sites S. Typically, $S \subset \mathbf{Z}^d$ with $d > 1$. These families are therefore known as *random fields*. The results will also apply to the special case $d = 1$. When $d > 1$, the complete lack of order on S causes difficulties.

Let us then consider X, a random field on S. Each marginal distribution X_s takes its values x_s from a discrete and finite set of states E_s. One representation of X is therefore a field $x = \{x_s; \ s \in S, \ x_s \in E_s\}$. We use $E = \prod_s E_s$ to denote the set of configurations of the field on S. Note that the sets E_s are not necessarily the same from one site to the next. This generalization will prove to be an important practical advantage. When E_s is independent of s, x is an image in the normal sense, provided that the configurations are visually comprehensible. We assume that the discrete probability distribution of X is such that any representation x of X has nonzero probability: $P[X = x] > 0, \ \forall x \in E$. We will use the notation $P(x)$ instead of $P[X = x]$.

5.1.1 Finite Markov Fields

The concept of Markov fields was first introduced by Levy [128]. The classic formalization goes back to Besag [21], but the interest of these fields was brought to light thanks to the work of Geman and Geman [84]. We now give a series of definitions leading to the definition of Markov fields.

Definition 5.1. System of neighborhoods and cliques.

   ○ *Every site $s \in S$ has an associated set $N_s \subset S$ of neighbors such that:*
      (i) $s \notin N_s$,
      (ii) $t \in N_s \ \Leftrightarrow \ s \in N_t$.
      $\mathcal{N} = \{N_s, \ s \in S\}$ *is called a neighborhood system.*

   ○ *A subset c of S is a clique if all its elements are mutual neighbors:*
      $s, t \ \in c \Leftrightarrow \ t \in N_s$, or if c is a singleton: $c = \{s\}$. We write C to denote the set of cliques associated with the neighborhood system $\mathcal{N}$.

Note that the pair $(S, \mathcal{N})$ constitutes a symmetric graph with the sites S as its vertices and the pairs (s, t) with $t \in N_s$ as its edges.

Definition 5.2. Local specifications.
The local conditional probabilities associated with the neighborhood system

$$P[X_s = x_s \mid \{X_t = x_t, \ t \in N_s\}] \ \dot{=} \ P[x_s \mid \{x_t, \ t \in N_s\}]$$

are called the local specifications.

Notation The following notation will be used from now on:

$$\breve{x}_s \doteq \{x_t, \ t \neq s\},$$
$$N_s(x) \doteq \{x_t, \ t \in N_s\}.$$

Definition 5.3. Markov field.
A random field X is a Markov field if

$$P[x_s \mid \{x_t, \ t \neq s\}] \ = \ P[x_s \mid \{x_t, \ t \in N_s\}].$$

5.1.2   Gibbs Fields

We now define a second type of random field, characterized by the analytical expression of the probability distribution. This type of random field is well known in statistical physics.

Definition 5.4. Potential and neighborhood potential.

o *Let there be a family V of functions concerning the restriction of x to subsets of S: $\{V_A(x), \ A \subset S\}$, where V_A applies to $\{x_s, s \in A\}$ only. The following function is called the energy of potential V:*

$$U(x) \ = \ \sum_{A \subset S} V_A(x).$$

o *When the subsets A are limited to the cliques, the energy is written as*

$$U(x) \ = \ \sum_{c \in C} V_c(x),$$

where V_c is called the neighborhood potential.

So for example, in the energy U_1 contained in (5.1), the local variation $(2g(t_i) - g(t_{i-1}) - g(t_{i+1}))^2$ is the potential of the clique $c = \{t_i, t_{i-1}, t_{i+1}\}$.

For all s, let us choose a reference value in E_s, to be written as 0_s. We emphasize that 0_s is merely the notation, and does not necessarily signify the value 0, (according to the encoding, E_s may not contain 0). The potential V is said to be canonical if $V_A(x) = 0$ when A contains a site s such that $x_s = 0_s$. This condition lets us have a unique potential representation. It does not have much practical utility, and tends to lead to unintelligible expressions (see Section 5.3.1).

Definition 5.5. Gibbs distribution.
A random field X is a Gibbs field of energy U if it is governed by the following probability distribution called a Gibbs distribution:

$$P[X = x] \ = \ \frac{1}{Z} \exp -U(x),$$
$$Z \ = \ \sum_x \exp -U(x).$$

In this definition, $\mathcal{Z}$ is the normalization constant (normalizing to 1) that makes P a probability. Note that $\mathcal{Z}$ is almost never accessible, either analytically or numerically, because of the huge size of E; for example $|E| = 256^{512^2}$ for a 512×512 image and grayscale encoding on 256 values. This distribution relates to statistical physics for the following reason.

Property 5.1. *On all probability distributions P such that $\mathrm{E}(U)$ is constant,*

$$\sum_x U(x)P(x) = a,$$

the Gibbs distribution

$$\Pi_T(x) = \frac{1}{\mathcal{Z}_T} \exp -\frac{1}{T} U(x), \quad \text{where} \quad \mathcal{Z}_T = \sum_x \exp -\frac{1}{T} U(x),$$

is the "most disordered" distribution, i.e., the one that maximizes the entropy

$$\hbar(P) = -\sum_x P(x) \log P(x).$$

In statistical physics, $\mathcal{Z}_T$ is called a partition function, and T is the temperature.

Proof:
From the expression for the Gibbs distribution $\Pi_T(x)$ we obtain $-U(x) = T[\log \mathcal{Z}_T + \log \Pi_T(x)]$. With this new expression, the expectation of U is written as

$$\mathrm{E}(U) = \sum_x U(x)P(x) = -T \log \mathcal{Z}_T - T \sum_x P(x) \log \Pi_T(x).$$

With the constraint $\mathrm{E}(U) = a$, we therefore obtain

$$-\sum_x P(x) \log \Pi_T(x) = \log \mathcal{Z}_T + \frac{a}{T}. \tag{5.2}$$

We can now use a fundamental lemma that is well known in information theory:

$$\hbar(P) = -\sum_x P(x) \log P(x) \leq -\sum_x P(x) \log Q(x),$$

which is true for any distribution Q, and where the inequality is strict for $P \neq Q$. (The proof of this lemma is a direct consequence of Jensen's inequality: $\mathrm{E}(\log X) \leq \log \mathrm{E}(X)$ applied to $X = Q/P$.) By setting $Q = \Pi_T$, we obtain from this lemma and the inequality (5.2) an upper threshold for the entropy: $\hbar(P) \leq \log \mathcal{Z}_T + \frac{a}{T}$, which is reached for $P = \Pi_T$. $\square$

5.2   Markov–Gibbs Equivalence

We will now establish the equivalence between Markov fields and Gibbs fields. The basic tool for this proof is the following lemma.

Lemma 5.1. (Möbius inversion formula).
Let Λ be a finite set. Let ϕ and ψ be functions with real values defined on the subsets of Λ. We then have the following equivalence:

$$\phi(A) = \sum_{B \subset A} (-1)^{|A-B|} \psi(B), \quad \forall A \subset \Lambda, \qquad (5.3)$$

$$\Leftrightarrow \ \psi(A) = \sum_{B \subset A} \phi(B), \quad \forall A \subset \Lambda, \qquad (5.4)$$

where $|A - B|$ denotes the cardinality of $A - B$.

Proof:
We will prove that $\psi(A) = \sum_{B \subset A} \phi(B)$ from the first equation. The same type of proof can be used for the reciprocal case. We have

$$\sum_{B \subset A} \phi(B) = \sum_{B \subset A} \left[\sum_{D \subset B} (-1)^{|B-D|} \psi(D) \right],$$

$$= \sum_{D \subset A} \psi(D) \left[\sum_{E \subset A-D} (-1)^{|E|} \right],$$

$$= \psi(A),$$

because according to the following lemma, we have $\left[\sum (-1)^{|E|} \right] \neq 0$ only for $A - D = \emptyset$.
Lemma:

$$\sum_{B \subset A} (-1)^{|A-B|} = \begin{cases} 0, & \text{if } A \neq \emptyset, \\ 1, & \text{otherwise.} \end{cases}$$

□

We will establish that any random field is a Gibbs field.

Theorem 5.1. *If X is a random field such that $P(x) > 0$, $\forall x \in E$, then it is a Gibbs field whose potential is given by*

$$V_A(x) = - \sum_{B \subset A} (-1)^{|A-B|} \log P(x^B), \qquad (5.5)$$

where $x_s^B = x_s$ if $s \in B$ and 0 otherwise.

Proof:
We must prove that $P(x) = \frac{1}{Z} \exp{-U(x)}$ with V given by (5.5). Using

the above lemma for $A \neq \emptyset$, we can rewrite V_A as:

$$V_A(x) = -\sum_{B \subset A} (-1)^{|A-B|} \log P(x^B) + \log P(\mathbf{0}) \sum_{B \subset A} (-1)^{|A-B|}$$

because the added term is zero. Here, $\mathbf{0}$ denotes the reference field $\{0_s, \ s \in S\}$. We therefore have

$$V_A(x) = -\sum_{B \subset A} (-1)^{|A-B|} \log \frac{P(x^B)}{P(\mathbf{0})},$$

which gives the following expression if we set $\phi(A) = V_A(x)$ and $\psi(B) = -\log \left(P(x^B)/P(\mathbf{0}) \right)$:

$$\phi(A) = \sum_{B \subset A} (-1)^{|A-B|} \psi(B).$$

This is one of the expressions of the Möbius formula. After inversion, this gives $\psi(B) = \sum_{A \subset B} \phi(A)$. For the case where $B = S$, we have $\psi(S) = \sum_{A \subset S} \phi(A)$ and $\psi(S) = -\log(P(x)/P(\mathbf{0}))$. Finally, we obtain

$$-\log \frac{P(x)}{P(\mathbf{0})} = \sum_{A \subset S} V_A(x),$$

which gives the desired expression with $\mathcal{Z}^{-1} = P(\mathbf{0})$:

$$P(x) = P(\mathbf{0}) \exp - \sum_{A \subset S} V_A(x) = P(\mathbf{0}) \exp -U(x).$$

□

We now show the equivalence with Markov fields.

Theorem 5.2. *Let X be a Gibbs field of potential V. If V is a neighborhood potential $\mathcal{N}$, it is a Markov field for this same neighborhood system.*

Proof:
Assume that the distribution of X is

$$P(x) = \frac{1}{\mathcal{Z}} \exp -U(x) \quad \text{with} \quad U(x) = \sum_{c \in C} V_c(x).$$

We need to prove (see Definition 5.3) that

$$P[x_s \mid \{x_t, \ t \neq s\}] = P[x_s \mid \{t \in N_s\}].$$

To do this, we calculate the following ratio for $x_s \neq 0_s$:

$$\frac{P(x_s \mid \{x_t, \ t \neq s\})}{P(0_s \mid \{x_t, \ t \neq s\})} = \frac{P(x) \, / \, P(\{x_t, \ t \neq s\})}{P(x_{/s}) \, / \, P(\{x_t, \ t \neq s\})},$$

where $x_{/s}$ is the field x in which x_s has been set to the reference value 0_s. With arbitrary site numbering, we have

$$x = (\dots, x_{s-1}, x_s, x_{s+1}, \dots),$$
$$x_{/s} = (\dots, x_{s-1}, 0_s, x_{s+1}, \dots).$$

The ratio can also be written as

$$\frac{P(x)}{P(x_{/s})} = \frac{\frac{1}{z}\exp{-U(x)}}{\frac{1}{z}\exp{-U(x_{/s})}} = \exp{-\left[U(x) - U(x_{/s})\right]}.$$

Taking the first and last expressions of this expansion, we obtain

$$P\left(x_s \mid \{x_t, \ t \neq s\}\right) = \frac{1}{z}\exp{-\left[U(x) - U(x_{/s})\right]}, \qquad (5.6)$$

where $z^{-1} = P\left(0_s \mid \{x_t, \ t \neq s\}\right)$ is the normalization constant. Finally, we note that in the energy difference, written as:

$$U(x) - U(x_{/s}) = \sum_{c \in C}\left[V_c(x) - V_c(x_{/s})\right],$$

only the potential differences on the cliques that include s need to be calculated. The other differences are zero:

$$\forall c \in C : s \notin c, \text{ we have } V_c(x) = V_c(x_{/s}), \text{ and therefore}$$
$$U(x) - U(x_{/s}) = \sum_{c \in C : s \in c}\left[V_c(x) - V_c(x_{/s})\right]. \qquad (5.7)$$

But $\{c \in C : s \in c\} = N_s$. This means that the value at site s is conditioned by $\{x_t, \ t \in N_s\}$ only, and the Markov property is proven. $\quad\square$

This proof is important because it gives us an analytical method to determine the local specifications of a Markov field from the overall probability over x. The formula (5.6) will be used frequently. This is summed up in the following corollary.

Corollary 5.1. *The local specifications of a Markov field of energy U are given by*

$$P[x_s \mid N_s(x)] = \frac{1}{z}\exp{-\left[U(x) - U(x_{/s})\right]} \qquad (5.8)$$
$$\text{with} \quad z = \sum_{x_s \in E_s}\exp{-\left[U(x) - U(x_{/s})\right]}.$$

Equation (5.7) can be rewritten as

$$U(x) - U(x_{/s}) \doteq \mathbb{U}(x_s, N_s(x)) - \mathbb{U}(0_s, N_s(x)),$$

where $\mathcal{U}(x_s, N_s(x))$ represents the energy of the restricted configuration $(x_t, \ t \in \{s, N_s\})$:

$$\mathcal{U}(x_s, N_s(x)) \doteq \sum_{c \in \mathcal{C}: \, s \in c} V_c(x).$$

We therefore have

$$P[x_s \mid N_s(x)] = \frac{\exp -\mathcal{U}(x_s, N_s(x))}{\sum_{\lambda \in E_s} \exp -\mathcal{U}(\lambda, N_s(x))} \propto \exp -\left(\sum_{c \in \mathcal{C}: \, s \in c} V_c(x) \right). \quad (5.9)$$

5.3   Examples

5.3.1   Bending Energy

Let us take the term U_1 from (5.1) and rewrite it as

$$U_1(g) = \sum_i \big(2g(t_{i+1}) - g(t_i) - g(t_{i+2})\big)^2.$$

Here $x \equiv g$ with $S = \{t_1, \ldots, t_n\}$ and $t_i \in \mathbb{R}$. Without taking into account the boundary conditions of S, we obtain

$$U_1(g) = \sum_i \big(6g(t_i)^2 - 4g(t_i)g(t_{i+1}) + g(t_i)g(t_{i+2})\big).$$

Here $U_1(g)$ is a neighborhood potential energy. In the first expression, the cliques are the triplets $c_i = \{t_i, t_{i+1}, t_{i+2}\}$, and the potential is $V_{c_i}(x) = \big(2g(t_{i+1}) - g(t_i) - g(t_{i+2})\big)^2$. This potential is not canonical. In the second expression the cliques are the singletons $c_i^{(1)} = \{t_i\}$ and the pairs $c_i^{(2)} = \{t_i, t_{i+1}\}$ and $c_i^{(3)} = \{t_i, t_{i+2}\}$, and the potentials are $V_{c_i^{(1)}} = 6g(t_i)^2$, $V_{c_i^{(2)}} = -4g(t_i)g(t_{i+1})$, $V_{c_i^{(3)}} = g(t_i)g(t_{i+2})$. This potential is canonical for the reference value $0_{t_i} = 0$. In both cases the neighborhood of t_i is $N_{t_i} = \{t_{i-2}, t_{i-1}, t_{i+1}, t_{i+2}\}$. In this example we note that with the canonical expression we lose the natural interpretation of the energy, i.e., bending. This gives a probabilistic interpretation of the discrete nonparametric spline model: g can be seen as the realization of a Markov field X with the Gibbs distribution

$$P[X = g] = \frac{1}{\mathcal{Z}} \exp -\alpha U_1(g),$$

with $\alpha > 0$. We can therefore see that the curves g with a high degree of bending are less likely than those with a little bending. This is governed by the parameter α.

5.3.2 Bernoulli Energy

Let X be a binary Markov field encoded on $E_s = \{0, 1\}$ and defined on $S \subset \mathbf{Z}^2$. The neighbors of s are the four nearest sites: $t \in N_s$ if and only if $\|s - t\| = 1$, where $\|\cdot\|$ denotes the Euclidean norm. The cliques are therefore the singletons s and the pairs such that $\|s - t\| = 1$. These pairs are written as $< s, t >$. The field X is called a Bernoulli–Markov field if it is governed by the following Gibbs distribution:

$$P[X = x] = \frac{1}{\mathcal{Z}} \exp -U(x),$$

$$U(x) = \alpha_1 \sum_s x_s + \alpha_2 \sum_{<s,t>} x_s x_t.$$

The potential is defined by the functions $V_s(x) = \alpha_1 x_s$ and $V_{<s,t>}(x) = \alpha_2 x_s x_t$. This field X is equivalent to the field Y encoded on $\{-1, +1\}$ if we consider the transformation $x_s \longmapsto y_s = 2x_s - 1$ of $\{0, 1\} \longmapsto \{-1, +1\}$. With this other encoding, the Markov field Y is known in statistical physics as the Ising model. The local specifications of X are Bernoulli's laws

$$P[X_s = 1 \mid N_s(x)] = \frac{\exp -\delta_s}{1 + \exp -\delta_s}, \qquad (5.10)$$

$$\delta_s = \alpha_1 + \alpha_2 \sum_{t \in N_s} x_t. \qquad (5.11)$$

In fact, from $P[X_s = x_s \mid N_s(x)] = \frac{1}{z} \exp - [U(x) - U(x_{/s})]$ we can immediately obtain

$$U(x) - U(x_{/s}) = \begin{cases} \alpha_1 + \alpha_2 \sum_{t \in N_s} x_t & \text{if } x_s = 1, \\ 0, & \text{if } x_s = 0, \end{cases}$$

$$\text{and} \quad z = 1 + \exp \left[-\alpha_1 - \alpha_2 \sum_{t \in N_s} x_t \right].$$

This leads to the stated Bernoulli law. $\qquad \square$

   There are many other binary field energy expressions that can also be called Bernoulli energies. The above isotropic field thus becomes nonisotropic with the definition

$$U(x) = \alpha_1 \sum_s x_s + \alpha_{2,h} \sum_{<s,t>_h} x_s x_t + \alpha_{2,v} \sum_{<s,t>_v} x_s x_t,$$

when $\alpha_{2,h} \neq \alpha_{2,v}$. These two parameters rule the local horizontal and vertical interactions of the random field over its horizontal and vertical neighborhoods $< \cdot, \cdot >_h$ and $< \cdot, \cdot >_v$, respectively.

Illustration:
Let us consider the case where $\alpha_2 = -\frac{1}{2}\alpha_1 = \alpha$, for which $\delta_s = \alpha(-2 +$

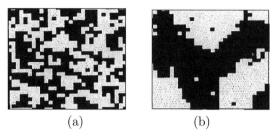

(a)   (b)

Figure 5.1. Bernoulli fields: (a) $\alpha = -1$, (b) $\alpha = -4$.

$\sum_{t \in N_s} x_t$). We set $n_s = \sum_{t \in N_s} x_t$. Because of the encoding $\{0, 1\}$, n_s is the number of neighboring sites of s at the value 1. With this parametrization, the local specifications satisfy $P[X_s = 1 \mid n_s] = P[X_s = 0 \mid 4 - n_s]$. The role of α is illustrated in the following table:

specifications	$\alpha = 0$	$\alpha = -1$	$\alpha = -4$
$P(X_s = 1 \mid n_s = 0)$	0.5	0.12	0.0
$P(X_s = 1 \mid n_s = 1)$	0.5	0.27	0.02
$P(X_s = 1 \mid n_s = 2)$	0.5	0.50	0.50
$P(X_s = 1 \mid n_s = 3)$	0.5	0.73	0.98
$P(X_s = 1 \mid n_s = 4)$	0.5	0.88	1.00

Figure 5.1 gives simulations of this field X. We note that a strongly negative α favors fields consisting of wide gray ($x_s = 1$) and black ($x_s = 0$) regions.

5.3.3   Gaussian Energy

As before, we assume that $S \subset \mathbf{Z}^2$ and that the neighborhood system is the system of the four nearest neighbors. We use E_s to denote a discrete interval that is symmetric about 0, and in this case 0 is the reference value 0_s. A Markov field with this neighboring system is called a centered Gaussian field if it is governed by the following Gibbs distribution:

$$P[X = x] = \frac{1}{Z} \exp -U(x),$$
$$U(x) = \alpha_1 \sum_s x_s^2 + \alpha_2 \sum_{<s,t>} x_s x_t, \qquad (5.12)$$

with $\alpha_1 > 0$. The potential is defined by the functions: $V_s(x) = \alpha_1 x_s^2$ and $V_{<s,t>}(x) = \alpha_2 x_s x_t$. Let us rewrite (5.12) in matrix form as $U(x) = x'Qx$ with $Q_{ss} = \alpha_1$, $Q_{st} = \alpha_2/2$ if s and t are neighbors, and $Q_{st} = 0$ otherwise. A necessary and sufficient condition for $P[X = x]$ to be a nondegenerate Gaussian distribution is that Q be a positive definite matrix. Note that the expression (5.12) is a special case of the more general model $U(x) = \alpha_1 \sum_s x_s^2 + \sum_{<s,t>} \alpha_{st} x_s x_t$, (see [21]).

Proposition 5.1. *The local specifications* $P(x_s \mid N_s(x))$ *of the Markov random field (5.12) are Gaussian distributions* $LG\big(\mu_s = \gamma \sum_{t \in N_s} x_t,\ \sigma_s^2\big)$ *with*

$$\gamma = -\frac{\alpha_2}{2\alpha_1}, \quad \sigma_s^2 = \frac{1}{2\alpha_1}.$$

This means that

$$\mathrm{E}[X_s \mid N_s(x)] = \gamma \sum_{t \in N_s} x_t \quad \text{and} \quad \mathrm{Var}[X_s \mid N_s(x)] = \frac{1}{2\alpha_1}.$$

The value γ is interpreted as a spatial autocorrelation factor, which satisfies $|\gamma| < \frac{1}{4}$ if the stationarity assumption for X is required. To favor configurations where neighboring x_s are close values, $\alpha_2 < 0$ must be satisfied. This term is responsible for the microtexture effect seen in simulated x fields.

Proof:
According to the corollary, we have $P[x_s \mid N_s(x)] \propto \exp - \big[U(x) - U(x_{/s})\big]$. This is a basic calculation. We have

$$U(x) - U(x_{/s}) = \alpha_1 x_s^2 + \alpha_2 x_s \sum_{t \in N_s} x_t = \alpha_1 \left[x_s^2 - 2\gamma x_s \sum_{t \in N_s} x_t \right],$$

$$= \alpha_1 \left[x_s - \gamma \sum_{t \in N_s} x_t \right]^2 - \alpha_1 \left[\gamma \sum_{t \in N_s} x_t \right]^2.$$

Because the second quadratic term is independent of x_s, we have

$$P[x_s \mid N_s(x)] \propto \exp - \left[\alpha_1 (x_s - \gamma \sum_{t \in N_s} x_t)^2 \right],$$

which is, up to a normalizing constant, the Gaussian distribution $LG\left(\mu_s = \gamma \sum_{t \in N_s} x_t,\ \sigma_s^2 = \frac{1}{2\alpha_1}\right)$. This assumes that E_s is a sufficiently fine discretization for us to approximate the discrete distribution with a continuous distribution. $\square$

5.4 Consistency Problem

In the above examples we started by defining an energy in order to then determine the local specifications. We have seen that for certain energies (Examples 2 and 3), the local specifications can be better interpreted than the joint probability distribution obtained from the energy. In actual applications, a Markov model can be built either in terms of global energy or in terms of local specifications. A global energy always defines a joint probability distribution, but a set of local specifications does not necessarily do so.

Proposition 5.2. *A family of local conditional probabilities* $\{P[x_s \mid N_s(x)],\ s \in S\}$ *is not generally "consistent" so as to define a joint probability* $P(x)$.

We will examine the problem of consistency in the case of Besag automodels. These Markov random fields have the energy:

$$U(x) = \sum_s V_{\{s\}}(x) + \sum_{<s,t>} V_{<s,t>}(x).$$

We will give conditions that are necessary for local specifications to be consistent.

Proposition 5.3. *We use* $\star$ *to denote* $N_s(x)$.
(i) If the local specifications to a Markov random field X *belong to an exponential family, i.e.,*

$$\log P[x_s \mid \star] = A_s(\star)B_s(x_s) + C_s(x_s) + D_s(\star), \qquad (5.13)$$

with $B_s(0_s) = C_s(0_s) = 0$, *then there exist parameters* α_s *and* β_{st} *with* $\beta_{st} = \beta_{ts}$ *such that*

$$A_s(\star) = \alpha_s + \sum_{t \in N_s} \beta_{st} B_t(x_t), \qquad (5.14)$$

and

$$\left. \begin{array}{l} V_{\{s\}}(x) = \alpha_s B_s(x_s) + C_s(x_s), \\ V_{<s,t>}(x) = \beta_{st} B_s(x_s) B_t(x_t). \end{array} \right\} \qquad (5.15)$$

(ii) Reciprocally, conditional probabilities that satisfy (5.13) and (5.14) are the specifications of a Markov random field whose potential is given by (5.15).

Thus, taking a set of local specifications is not enough for a coherent definition of the joint probability $P(x)$. To be consistent, the local specifications must satisfy certain constraints; in this case, they must satisfy the constraints (5.14).

Proof:
(i) Let X be a Markov random field of energy U satisfying (5.13). Let us establish (5.15). We have

$$U(x) - U(x_{/s}) = \log \frac{P(x_s \mid \star)}{P(0_s \mid \star)},$$

which can also be written as

$$V_{\{s\}}(x) + \sum_t V_{<s,t>}(x)$$

$$= [A_s(\star)B_s(x_s) + C_s(x_s) + D_s(\star)] - [A_s(\star)B_s(0_s) + C_s(0_s) + D_s(\star)],$$
$$= A_s(\star)B_s(x_s) + C_s(x_s).$$

For a neighborhood that is identically zero, i.e., $x_t = 0$, $\forall t \in N_s$, we write $A_s(\mathbf{0})$ instead of $A_s(\star)$. In this case, we obtain

$$V_{\{s\}}(x) = A_s(\mathbf{0})B_s(x_s) + C_s(x_s) \doteq \alpha_s B_s(x_s) + C_s(x_s),$$

where we have set $\alpha_s = A_s(\mathbf{0})$. We have therefore established the expression for $V_{\{s\}}(x)$ given in (5.15). For the expression for $V_{<s,t>}$, let us set $s = 1$ and $t = 2$ to simplify the notation. The expressions $U(x) - U(x_{/1})$ and $U(x) - U(x_{/2})$ give

$$V_{\{1\}}(x) + V_{<1,2>}(x) = A_1(0, x_2, 0, \dots)B_1(x_1) + C_1(x_1),$$
$$V_{\{2\}}(x) + V_{<1,2>}(x) = A_2(x_1, 0, \dots)B_2(x_2) + C_2(x_2).$$

Using the expression for $V_{\{s\}}(x)$ established above, these two expressions give

$$V_{<1,2>}(x) = [A_1(0, x_2, 0, \dots) - A_1(\mathbf{0})] B_1(x_1,)$$
$$V_{<1,2>}(x) = [A_2(x_1, 0, \dots) - A_2(\mathbf{0})] B_2(x_2).$$

From this, we deduce

$$A_1(0, x_2, 0, \dots) - A_1(\mathbf{0}) = \frac{A_2(x_1, 0, \dots) - A_2(\mathbf{0})}{B_1(x_1)} B_2(x_2) \doteq \beta_{1,2} B_2(x_2),$$

$$A_2(x_1, 0, \dots) - A_2(\mathbf{0}) = \frac{A_1(0, x_2, 0, \dots) - A_1(\mathbf{0})}{B_2(x_2)} B_1(x_1) \doteq \beta_{2,1} B_1(x_1).$$

If we write these values into $V_{<1,2>}(x)$, we obtain $\beta_{1,2} = \beta_{2,1}$ and therefore $V_{<1,2>}(x) = \beta_{1,2} B_1(x_1)B_2(x_2)$, which completes the proof of (5.15). We deduce (5.14) directly from these results.

(ii) Let X be a random field of energy U: $P(x) \propto \exp -U(x)$. We must show that U is a neighborhood energy when the conditional probabilities satisfy (5.13) and (5.14). We again have

$$U(x) - U(x_{/s}) = A_s(\star)B_s(x_s) + C_s(x_s),$$

as a consequence of (5.13). Taking the expression (5.14) of A_s, we obtain

$$U(x) - U(x_{/s}) = [\alpha_s B_s(x_s) + C_s(x_s)] + \left[B_s(x_s) \sum_{t \in N_s} \beta_{st} B_t(x_t) \right].$$

The first term in square brackets depends on the site s only, and the second depends on the sites $\{t \in N_s\}$ only. Consequently, U is a neighboring energy, and we can therefore identify these terms with the potential defined by (5.13).   □

The Bernoulli energy and the Gaussian energy discussed in the previous section are special cases of auto-models. Let us look at another example.

EXAMPLE: Binomial energy.
We wish to build a Markov random field based on its local characteristics. Here we are concerned with the case where the conditional random variables $(X_s \mid \star)$ are governed by binomial distributions $\mathcal{B}(n, \theta_s(\star))$. Using $\mathcal{C}_n^p$ to denote the number of combinations of p in n, these distributions are as follows:

$$
\begin{aligned}
\log P(x_s \mid \star) &= \log \left[\mathcal{C}_n^{x_s} \, \theta_s(\star)^{x_s} \, (1 - \theta_s(\star))^{n - x_s} \right], \\
&= \log \mathcal{C}_n^{x_s} + x_s \log \frac{\theta_s(\star)}{1 - \theta_s(\star)} + n \log(1 - \theta_s(\star)), \\
&\doteq C_s(x_s) + B_s(x_s) A_s(\star) + D_s(\star),
\end{aligned}
$$

with $B_s(x_s) = x_s$, and $A_s(\star) = \log \dfrac{\theta_s(\star)}{1 - \theta_s(\star)}$.

For the local specifications to define a Markov random field, the condition (5.14) must be satisfied. Here this condition is

$$
A_s(\star) = \alpha_s + \sum_{t \in N_s} \beta_{st} x_t.
$$

Conclusion: For the family of random variables $(X_s \mid \star)$ governed by binomial distributions $\mathcal{B}(n, \theta_s(\star))$ to be the local specifications of a Markov random field, $\theta_s(\star)$ must be expressed as follows:

$$
\theta_s(\star) = \frac{\exp A_s(\star)}{1 + \exp A_s(\star)},
$$

$$
\text{with } A_s(\star) = \alpha_s + \sum_{t \in N_s} \beta_{st} x_t.
$$

Note that in the case $n = 2$ we again find the Bernoulli model examined above.

6

Bayesian Estimation and Inverse Problems

These concepts of Bayesian estimation and inverse problems have already been introduced (Section 1.1). Here they are described more specifically in the Markov context, which will be one of the most important formalisms in Part III.

6.1 Principle

In the preceding chapter we defined Markov models for fields X that were understood to be directly observable. In fact, in actual image analysis situations, the representations of X are not directly observable. This is true in particular for fields that correspond to virtual realities that are not accessible to measuring devices. Let us give a heuristic definition of this concept:

Definition 6.1. *A random field X is said to be hidden if it is not "physically" observable.*

To illustrate this situation, let us look again at the example of the approximation spline obtained by minimizing the discrete energy (2.7):

$$U(g \mid y) = \alpha U_1(g) + U_2(y \mid g),$$
$$= \alpha \sum_i \big(2g(t_i) - g(t_{i-1}) - g(t_{i+1})\big)^2 + \sum_i \big(y_i - g(t_i)\big)^2.$$

Here, the occurrence of the one-dimensional field X is a discretized curve g. In the preceding chapter we saw that $U_1(g)$ could be interpreted as a potential energy, making X a Markov random field. In turn, y is interpreted as a degraded version of g, where the level of degradation is quantified by the term $U_2(y \mid g)$. The approximation amounts to determining a specific occurrence of the hidden field X, for example $\widehat{g}$, based on y, and minimizing the energy U. We have now processed an inverse problem. In this example we can see that this consists in building an energy that sets up a competition between a term reflecting the prior knowledge on g, in this case bending, and a term quantifying the difference between g and y, and then minimizing with respect to g the resulting energy U. In more general terms, this approach is stated as follows.

Method 6.1.
(i) *The definition of an inverse problem is summarized by the ordered triplet* $\left(y^O, X, U(x \mid y, \theta)\right)$.

1. *The first element is an observed field y^O on a grid $\widetilde{S}$.*

2. *The second element is the definition of a hidden random field on a grid S, where S is not necessarily identical to $\widetilde{S}$.*

3. *The third element is the definition of an energy expressed as*

$$
\begin{aligned}
U(x \mid y, \ \theta) &= U_1(x, \theta_1) \ + \ U_2(y \mid x, \theta_2), \\
&= \sum_{i=1}^{q_1} U_{1,i}(x, \theta_{1,i}) + \sum_{i=1}^{q_2} U_{2,i}(y \mid x, \theta_{2,i}). \quad (6.1)
\end{aligned}
$$

   The energies U_1 and U_2 are neighborhood potential energies and the $\theta_{i,j}$ are weighting parameters.

(ii) *Solving this inverse problem involves calculating the minimum with respect to x of a cost function defined according to $U\left(x \mid y^O, \ \theta\right)$.*

EXAMPLES.
(1) Let us illustrate the second point concerning the hidden field X. We consider the situation where X represents the "virtual" contours in an image y^O that is observed on a grid $\widetilde{S} \subset \mathbb{Z}^2$. This means that X is a binary field defined on a grid S, obtained by placing a site s in the middle of every edge $[\widetilde{s}, \widetilde{t}]$ so that $\|\widetilde{s} - \widetilde{t}\| = 1$. On a 3×3 grid $\widetilde{S}$, with $\circ$ denoting the sites of $\widetilde{S}$ and $+$ denoting the sites of S, we can represent this with the following diagram:

```
o   +   o   +   o
  +       +       +
o   +   o   +   o
  +       +       +
o   +   o   +   o
```

For any site of S we set $E_s = \{0,1\}$. The presence of a contour at s is signified by $x_s = 1$, and the absence of a contour is signified by $x_s = 0$. On $\widetilde{S}$ we have $E_{\widetilde{s}} = \{0,1,\ldots,2^m - 1\}$ if y^O is encoded on m bits. We must emphasize that the definition of a hidden field is not always trivial as in this example. Sometimes there are several possible formulations for a problem.

(2) Let us illustrate the third point using the Gaussian energy (5.12), which in this case is expressed as

$$U_1(x, \theta_1) = \theta_{1,1} \sum_s x_s^2 + \theta_{1,2} \sum_{<s,t>} x_s x_t.$$

The first term of the energy is a variance term, while the second is a spatial correlation term. More generally, the breakdown (6.1) corresponds to a separation of the functionalities, where each term $U_{j,i}$ has its own role.

Let us interpret the energy $U(x \mid y)$. [1] The energy $U_1(x)$ reflects the prior knowledge on x. It is called the prior energy or internal energy. With respect to this energy, X is a Markov random field with the prior distribution

$$P[X = x] = \frac{1}{\mathcal{Z}} \exp -U_1(x).$$

The energy $U_2(y \mid x)$ quantifies the difference between y and x. It is called the fidelity energy or external energy. It leads to the following distribution of the conditional field $(Y \mid X = x)$:

$$P[Y = y \mid X = x] = \frac{1}{\kappa} \exp -U_2(y \mid x).$$

In general, κ should be dependent on x. In practice, however, we will choose a distribution such that it is not, for example, a distribution that depends on $x - y$ only. In real applications, this distribution often results from the physics of the problem. It could be, for example, the distribution of a measured noise whose occurrences on the sites s are independent. In

---

[1] For the remainder of this chapter θ is omitted from the notation because it has no further relevance to our arguments.

this case, the simplest expression for the distribution is

$$P[Y = y \mid X = x] = \prod_s P[Y_s = y_s \mid X_s = x_s],$$

or, more generally,

$$P[Y = y \mid X = x] = \prod_s P[Y_s = y_s \mid X = x]. \tag{6.2}$$

This latter distribution is typical of tomography problems (Chapter 12). It is unusual to introduce dependencies on $Y \mid X = x$ in considering a Markov-type structure, and we will therefore assume that the field $Y \mid X = x$ is independent in the sense of (6.2).

By applying the Bayes formula, we obtain

$$\begin{aligned}
P[X = x, Y = y] &= P[X = x]P[Y = y \mid X = x] \\
&= \frac{1}{Z_K} \exp -[U_1(x) + U_2(y \mid x)] \\
&\doteq \frac{1}{Z_K} \exp -U(x \mid y). \tag{6.3}
\end{aligned}$$

The pair (X, Y) is therefore a Markov random field of energy $U(x \mid y)$. The posterior distribution is

$$\begin{aligned}
P[X = x \mid Y = y] &= \frac{1}{P[Y = y]} P[X = x, Y = y] \\
&= \frac{1}{P[Y = y]} \frac{1}{Z_K} \exp -U(x \mid y) \\
&\propto \frac{1}{Z_K} \exp -U(x \mid y). \tag{6.4}
\end{aligned}$$

The conditional field $(X \mid Y = y)$ is therefore a Markov random field of energy $U(x \mid y)$, just like (X, Y). In general, we write $U(x \mid y)$ instead of $U(x, y)$ because the posterior distribution plays an essential role in the Bayesian context. The energy $U(x \mid y)$ is therefore called a posterior energy.

The term "inverse problem" comes from the fact that in certain situations, an operator $\mathcal{O}$ connecting the field X to the field Y can be found (Section 1.1.2):

$$Y = \mathcal{O}(X, W),$$

where W denotes a random field that is partially responsible for the nonobservability of X. In this situation, W is likened to an observation noise. In image analysis, the operators $\mathcal{O}$ are often nonlinear. To illustrate our assertions, let us consider a special case that is easy to formalize: the case where every occurrence of X is degraded by an additive noise w, the occurrence of a random field W. This corresponds to the degradation model

$$Y = \mathcal{O}(X, W) = X + W,$$

with $(X + W)_s \dot{=} X_s + W_s$. The field before degradation is X, and the observable field after degradation is Y. This is the simplest model found in image restoration. We can see that it contains in particular the spline approximation model (2.1). It allows us to define the energy term U_2. To do this, we simply need to take a probability distribution for the random field $W = Y - X$:

$$P[W = w] \propto \exp -R(w), \tag{6.5}$$

where R is a neighborhood energy. Let us also note that X is assumed to be a Markov random field with the following prior distribution:

$$P[X = x] \quad \propto \quad \exp -U_1(x). \tag{6.6}$$

We can then state the following result.

Proposition 6.1. *If the fields X and W expressed respectively as (6.5) and (6.6) are independent, then the energy*

$$U(x \mid y) = U_1(x) + R(y - x).$$

is the energy of the conditional Markov random field $(X \mid Y = y)$.

Proof:
Starting with the Bayes formula, we obtain

$$
\begin{aligned}
P[X = x \mid Y = y] &= P[Y = y \mid X = x] P[X = x] / P[Y = y] \\
&\propto P[X + W = y \mid X = x] P[X = x] \\
&= P[W = y - x \mid X = x] P[X = x] \\
&= P[W = y - x] P[X = x],
\end{aligned}
$$

where the last line comes from the independence between W and X. Finally,

$$
\begin{aligned}
P[X = x \mid Y = y] &\propto \exp -[U_1(x) + R(y - x)] \\
&\propto \exp -U(x \mid y).
\end{aligned}
$$

Because U_1 and R derive from a neighborhood potential, the same applies to their sum, and therefore $(X \mid Y = y)$ is a Markov random field.   □

If we assume that the random variables W_s are mutually independent and governed by the Gaussian distribution $LG(0, \sigma^2)$,

$$P[W = w] = \prod_s P(w_s) \propto \exp -\frac{1}{2\sigma^2} \sum_s w_s^2,$$

we then have

$$R(w) = \frac{1}{2\sigma^2} \sum_s w_s^2.$$

This gives

$$U_2(y \mid x) \equiv R(y - x) = -\frac{1}{2\sigma^2} \sum_s (y_s - x_s)^2.$$

Therefore, the term U_2 of the spline energy (2.7) can be interpreted as the result of a Gaussian degradation.

6.2   Cost Functions

We will now discuss in greater detail the concept of cost function introduced in point *(ii)* of the method 6.1. Following on from point *(i)*, let us note that for a fixed value of y, X is a Markov random field of energy $U(x \mid y)$. Let us assume an "ideal" field written as x^* that is a solution to the inverse problem (Section 1.1.1), assuming also that we have an estimate $\tilde{x} = \phi(y)$ of x^*, where ϕ is a deterministic function of y.

The Bayesian estimation method proceeds as follows. First, we take an elementary cost function $\Gamma : E \times E \to \mathsf{R}^+$, where $E = \prod_s E_s$ is the set of occurrences of X such that

$$\Gamma(\tilde{x}, x^*) \geq 0 ,$$
$$\Gamma(\tilde{x}, x^*) = 0 \Leftrightarrow \tilde{x} = x^*.$$

The function $\Gamma(\tilde{x}, x^*)$ is the elementary cost of the error involved in replacing x^* by $\tilde{x}$. Because x^* is unknown, we have to work with a cost function that is averaged with respect to the posterior distribution $P(x \mid y)$.

Definition 6.2. *Given an elementary cost function Γ, a posterior distribution $P(x \mid y)$, and an observed value y^O of Y, we call the following average cost function a Bayesian cost function:*

$$\begin{aligned}\bar{\Gamma}(\tilde{x}) &= \mathbf{E}\left[\Gamma(\tilde{x}, X) \mid Y = y^O\right] \\ &= \sum_{x \in E} \Gamma(\tilde{x}, x)\, P\left[X = x \mid Y = y^O\right], \quad \forall\, \tilde{x} \in E.\end{aligned}$$

The Bayesian estimate $\hat{x}$ of x^* is therefore

$$\hat{x} = \arg\min_{x \in E} \bar{\Gamma}(x).$$

6.2.1 Cost Function Examples

A Bayesian estimate, then, depends on the choice of elementary cost function Γ. We now give detailed estimates for various functions Γ.

EXAMPLE 1.
We choose

$$\Gamma(\tilde{x}, x) = \begin{cases} 1, & \text{if } \tilde{x} \neq x, \\ 0, & \text{if } \tilde{x} = x. \end{cases}$$

This is a severe cost function; just one erroneous site can result in the maximum cost. The average cost function is

$$\begin{aligned} \bar{\Gamma}(\tilde{x}) &= \sum_{x \in E} \Gamma(\tilde{x}, x) P\left[X = x \mid Y = y^O\right] \\ &= \sum_{x \neq \tilde{x}} P\left[X = x \mid Y = y^O\right] \\ &= 1 - P\left[X = \tilde{x} \mid Y = y^O\right]. \end{aligned}$$

The Bayesian estimation is therefore

$$\hat{x} = \arg\min_{x \in E} \bar{\Gamma}(x) = \arg\max_{x \in E} P\left[X = x \mid Y = y^O\right].$$

This is called the estimation of "maximum a posteriori likelihood", or MAP, as seen in Chapter 2. One of the advantages of the statistical approach is that it allows us to define other forms of estimation, as illustrated in the following examples.

EXAMPLE 2.
For a less severe cost function than the one shown above, we can choose

$$\Gamma(\tilde{x}, x) = \sum_{s \in S} \mathbf{1}[\tilde{x}_s \neq x_s].$$

This function can be written using local elementary cost functions

$$\Gamma(\tilde{x}, x) = \sum_{s \in S} \Gamma_s(\tilde{x}_s, x_s),$$

where Γ_s is defined almost as in the first example:

$$\Gamma_s(\tilde{x}_s, x_s) = \begin{cases} 1, & \text{if } \tilde{x}_s \neq x_s, \\ 0, & \text{if } \tilde{x}_s = x_s. \end{cases}$$

The average cost function is

$$
\bar{\Gamma}(\tilde{x}) \;=\; \sum_{x \in E} \left[\sum_{s} \Gamma_s(\tilde{x}_s, x_s) \right] P\left[X = x \mid Y = y^O \right]
$$

$$
\;=\; \sum_{s} \left[\sum_{x \in E} \Gamma_s(\tilde{x}_s, x_s) \, P\left[X = x \mid Y = y^O \right] \right]
$$

$$
\;=\; \sum_{s} \mathbf{E}\left[\Gamma_s\left(\tilde{x}_s, X_s \right) \mid Y = y^O \right]
$$

$$
\;=\; \sum_{s} 1 - P\left[X_s = \tilde{x}_s \mid Y = y^O \right].
$$

The Bayesian estimation $\hat{x} = \arg\min_{x \in E} \bar{\Gamma}(x)$ is therefore obtained in this case according to

$$
\hat{x}_s \;=\; \arg\max_{x_s \in E_s} P\left[X_s = x_s \mid Y = y^O \right] , \quad \forall s \in S.
$$

This is similar to the MAP estimation but is performed locally. It is called the "maximum posterior marginal likelihood" estimation, or MPM.

EXAMPLE 3.
When $E_s \subset \mathsf{R}$, the above cost function does not take the size of error into account. To remedy this problem, we can choose

$$
\Gamma(\tilde{x}, x) \;=\; \sum_{s \in S} (\tilde{x}_s - x_s)^2 .
$$

This function can also be written using local elementary cost functions Γ_s: $\Gamma(\tilde{x}, x) = \sum_{s \in S} \Gamma_s(\tilde{x}_s, x_s)$, with $\Gamma_s(\tilde{x}_s, x_s) = (\tilde{x}_s - x_s)^2$. The estimation $\hat{x} = \arg\min_{x \in E} \bar{\Gamma}(x)$ is then defined for all s by

$$
\hat{x}_s \;=\; \arg\min_{x_s \in E_s} \mathbf{E}\left[(X_s - x_s)^2 \mid Y = y^O \right]
$$

$$
\;=\; \arg\min_{x_s \in E_s} \left(\mathbf{E}\left[X_s \mid Y = y^O \right] - x_s \right)^2 .
$$

This means that $\hat{x}_s$ is the value of E_s that is nearest to $\mathbf{E}\left[X_s \mid Y = y^O \right]$. This is called the "threshold posterior mean", or TPM estimation, also known as mean field estimation.

Example of an application. We again consider a special case that is simple to formalize. Let X be a binary field encoded with $E_s = \{-1, +1\}$. Every occurrence of X is degraded by a multiplicative noise w from a random field W. This is expressed by the degradation model

$$
Y = \mathcal{O}(X, W) = XW,
$$

where $(XW)_s \doteq X_s W_s$. The random variables W_s are independent and governed by the distribution $P[W_s = -1] = e^{\beta} / \left(e^{\beta} + e^{-\beta} \right)$ where $\beta \in \mathsf{R}$, which gives

$$P[W_s = w_s] = \frac{e^{-\beta w_s}}{e^{\beta} + e^{-\beta}}.$$

We assume that X is governed by the Ising model (Section 5.3.2) with energy $U_1(x) = \alpha \sum_{<s,t>} x_s x_t$:

$$P[X = x] = \frac{1}{Z} \exp - \left[\alpha \sum_{<s,t>} x_s x_t \right],$$

where $\alpha < 0$, X and W are independent. As in the additive case, the degradation model lets us calculate the conditional distribution. If we write $\left(\frac{Y}{X}\right)_s \dot{=} \frac{Y_s}{X_s}$, we have

$$
\begin{aligned}
P[X = x \mid Y = y] &= P[Y = y \mid X = x]\, P[X = x]/P[Y = y] \\
&\propto P[XW = y \mid X = x]\, P[X = x] \\
&= P\left[W = \frac{y}{x} \mid X = x\right] P[X = x] \\
&= P\left[W = \frac{y}{x}\right] P[X = x] \\
&= \frac{1}{(e^{\beta} + e^{-\beta})^{|S|}} \frac{1}{Z} \exp - \left[\sum_s \frac{y_s}{x_s} \beta + \alpha \sum_{<s,t>} x_s x_t \right].
\end{aligned}
$$

This means that $(X \mid Y = y)$ is a Markov random field with energy

$$U(x \mid y) = \alpha \sum_{<s,t>} x_s x_t + \beta \sum_s x_s y_s. \tag{6.7}$$

The Bayesian estimation of the MAP is therefore

$$
\begin{aligned}
\widehat{x} &= \arg\max_x P\left[X = x \mid Y = y^O\right] \\
&= \arg\min_x \left\{ \alpha \sum_{<s,t>} x_s x_t + \beta \sum_s x_s y_s^O \right\},
\end{aligned}
$$

and the Bayesian estimation of the MPM is

$$\widehat{x}_s = \begin{cases} +1 & \text{if } P\left[X_s = 1 \mid Y = y^O\right] \geq \frac{1}{2}, \\ -1 & \text{if } P\left[X_s = 1 \mid Y = y^O\right] < \frac{1}{2}. \end{cases}$$

6.2.2 Calculation Problems

The use of the estimation methods described above gives rise to some very difficult calculation problems. Estimating the MAP requires the following calculation:

$$\max_{x \in E} P[X = x \mid Y = y].$$

Estimating the MPM requires the following calculation

$$\max_{x_s \in E_s} P[X_s = x_s \mid Y = y].$$

Estimating the TPM requires the following calculation

$$\mathrm{E}[X_s \mid Y = y] = \sum_{x_s \in E_s} x_s P[X_s = x_s \mid Y = y].$$

In view of the huge size of E, it is impossible to calculate these expressions directly and exactly. We therefore seek to approximate them using Monte Carlo methods, as described in the following chapter.

Let us, however, provide a few introductory guidelines. Let us assume that we can simulate the conditional field $(X \mid Y = y)$. We write $\{x^{(1)}, \ldots, x^{(N)}\}$ to denote a series of occurrences of this random field obtained according to the Gibbs distribution $P(x \mid y) \propto \exp -U(x \mid y)$. We can then perform an approximate calculation of the estimates of the MPM and the TPM using

$$P[X_s = \lambda \mid Y = y] \approx \frac{1}{N} \sum_{i=1}^{N} \mathbf{1}\left[x_s^{(i)} = \lambda\right],$$

$$\mathrm{E}[X_s \mid Y = y] \approx \frac{1}{N} \sum_{i=1}^{N} x_s^{(i)}.$$

The first approximation is the frequency of observation of the value $\lambda \in E$ at the site s, and the second is the empirical mean at the site s. It is not so simple to perform an approximate calculation of the MAP. The first two approximations are justified by the law of large numbers, which is stated as follows:

Theorem 6.1. *Let* $\{X^{(i)}, i \geq 1\}$ *be a series of independent random variables with real values that all follow the same distribution as a random variable* X *with finite variance and expectation. Let us set* $\bar{X}_N = (X^{(1)} + \cdots + X^{(N)})/N$. *We then have the convergence*

$$P\left[\lim_{N \to \infty} \bar{X}_N = \mathrm{E}[X]\right] = 1.$$

7
High-Dimensionality Simulation and Optimization

Two types of problem arise in the wake of the preceding chapter. With the expression for the Gibbs distribution $P[X = x \mid Y = y] = \frac{1}{Z} \exp -U(x \mid y)$ of a Markov random field X, these problems are

- (a1) To simulate $[X \mid Y = y]$ according to this distribution,
  (a2) To calculate $\mathrm{E}[f(X) \mid Y = y] = \sum_{x \in E} f(x) P[X = x \mid Y = y]$,

- (b) To minimize $U(x \mid y)$.

The difficulties are caused by the very large size of E. In this chapter, because the observed value y is fixed, we omit the conditioning by $Y = y$ by writing simply $P(X = x)$, $\mathrm{E}[f(X)]$, and $U(x)$ whether or not there is conditioning.

We first describe *stochastic* algorithms that can handle the stated problems. The main drawback of these algorithms is their cost in computer time. Second, we present deterministic algorithms that reduce the calculation time, but to the detriment of accuracy. The stochastic algorithms presented relate to two dynamics: Metropolis and the Gibbs sampler. The first technique goes beyond the scope of Markov random fields. We will present this technique in the context of images, but the results can also be used in other situations. In particular, E, a *finite* set, will be an abstract set that can designate entities other than field configurations. This applies to combinatorial optimization, for example, where E is a finite set that does not necessarily have a product structure. Readers with little experience of stochastic techniques can start reading this chapter at Section 7.4. Sections 7.1 and 7.2 are an overview of the best known stochastic algorithms for simulation and optimization, while Section 7.3 gives the elements required to prove that they converge.

7.1 Simulation

7.1.1 Homogeneous Markov Chain

Finite Markov chains [116] provide the mathematical context for dealing with problem (a). A finite Markov chain is a series of random variables $\{Z_n, \ n \geq 0\}$ that are time-indexed and have values in E, satisfying the Markov property

$$P[Z_n = x(n) \mid \{Z_k = x(k), \ k < n\}] \ = \ P[Z_n = x(n) \mid Z_{n-1} = x(n-1)],$$

whatever the configurations $\{x(k), k \leq n\}$. In this section we deal only with *homogeneous chains* (stationary in time). Homogeneity means that the probability of going in a single instant from a configuration x to a configuration y is independent of the instant under consideration:

$$P[Z_n = y \mid Z_{n-1} = x] \doteq p_{xy}, \quad \forall x, y \in E.$$

All of these transition probabilities are arranged in a matrix $\mathcal{P}$:

$$\mathcal{P} \doteq (p_{xy}).$$

This matrix is said to be stochastic because it satisfies $\mathcal{P} \geq 0$ and $\sum_y p_{xy} = 1$ for all x. We also write

$$p_{xy}^{(n)} \doteq P[Z_n = y \mid Z_0 = x],$$

to denote the probability of transition in n instants and $\mathcal{P}^{(n)}$ to denote the associated transition matrix. We also write

$$\Pi^{(n)}(x) \doteq P[Z_n = x],$$

for the probability of the configurations at the instant n. The distribution $\Pi^{(n)}$ is considered as a *line* vector. We clearly have

$$\Pi^{(n)}(y) \ = \ \sum_x \Pi^{(n-1)}(x) \, p_{xy},$$

and, in vector form $\Pi^{(n)} = \Pi^{(n-1)}\mathcal{P}$ and from nearest neighbor to nearest neighbor,

$$\Pi^{(n)} = \Pi^{(0)}\mathcal{P}^n.$$

We therefore have $\mathcal{P}^n = \mathcal{P}^{(n)}$. This means that a homogeneous Markov chain is completely determined when $\Pi^{(0)}$ and $\mathcal{P}$ are given.

Our next question concerns the asymptotic behavior of $\Pi^{(n)}$. In certain situations we can see that the vector $\Pi^{(n)}$ converges to a vector Π that is independent of $\Pi^{(0)}$. The Markov chain is then said to be ergodic, and the limit distribution Π is called the *equilibrium distribution*. Note that if a distribution Π satisfies $\Pi\mathcal{P} = \Pi$, then $\Pi\mathcal{P}^n = \Pi$, such that it appears as a possible equilibrium distribution. To obtain this limit, we must restrict the family of Markov chains. In the finite case, this means considering the chains that are *regular*, i.e., such that there is an r for which $\mathcal{P}^r > 0$. This type of chain has the property of *irreducibility*: For any pair (x, y), it is possible to obtain y from x in a finite number of transitions.

Theorem 7.1. *A homogeneous, finite, regular Markov chain with transition matrix $\mathcal{P}$ has a unique equilibrium distribution $\Pi > 0$ such that*

$$\lim_{n \to \infty} \Pi^{(0)}\mathcal{P}^n = \Pi, \tag{7.1}$$

where this distribution satisfies the invariance property with respect to $\mathcal{P}$:

$$\Pi = \Pi\,\mathcal{P}. \tag{7.2}$$

For the proof of this theorem, refer to [76, 182] and Section 7.3.

7.1.2 Metropolis Dynamic

Our purpose here is to simulate a distribution $\Pi(x)$ on finite E whose expression is given, where $\Pi(x) > 0$ for all $x \in E$. For this, we need to build a Markov chain Z with equilibrium distribution $\Pi(x)$. The construction of the chain $Z_1, Z_2, \ldots, Z_n, \ldots$ will therefore give configurations $x(1), x(2), \ldots, x(n), \ldots$ which will be governed by the following distribution for large n:

$$\lim_{n \to \infty} P[Z_n = x \mid Z_0 = x(0)] = \Pi(x).$$

This amounts to finding a regular matrix $\mathcal{P}$ that satisfies the invariance property (7.2). In the limiting case we will therefore obtain the equilibrium distribution (7.1) thanks to the property of uniqueness.

The Metropolis dynamic that is used to apply this idea comes from Metropolis et al. [135] (also refer to [105]). To build the Markov chain Z, we start by taking an initial transition matrix Q that is symmetric and irreducible on E. This is called an *exploration* matrix, and it is used to generate at each instant n a candidate configuration for Z_n. We then set

$$p_{xy} = \begin{cases} q_{xy} & \text{if } \Pi(y) \geq \Pi(x) \text{ and } x \neq y, \\ q_{xy}\frac{\Pi(y)}{\Pi(x)} & \text{if } \Pi(y) < \Pi(x) \text{ and } x \neq y, \\ 1 - \sum_{y:y \neq x} p_{xy} & \text{if } y = x. \end{cases} \qquad (7.3)$$

Therefore, $\mathcal{P} = (p_{xy})$ is a regular stochastic matrix. Note that when $\Pi(x)$ is a Gibbs distribution, the expression is independent of $\mathcal{Z}$ because p_{xy} involves only the fraction $\Pi(y)/\Pi(x)$, which is independent of $\mathcal{Z}$.

Lemma 7.1. *The vector Π satisfies the invariance property $\Pi = \Pi\mathcal{P}$ with respect to the transition matrix $\mathcal{P} = (p_{xy})$ defined in (7.3).*

Proof:
We start by showing that $\mathcal{P}$ is *reversible* in the following sense:

$$\Pi(x)p_{xy} = \Pi(y)p_{yx}. \qquad (7.4)$$

If we look, for example, at the expression for p_{xy} when $\Pi(y) < \Pi(x)$, we obtain

$$\begin{aligned} \Pi(x)p_{xy} &= \Pi(x)q_{xy}\frac{\Pi(y)}{\Pi(x)} \\ &= q_{yx}\Pi(y) \quad \text{(because } Q \text{ is symmetric)} \\ &= \Pi(y)p_{yx} \quad \text{(according to the first line of (7.3)).} \end{aligned}$$

This demonstrates (7.4). We therefore have the expected result

$$(\Pi\mathcal{P})(x) = \sum_{y}\Pi(y)p_{yx} = \sum_{y}\Pi(x)p_{xy} = \Pi(x),$$

where the second equality is true thanks to reversibility. $\qquad\square$

Finally, by applying Theorem 7.1, we obtain

$$\lim_{n\to\infty} P[Z_n = y \mid Z_0 = x] = \Pi(y).$$

The algorithm associated with the exploration matrix Q of E is as follows:

Algorithm 7.1. (Metropolis).

- *Step 0 :* Select $x(0)$ arbitrarily.

- *Step $(n + 1)$:* Transition $n \to (n + 1)$.
  Let $x(n)$ be the occurrence of Z_n at Step n.
  Take a $y \in E$ according to $q_{x(n),y}$.
    - If $\Pi(y) \geq \Pi(x(n))$ and $y \neq x(n)$, take $x(n + 1) = y$,
    - Otherwise,

$$x(n + 1) = \begin{cases} y & \text{with probability} \quad \frac{\Pi(y)}{\Pi(x(n))}, \\ x(n) & \text{with probability} \quad 1 - \frac{\Pi(y)}{\Pi(x(n))}. \end{cases}$$

We clearly have

$$P[Z_{n+1} = y \mid Z_n = x(n)] = p_{x(n),y},$$

as defined in (7.3), and therefore the equilibrium distribution is indeed (7.1). In practice, the second run of the algorithm consists in simulating a random number ξ according to the uniform distribution on $[0, 1]$, and then doing

$$x(n + 1) = \begin{cases} y & \text{if } \xi < \frac{\Pi(y)}{\Pi(x(n))}, \\ x(n) & \text{otherwise}, \end{cases}$$

or, for a Gibbs distribution of energy U,

$$x(n + 1) = \begin{cases} y & \text{if } \log \xi < -[U(y) - U(x(n))], \\ x(n) & \text{otherwise}. \end{cases} \tag{7.5}$$

7.1.3 Simulated Gibbs Distribution

- METROPOLIS DYNAMIC

Let us look at how this algorithm is used when X is a Markov random field indexed by a finite set of sites S, with energy U that comes from a potential V:

$$\Pi(x) = \frac{1}{\mathcal{Z}} \exp -U(x).$$

In this case, the transition probabilities are rewritten as follows when $x \neq y$:

$$p_{xy} = q_{xy} \exp -[U(y) - U(x)]_+ \tag{7.6}$$

$$\text{where } a_+ = \begin{cases} 0 & \text{if } a < 0, \\ a & \text{if } a \geq 0. \end{cases}$$

We write x to denote the running configuration $x(n)$. If the configurations x and y differ at only one site s, it is easy to calculate the ratio $\Pi(y)/\Pi(x)$

found in the algorithm (see (5.7)):

$$\frac{\Pi(y)}{\Pi(x)} = \exp - \left(\sum_{c \in C: \ s \in c} [V_c(y) - V_c(x)] \right).$$

We are therefore tempted to choose transitions that change the running configuration x at only one site s. To do this, a value at the site s is simply generated by a random selection that is uniform in E_s. This is the "local" exploration matrix $Q^{[s]}$:

$$q_{xy}^{[s]} = \begin{cases} \frac{1}{|E_s|} & \text{if } x_t = y_t, \ \ \forall t \neq s, \\ 0 & \text{otherwise.} \end{cases}$$

From this we obtain the "local" transition matrix $\mathcal{P}^{[s]}$ of elements $p_{xy}^{[s]} = q_{xy}^{[s]} \exp -[U(y) - U(x)]_+$. Changes are made according to a site visiting order, which we can label $(s_1, \ldots, s_{|S|})$, for example, alphabetical order when $S \subset \mathbf{Z}^2$. When all of S has been visited, we say that a *sweep* of S has been performed. A succession of these sweeps gives the Markov chain with values in E: $Z_0, (Z_1, \ldots, Z_{|S|}), \ldots, (Z_{(k-1)|S|+1}, \ldots, Z_{k|S|}), \ldots$ where each sequence $(Z_{(k-1)|S|+1}, \ldots, Z_{k|S|})$, $k \geq 1$, is generated by the local transition matrices: $\mathcal{P}^{[s_1]}, \ldots, \mathcal{P}^{[s_{|S|}]}$. This chain is not homogeneous. The extracted chain $(Z_0, Z_{|S|}, \ldots, Z_{k|S|}, \ldots)$, however, is homogeneous, and its transition matrix is $\mathcal{P} = \mathcal{P}^{[s_1]} \ldots \mathcal{P}^{[s_{|S|}]}$. When $\Pi = \Pi \, \mathcal{P}^{[s_j]}$ for all j, then $\Pi = \Pi \, \mathcal{P}$. If, in addition, $\mathcal{P}$ is irreducible, then Π is the unique equilibrium distribution. In summary, the simulation algorithm for a Gibbs distribution uses the Metropolis algorithm with Q replaced by $Q^{[s]}$ and where, after processing, we keep the sequence $\{x(k|S|), \ k \geq 1\}$ that results from the various sweeps.

• GIBBS SAMPLER

Here again, we need to simulate a distribution $\Pi(y)$, $y \in E$, whose expression is given. The dynamic of the Gibbs sampler is analogous to the Metropolis dynamic. At each instant of the Markov chain Z_n, we change the value of one site only. Let us then take $\{s_1, s_2, \ldots\}$, a path through the sites such that each site is visited an infinite number of times. We may consider any path, such as a random path, or we may consider the periodic path of the Metropolis dynamic, which proceeds by successive sweeps of S. At an instant n the visited site is s_n, and the transition from $(Z_{n-1} = x)$ to $(Z_n = y)$ occurs according to

$$y_s = \begin{cases} x_s & \text{if } s \neq s_n, \\ \lambda & \text{if } s = s_n, \end{cases}$$

where λ is the result of random selection according to the conditional distribution at the site s_n: $\Pi(y_{s_n} \mid \breve{y}_{s_n} = \breve{x}_{s_n})$, where $\breve{x}_s = \{x_t : t \neq$

s}. The Markov chain is not homogeneous. The generic element of its transition matrix $\mathcal{P}^{(n-1,n)}$ is

$$
\begin{aligned}
p_{xy}^{(n-1,n)} &= P^{(n-1,n)}[Z_n = y \mid Z_{n-1} = x] \\
&= \mathbf{1}[\breve{y}_{s_n} = \breve{x}_{s_n}] \, \Pi(y_{s_n} \mid \breve{y}_{s_n} = \breve{x}_{s_n}).
\end{aligned}
$$

and the generic element of the transition matrix of the kth sweep is

$$
\begin{aligned}
p_{xy}^{(k|S|,(k+1)|S|)} &= \prod_{j=1}^{|S|} p_{xy}^{(k|S|+j-1,k|S|+j)} \\
&= \prod_{j=1}^{|S|} \Pi(y_j \mid y_1, \ldots, y_{j-1}, x_{j+1}, \ldots, x_{|S|}) . \quad (7.7)
\end{aligned}
$$

The algorithm can be summarized as follows:

Algorithm 7.2. (Gibbs Sampler).

- *Step 0* : Select $x(0)$ arbitrarily.

- *Step n* : Transition $(n-1) \to n$.
  Let x be the occurrence of Z_{n-1} at Step $(n-1)$.
  Update site s_n by selecting $y \in E$ according to the distribution
  $$
  \left\{ p_{x,y}^{(n-1,n)} = \mathbf{1}[\breve{y}_{s_n} = \breve{x}_{s_n}] \, \Pi(y_{s_n} \mid \breve{y}_{s_n} = \breve{x}_{s_n}), \ y_{s_n} \in E_{s_n} \right\}.
  $$

The following theorem, proposed by Geman and Geman [84], ensures the convergence of this chain.

Theorem 7.2. *If* $\Pi(y) > 0$ *for all* $y \in E$ *and if each site is visited an infinite number of times, then*

$$
\lim_{n \to \infty} P[Z_n = y \mid Z_0 = x(0)] = \Pi(y).
$$

For the proof of this, refer to [182], or adapt the proof given in Section 7.3.

We thus have two sampling methods for a distribution Π. One of the main applications of this is to calculate the expectation $\sum_x f(x)\Pi(x)$ for an arbitrary function $f : \ E \mapsto \mathrm{R}$. For the Markov chains discussed above, the law of large numbers is written as follows:

$$
\lim_{n \to \infty} \frac{1}{n} \sum_{i=n_0}^{n} f(x^{(i)}) = \sum_x f(x)\Pi(x),
$$

where $x^{(i)}$ denotes the result obtained after the ith sweep. This provides the justification for the approximation of the MPM estimation introduced in the preceding chapter. The Gibbs sampler is the preferred tool in the case of Markov random fields, while the Metropolis dynamic applies more generally to combinatorial optimization because it does not assume any specific structure on E. These methods will also be used to deal with the problem $\min_{x \in E} U(x)$ in the following section.

7.2 Stochastic Optimization

We take the situation where $U : E \mapsto \mathbb{R}$ can have several local minima as well as several overall minima, which are reached for the following configurations

$$E_{\min} = \{\widetilde{x} \in E : U(\widetilde{x}) = \min_x U(x)\} .$$

The local minima present the primary difficulty. Any deterministic iterative procedure such as gradient descent (Section 7.4.2) can supply only one local minimum when sufficient knowledge of the landscape of minima is not available. The idea, therefore, is to use an iterative procedure that does not get stuck in the well of a local minimum, and that does this randomly. The desired procedure will be a stochastic gradient algorithm. To minimize $U(x)$, we use the following distribution:

$$\Pi_T(x) \;=\; \frac{1}{\mathcal{Z}_T} \exp -\frac{1}{T} U(x),$$

where T is a positive parameter. Note that $\Pi_T(x)$ becomes increasingly concentrated around $E_{\min}$ as T approaches 0. We could describe this by saying that as T decreases toward 0, the energetic landscape is brought into sharper relief, and the distribution gradually becomes concentrated around its modes, which are independent of T.

Proposition 7.1. *If Π_0 denotes the uniform distribution on $E_{\min}$, $\Pi_0(x) = \mathbf{1}[x \in E_{\min}] \, |E_{\min}|^{-1}$, we have*

$$\lim_{T \to 0} \Pi_T(x) \;=\; \Pi_0(x). \tag{7.8}$$

Moreover, for $x \in E_{\min}$, the function $\Pi_T(x)$ is an increasing function when $T \to 0$, and a decreasing function when $x \notin E_{\min}$.

Proof:

$$\Pi_T(x) \;=\; e^{-(U(x)-U_{\min})/T} \left[\sum_y e^{-(U(y)-U_{\min})/T} \right]^{-1}$$

$$=\; e^{-(U(x)-U_{\min})/T} \left[\sum_{y \notin E_{\min}} e^{-(U(y)-U_{\min})/T} + \sum_{y \in E_{\min}} 1 \right]^{-1} .$$

Therefore, when $T \to 0$, $\Pi_T(x) \to 0$ if $x \notin E_{\min}$, and otherwise $\Pi_T(x) = |E_{\min}|^{-1}$. In the latter case, we can see that $\Pi_T(x)$ increases monotonically toward $|E_{\min}|^{-1}$. $\qquad\qquad\qquad\qquad\qquad\square$

Note that for fixed T, the Metropolis dynamic or the Gibbs sampler dynamic generates a Markov chain such that

$$P[Z_n = x \mid Z_0 = x(0)] = \Pi_T(x). \qquad (7.9)$$

When T has a high value, the simulation tends to explore E broadly: When $T \to \infty$, $\Pi_T(x) \to |E|^{-1}$. On the other hand, for very small T, if we manage to generate a configuration $x \in E$, it will most probably be in $E_{\min}$. In this context, Kirkpatrick et al. [123] and Černy [34] independently introduced the *simulated annealing* algorithm for combinatory optimization problems. This algorithm was inspired by a process used in the field of physical chemistry, whereby cooling is applied to a system to take it to a crystalline, minimum energy state. The value of T is brought down very slowly toward 0 as we go along the Z chain so that

$$\lim_{n \to \infty} P[Z_n \in E_{\min} \mid Z_0 = x(0)] = 1, \quad \forall\, x(0) \in E.$$

Here, T represents the temperature, and a decrease in T signifies cooling. The temperature at an instant n is denoted by T_n, and a cooling diagram is what we call a series $\{T_n\}_{n \in \mathbb{N}}$ such that

$$T_n \geq T_{n+1} \quad \text{and} \quad \lim_{n \to \infty} T_n = 0.$$

The cooling procedure is tricky, however, because we cannot simply combine the results (7.8) and (7.9), since we do not necessarily have $\lim_{n \to \infty} P[Z_n = x \mid Z_0 = x(0)] = \Pi_0(x)$. For this, the cooling would have to satisfy certain assumptions, as described below.

● SIMULATED ANNEALING AND METROPOLIS DYNAMIC

Let us take another look at Section 7.1.3. Let Q be an irreducible and symmetric exploration matrix, and let $\{T_n\}_{n \in \mathbb{N}}$ be a cooling diagram. The transition matrix (7.6) is no longer homogeneous because it is dependent on n. We write it as $\mathcal{P}^{(n-1,n)}$. When $x \neq y$, it has the following element:

$$p_{xy}^{(n-1,n)} = q_{xy} \exp - \left[\frac{1}{T_n}(U(x) - U(y))_+ \right].$$

The simulated annealing algorithm is formally the Metropolis algorithm for variable temperature. Let us present it without considering the fact that U is derived from a potential.

Algorithm 7.3. (Simulated Annealing I).

- *Step 0* : Select $x(0)$ arbitrarily.

- *Step n* : Transition $(n-1) \rightarrow n$.
  Let $x(n-1)$ be the occurrence of Z_{n-1} at step $(n-1)$.
  Take one $y \in E$ according to the distribution
  $$\left\{ p^{(n-1,n)}_{x(n-1),y} \ , \ y \in E \right\}.$$

The basic problem here is to determine the conditions on the cooling diagram such that this algorithm minimizes U. These conditions will be given by the following theorem, credited to Hajek. First, we give several definitions that must be established before stating this theorem.

For all $x \in E$, we define its neighborhood $\mathcal{N}_x$ as $\mathcal{N}_x = \{y \in E, \ q_{xy} > 0\}$. With respect to these neighborhoods, we define the concept of a local minimum: A point $y \in E$ is a local minimum of U if $U(y) \leq U(x)$ for all $x \in \mathcal{N}_y$. We write E_{Lmin} for the set of these minima ($E_{\mathrm{min}} \subset E_{\mathrm{Lmin}}$). Two configurations x and y are said to connect with the height $\hbar$ if $y = x$ and $U(x) \leq \hbar$, or if there exists a sequence $x(1) = x, x(2), \ldots, x(n) = y$ such that $(U(x(j)) \leq \hbar, \ x(j+1) \in \mathcal{N}_{x(j)})$ for all $j = 1, \ldots, n$. Finally, the depth κ_x of a local minimum x is the smallest value $K > 0$ for which there exists $y \in E$ such that x and y connect to the height $U(x) + K$ with $U(y) < U(x)$.

Theorem 7.3. *Let us take the algorithm "Simulated Annealing I". We then have*

$$\lim_{n \to \infty} P[Z_n \in E_{\mathrm{min}} \mid Z_0 = x] = 1, \quad \forall \ x \in E,$$

if and only if $\sum_{n=1}^{\infty} \exp -(K/T_n) = \infty$, *where* $K = \max\{\kappa_x, x \in (E_{\mathrm{Lmin}} - E_{\mathrm{min}})\}$.

Proof of this theorem is given in [106, 7]. In this theorem, the cooling diagram of the form

$$T_n = \frac{C}{\log n} \quad \text{with} \quad C \geq K,$$

satisfies the condition. This condition and the condition for the theorem associated with the "Simulated Annealing II" algorithm presented below (page 134) are difficult to implement. Strictly observing this speed results in prohibitive calculation times. This is why we often violate the condition, for example with a diagram of the following type

$$T_n = T_0 \lambda^n, \quad \text{where} \quad 0.8 \leq \lambda \leq 0.99.$$

Another approach would be to parallelize these algorithms. For information on this subject, refer to [10].

EXAMPLE.
We again consider the index given in definition 4.1 (Section 4.3.1, page 88). Algorithm 4.3 is in fact a Metropolis version of a simulated annealing algorithm. This index is written as

$$I(a) = \sum_{l=1}^{N} I_l \, \mathbf{1}[a \in R_l].$$

The regions R_l are a partition of R^n on which $I(a)$ is piecewise constant. Maximizing I in a is equivalent to maximizing it on the simplex $E = \{x \in \{0,1\}^N : \sum_l x_l = 1\}$. We therefore set

$$I(x) = \sum_{l=1}^{N} I_l \, x_l \quad \text{and} \quad \Pi_T(x) = \frac{1}{\mathcal{Z}_T} \exp \frac{1}{T} I(x).$$

The exploration matrix $Q = (q_{x,x'})$ of the regions (or, equivalently, of E) is defined via the symmetry operation Ξ_k (4.18). Given a point $a \in R$, we say that R and R' are neighbors if there exists a hyperplane H_k such that $\Xi_k(a) \dot{=} a' \in R'$. This matrix is symmetric and irreducible because the R_l constitute a partition, and it is therefore always possible to find a path of finite length joining two non-neighboring regions. With this given, the Metropolis dynamic (7.5) at iteration n is as follows. It is expressed with respect to $I(a)$ instead of $I(x)$, because $I(x)$ is for the purpose of formalization only. Let $a(n)$ be the current value of a. The following value, a', is obtained not by random selection according to $q_{x(n),x'}$, but in an equivalent way by uniform random selection on the indices k of the hyperplanes H_k. We then have $a' = \Xi_k(a(n))$ if k is the result of this selection, and finally, as in (7.5),

$$a(n + 1) = \begin{cases} a' & \text{if } T_n \, log\xi < (I(a') - I(a(n))), \\ a(n) & \text{otherwise}, \end{cases}$$

which matches the Algorithm 4.3.

• SIMULATED ANNEALING AND GIBBS SAMPLER

Let us look again at the periodic path through the sites of S. In this case, $\{T_n\}_{n \in N}$ is chosen to be a constant value for each sweep of S. We write $T_{[k]}$ to denote the temperature at the kth sweep:

$$T_{[k]} \dot{=} T_{k|S|+1} = T_{k|S|+2} = \cdots = T_{(k+1)|S|}, \quad \forall \, k \geq 0.$$

The simulated annealing algorithm is formally the Gibbs sampler algorithm, but at variable temperature. We write $\Pi_T(y_s \mid \breve{y}_s = \breve{x}_s)$ for the conditional distribution at the site s. This algorithm is the following:

Algorithm 7.4. (Simulated Annealing II).

- *Step 0* : Select $x(0)$ arbitrarily.

- *Step n* : Transition $(n-1) \to n$.
  Let x be the occurrence of Z_{n-1} at step $(n-1)$.
  Update site s_n by taking a $y \in E$ according to the distribution
  $$\left\{ \mathbf{1}[\breve{y}_{s_n} = \breve{x}_{s_n}] \, \Pi_{T_n}(y_{s_n} \mid \breve{y}_{s_n} = \breve{x}_{s_n}), \; y_{s_n} \in E_{s_n} \right\}.$$

The remaining question is the choice of cooling diagram. The following theorem from Geman and Geman [84] provides a sufficient condition for convergence. A proof is given in Section 7.3.

Theorem 7.4. *Let us set* $\Delta = \max(U(x) - U(y))$, *where the* max *is calculated on the pairs* (x, y) *that are different at a single site. Let* $\{T_{[k]}\}_{k \in \mathbb{N}}$ *be a series that is decreasing toward* 0 *and such that for sufficiently large* k,

$$T_{[k]} \geq \frac{|S|\Delta}{\log k}. \tag{7.10}$$

Then, $\lim_{n \to \infty} P[Z_n \in E_{\min} \mid Z_0 = x(0)] = 1$, *for all* $x(0) \in E$.

7.3 Probabilistic Aspects

[1] This section provides a more general approach to the concepts of Markov chain convergence.

- **MARKOV CHAINS AND CONTRACTION COEFFICIENT**

Our formalism now includes nonhomogeneous chains. Let $\{Z_n, \; n \leq 0\}$ be a Markov chain such that

$$P[Z_n = y \mid Z_{n-1} = x] \doteq p_{xy}^{(n-1,n)}, \quad \forall x, y \in E.$$

This chain is described by the transition matrices $\{\mathcal{P}^{(n-1,n)}\}_{n=1}^{\infty}$ (which may or may not be dependent on n) and by a vector ψ representing the initial distribution. We write the following for all $k > n$:

$$\psi^{(n,k)} = \psi \mathcal{P}^{(n,n+1)} \mathcal{P}^{(n+1,n+2)} \dots \mathcal{P}^{(k-1,k)} = \psi \mathcal{P}^{(n,k)}.$$

This is the probability at instant k for an initialization ψ at instant n. The main question is about the asymptotic behavior of $\psi^{(0,k)}$ or $\psi^{(n,k)}$ when $k \to \infty$. When convergence occurs,

$$\lim_{k \to \infty} \psi \mathcal{P}^{(n,k)} = \Pi, \tag{7.11}$$

---

[1]This section can be omitted at the first reading.

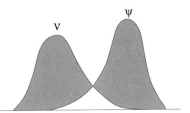

Figure 7.1. Distance L^1 between two probability distributions (shaded surface).

and this is independent of ψ (memory loss), the chain is said to be strongly ergodic. There can, however, be memory loss without convergence, in which case the chain is said to be weakly ergodic. The distance between probability distributions used in these concepts is based on the norm L^1:

$$\|\psi - \nu\| = \sum_{x \in E} |\psi(x) - \nu(x)|,$$

illustrated in Figure 7.1.

Definition 7.1. Ergodicity.

○ *A Markov chain is strongly ergodic if there is a probability vector* Π *such that for all* n,

$$\lim_{k \to \infty} \max_{\psi} \left\| \psi \mathcal{P}^{(n,k)} - \Pi \right\| = 0,$$

○ *It is weakly ergodic if for all* n,

$$\lim_{k \to \infty} \max_{\psi, \nu} \left\| \psi \mathcal{P}^{(n,k)} - \nu \mathcal{P}^{(n,k)} \right\| = 0.$$

It is difficult to use these definitions directly, so we tend to use the *contraction coefficient* δ credited to Dobrushin:

$$\delta(\mathcal{P}) = \frac{1}{2} \max_{x,y} \|p_{x.} - p_{y.}\|.$$

This coefficient belongs to $[0, 1]$ and has the following properties:

1. $\delta(\mathcal{P}) = 1 - \min_{xy} \left(\sum_u \min\{p_{xu}, p_{yu}\} \right) \leq 1 - |E| \min_{xy} p_{xy}$,

2. $\|\psi \mathcal{P} - \nu \mathcal{P}\| \leq \|\psi - \nu\| \delta(\mathcal{P})$,

3. $\delta(\mathcal{P}\mathcal{Q}) \leq \delta(\mathcal{Q})\delta(\mathcal{Q})$.

The name "contraction coefficient" is justified by the second property. These properties are useful for examining the convergence of stochastic algorithms in this chapter. With the second property, weak ergodicity is equivalent to $\lim_{k \to \infty} \delta(\mathcal{P}^{(n,k)}) = 0$ for all n, and with the third property we obtain a sufficient condition:

$$\forall\, n, \quad \prod_{k \geq n} \delta\left(\mathcal{P}^{(n,k)}\right) = 0. \tag{7.12}$$

Furthermore, a sufficient condition for strong ergodicity is given by the following theorem.

Theorem 7.5. *Let $\{\mathcal{P}^{(n-1,n)}\}$ be the series of transition matrices of a weakly ergodic Markov chain, and let $\{\Pi^{(n)}\}$ be the associated series of invariant vectors: $\Pi^{(n)}\mathcal{P}^{(n-1,n)} = \Pi^{(n)}$. If these vectors satisfy*

$$\sum_{n=1}^{\infty} \left\| \Pi^{(n)} - \Pi^{(n+1)} \right\| < \infty, \tag{7.13}$$

then the chain is strongly ergodic.

Because of the condition (7.13), the series $\{\Pi^{(n)}\}$ is a Cauchy series and is therefore convergent: $\lim_{n \to \infty} \Pi^{(n)} = \Pi$.

● CONVERGENCE EXAMPLES

The above concepts allow us to demonstrate the convergence of the stochastic algorithms described above [89, 182]. Let us look at two examples.

Proof (theorem 7.1):
Let us prove (7.1): $\lim_n \Pi^{(0)}\mathcal{P}^n = \Pi$, where $\mathcal{P}$ is a regular matrix, i.e., $\mathcal{P}^r > 0$. For this, we write

$$\left\| \Pi^{(0)}\mathcal{P}^n - \Pi \right\| = \left\| \Pi^{(0)}\mathcal{P}^n - \Pi\mathcal{P}^n \right\| \leq \left\| \Pi^{(0)} - \Pi \right\| \delta(\mathcal{P}^n) \leq 2\delta(\mathcal{P}^n).$$

To establish (7.1), it is then sufficient to show that $\delta(\mathcal{P}^n) \to 0$. For any matrix $\mathcal{P}$ that is not necessarily regular, property 3 above means that we have $\delta(\mathcal{P}^{n+1}) \leq \delta(\mathcal{P})\delta(\mathcal{P}^n) \leq \delta(\mathcal{P}^n)$. If, additionally, $\mathcal{P}^r > 0$, then for $n > r$, $\delta(\mathcal{P}^n) \leq \delta((\mathcal{P}^r)^k \mathcal{P}^{n-rk}) \leq \delta(\mathcal{P}^r)^k$, where k is the largest integer such that $rk \leq n$; according to property 1, $\mathcal{P}^r > 0$ implies $\delta(\mathcal{P}^k) < 1$, which proves that $\delta(\mathcal{P}^n)$ converges to 0. □

Proof (theorem 7.4):
We apply Theorem 7.5. The transition matrix $\mathcal{P}^{(n-1,n)}$ has the following element:

$$p_{xy}^{(n-1,n)} = \mathbf{1}_{[\breve{y}_{s_n} = \breve{x}_{s_n}]}\, \Pi_{T_n}(y_{s_n} \mid \breve{y}_{s_n} = \breve{x}_{s_n}).$$

At the kth sweep, we write $\mathcal{P}^{[k]}$ to denote the transition matrix (7.7),

$$\mathcal{P}^{[k]} \doteq \mathcal{P}^{(k|S|,(k+1)|S|)},$$

and $\Pi^{[k]}$ for the vector associated with the distribution $\Pi_{T_{[k]}}$. We need to check that the following conditions are satisfied: (i) The invariance of $\Pi^{[k]}$, (ii) The inequality (7.13), (iii) Weak ergodicity (7.12).

Proof of (i). Each component $\mathcal{P}^{(n,n+1)}$ of $\mathcal{P}^{[k]}$ is reversible: $\Pi_{T_{[k]}}(x)P_{xy}^{(n,n+1)}$ $= \Pi_{T_{[k]}}(y)P_{yx}^{(n,n+1)}$. As in Lemma 7.1, we deduce from this the invariance $\Pi^{[k]}\mathcal{P}^{(n,n+1)} = \Pi^{[k]}$ and from this, the invariance with respect to $\mathcal{P}^{[k]}$:

$$\Pi^{[k]}\mathcal{P}^{[k]} \;=\; \Pi^{[k]}.$$

Proof of (ii). The inequality $\sum_k \left\| \Pi^{[k]} - \Pi^{[k+1]} \right\| < \infty$ is the result of any increase or decrease of $\Pi^{[k]}(x)$ expressed in Proposition 7.1. Proof of this result comes from the following property:

$$0 \le \sum_k \left\| \Pi^{[k]} - \Pi^{[k+1]} \right\| = 2 \sum_x \sum_k \left(\Pi^{[k+1]}(x) - \Pi^{[k]}(x) \right)_+ .$$

Because the function is monotonic, there exists a k_0 such that either $\left(\Pi^{[k+1]}(x) - \Pi^{[k]}(x) \right)_+ = 0$ for all $k \ge k_0$, or $\left(\Pi^{[k+1]}(x) - \Pi^{[k]}(x) \right)_+ =$ $\Pi^{[k+1]}(x) - \Pi^{[k]}(x)$, for which $\sum_{k_0}^K = \left(\Pi^{[k+1]}(x) - \Pi^{[k]}(x) \right)_+ = \Pi^{[K+1]}(x) - \Pi^{[k_0]}(x) \le 1$, for all K.

Proof of (iii). We set $K_{[k]} \doteq \min_{\lambda \in E, s \in S} \Pi_{T_{[k]}}(\lambda_s \mid \check{\lambda}_s)$. We also have

$$\Pi_{T_{[k]}}(y_s \mid \check{y}_s = \check{x}_s) \;=\; \frac{\Pi_{T_{[k]}}(y_s, \check{y}_s = \check{x}_s)}{\Pi_{T_{[k]}}(., \check{y}_s = \check{x}_s)}$$

$$= \left[\sum_{\lambda \in E_s} \exp - \frac{1}{T_{[k]}} [U(\lambda, \check{y}_s = \check{x}_s) - U(y_s, \check{y}_s = \check{x}_s)] \right]^{-1}$$

$$\ge \frac{1}{|E_s|} \exp - \frac{\Delta}{T_{[k]}},$$

where Δ is defined in Theorem 7.4. Property 1 above results in the inequality $\delta(\mathcal{P}^{[k]}) \le 1 - |E| \, K_{[k]}^{|S|}$, which with the previous lower bound of $\Pi_{T_{[k]}}$, gives

$$\delta(\mathcal{P}^{[k]}) \le 1 - c \exp - \frac{\Delta |S|}{T_{[k]}}, \tag{7.14}$$

where $c = |E|/|E_s|^{|S|}$. Finally, according to (7.12), the weak ergodicity condition is satisfied if $\prod_{k \ge n} \delta(\mathcal{P}^{[k]}) = 0$ or, equivalently,

$$\sum_{k \ge 1} \left(1 - \delta(\mathcal{P}^{[k]}) \right) = \infty,$$

(since for two series such that $0 \leq a_n \leq b_n \leq 1$, we have $\sum_n a_n = \infty \Rightarrow \prod_n (1 - b_n) = 0$). If we set $T_{[k]} = \gamma \log k$, from (7.14) we obtain

$$1 - \delta(\mathcal{P}^{[k]}) \geq c\, k^{-\Delta|S|\gamma}.$$

We can therefore be sure that the series diverges if $\Delta|S|\gamma > 1$, i.e., if $T_{[k]} \geq |S|\Delta / \log k$. $\square$

This type of proof also applies to the convergence of the Gibbs sampler, even though its chain is homogeneous. In that case, we only need to demonstrate the invariance $\Pi \mathcal{P}^{[k]} = \Pi$ for all k (which therefore implies (7.13)), and then the weak ergodicity, to obtain the strong ergodicity.

7.4   Deterministic Optimization

7.4.1   ICM Algorithm

Once again we are concerned with minimizing the function $U(x)$, $x \in E$, but our goal is less ambitious than before, because we are only looking for a local minimum of $U(x)$. In practical terms, we have an initial configuration $x(0)$ that is near to a local minimum that is of interest for the application under consideration. We will now introduce the *iterated conditional mode* (ICM) method, as given by Besag (1988). The optimization problem is

Given $x(0)$, locally maximize
$$P[X = x] = \tfrac{1}{Z} \exp -U(x).$$

The associated algorithm is deterministic and iterates sweeps of S while updating at each site of S, as was seen above in the case of stochastic algorithms. The ICM algorithm starts from the following breakdown

$$P[X = x] = P\big[X_s = x_s \mid \check{X}_s = \check{x}_s\big]\, P\big[\check{X}_s = \check{x}_s\big],$$

where again we write $\check{x}_s = \{x_t,\ t \neq s\}$. Note that maximizing the conditional probability $P\big[X_s = x_s \mid \check{X}_s = \check{x}_s\big]$ at x_s increases $P[X = x]$. This maximization is merely the search for the conditional mode at s; hence the name of this method. This search can, however, be costly if the conditioning $\check{x}_s$ is unknown. Our first reflex, then, is to replace it with a conditioning taken from a running configuration, which gives the following algorithm.

Algorithm 7.5. (ICM algorithm).

- *Step 0* : Let $x(0)$ be a configuration of interest.
- *Step n* : Transition $(n-1) \to n$.
  Let s_n be the site visited in Step n. Calculate

$$x_s(n) = \begin{cases} x_s(n-1) & \text{if} \quad s \neq s_n, \\ \arg\max_{x_s \in E_s} P\big(x_s \mid \breve{x}_s(n-1)\big) & \text{if} \quad s = s_n. \end{cases}$$

Proposition 7.2. *The series of configurations $\{x(n)\}_{n \in \mathsf{N}}$ of the ICM algorithm has an increasing probability.*

Proof:
The proof is simple. We set $s = s_n$. We simply need to show that $P[X = x(n)] \geq P[X = x(n-1)]$ for all $n > 1$. As above, we have the breakdown at the iteration $(n-1)$:

$$P[x(n-1)] = P\big[x_s(n-1) \mid \breve{x}_s(n-1)\big] P\big[\breve{x}_s(n-1)\big].$$

When going from $(n-1)$ to n, we have $\breve{x}_s(n-1) = \breve{x}_s(n)$, and therefore, at the iteration n, the breakdown is

$$P[x(n)] = P\big[x_s(n) \mid \breve{x}_s(n-1)\big] P\big[\breve{x}_s(n-1)\big].$$

But the algorithm gives us

$$P\big[x_s(n) \mid \breve{x}_s(n-1)\big] = \max_{x_s} P\big[x_s \mid \breve{x}_s(n-1)\big],$$

and the stated property is therefore satisfied. □

The algorithm converges toward a local maximum that is strongly dependent on the initial configuration $x(0)$. In practice, this procedure should be using sparingly, depending on the context. We should also emphasize that experience has shown that this algorithm converges very rapidly, in just a few sweeps. We will now illustrate the ICM on an inverse problem example.

EXAMPLE.
We consider a Markov random field X with the following local specifications

$$P\big[X_s = \lambda \mid \breve{X}_s = \breve{x}_s\big] \propto \exp\big[\alpha K_s(\lambda, \breve{x}_s)\big], \tag{7.15}$$

where $\lambda \in E_s = \{\lambda_1, \ldots, \lambda_m\}$ for all s. We write $K_s(\lambda, \breve{x}_s)$ for the number of neighbors of s at value λ:

$$K_s(\lambda, \breve{x}_s) = \sum_{t \in N_s} \mathbf{1}_{[x_t = \lambda]},$$

and α is a positive parameter assumed to be known. This means that a large $K_s(\lambda, \breve{x}_s)$ strengthens the probability of the value λ at site s. We should, however, check that these specifications do indeed connect.

In this example X is a hidden field, because only a field Y can be observed instead of X:

$$Y_s = \mu(X_s) + W_s,$$

where $\mu(\lambda)$ is a known real function and W_s is an independently and identically distributed Gaussian noise: $W_s \sim LG(0, \sigma^2)$. Taking the results found in Chapter 5, we know that the posterior distribution is

$$P[X = x \mid Y = y] \propto \exp -[U(x) + R(y - \mu(x))]$$
$$\propto \exp -U(x \mid y),$$

where the function μ applies in this case to each coordinate of x, and where $R(w) = \frac{1}{2\sigma^2} \sum_s w_s^2$. We again write $P_y[X = x] = P[X = x \mid Y = y]$ for this distribution. To solve the inverse problem in the MAP sense, we would have to find an overall maximum of $P_y[X = x]$. Let us assume, however, that we have an initial solution $x(0)$ for our inverse problem, and then implement the ICM algorithm. To do this, we determine the expression for the local specifications as indicated by Corollary 5.1 (page 105).

$$P_y[X_s = \lambda \mid \check{X}_s = \check{x}_s]$$
$$\propto \exp -[U(x \mid y) - U(x_{/s} \mid y)]$$
$$\propto \exp -[U(x) - U(x_{/s}) + R(y - \mu(x)) - R(y - \mu(x_{/s}))]$$
$$\propto \exp \left[\alpha K_s(\lambda, \check{x}_s) - \frac{1}{2\sigma^2}(y_s - \mu(\lambda))^2 + \frac{1}{2\sigma^2} y_s^2 \right],$$

where we have assumed that the reference value 0_s is such that $\mu(0_s) = 0$.

At iteration n, the ICM algorithm calculates $\max_\lambda P_y\left[X_{s_n} = \lambda \mid \check{X}_{s_n} = \check{x}_{s_n}(n-1)\right]$, or equivalently,

$$\max_\lambda \left\{ \alpha R_{s_n}(\lambda, \check{x}_{s_n}) - \frac{1}{2\sigma^2}(y_{s_n} - \mu(\lambda))^2 \right\}.$$

Now we must select the configuration $x(0)$, which is left to the user's choice. We experiment with the configuration that corresponds to the maximum likelihood without prior knowledge:

$$x(0) = \arg\max_x \{\exp -R(y - \mu(x))\},$$

$$\text{or } x_s(0) = \arg\max_\lambda \left\{ -\frac{1}{2\sigma^2}(y_s - \mu(\lambda))^2 \right\}, \quad \forall s \in S.$$

J. Besag describes an experiment performed with a (120×120) image x^* encoded on a scale of six values: $\mu(\lambda) = 0, \ldots, 5$. This image was created artificially and includes well structured regions. It was then degraded to an image y by a noise w of covariance $\sigma^2 = 0.6$. It is encoded on a scale of 256 gray levels. The configuration $x(0)$ has 35% of sites that are erroneous compared to x^*. On the other hand, the ICM algorithm with $\alpha = 1.5$, after six sweeps, gives a configuration $x(6)$ that has an error rate of only 1.7%. If the result is satisfactory to the user, the local optimum is kept without any concern about whether or not it is global. This approach is ineffective, however, when no satisfactory initial configuration is available.

This experiment illustrates the importance of the gain obtained using the regularization given by $U_1(x)$. Without the regularization, the configuration $x(0)$ is calculated independently, site by site. This results in a high rate of incorrect decisions. With the regularization, however, the decisions interact because the prior information contained in $U_1(x)$ favors configurations consisting of well-structured regions. This is a typical example that encompasses the whole philosophy of the Bayesian–Markov approach.

7.4.2 Relaxation Algorithms

Deterministic optimization methods are usually of the *gradient descent* type. They are described in detail in several reference works, such as [131, 92]. These methods seek to optimize nonlinear functions $U(x)$ of the class $\mathcal{C}^1$ or $\mathcal{C}^2$ on R^n. A special role is played by convex and concave functions in maximization and minimization, respectively.

A function f defined on a convex set C is said to be convex function if for every $x, y \in C$ and $0 \le \lambda \le 1$,

$$f(\lambda x + (1 - \lambda)y) \le \lambda f(x) + (1 - \lambda)f(y). \tag{7.16}$$

The function is said to be concave if the inequality is reversed. In this definition C is a convex set if for all x, y in C, the line segment joining x and y is also in C.

Outside this class of functions, these methods seek a local optimum only. In minimization, the gradient descent method forms the basis of many methods. It is based on a linear approximation of U given by the Taylor expansion

$$U(x + \epsilon\wp) \approx U(x) + \epsilon\,\wp'\nabla U(x), \tag{7.17}$$

where $\nabla U = \frac{\mathrm{d}}{\mathrm{d}x}U$ is the gradient of U, $\wp$ a direction in R^n, and ϵ a positive scalar value. If we want $\wp$ to be a direction of descent along the surface defined by U in R^{n+1}, i.e., $U(x + \epsilon\wp) < U(x)$, we must choose $\wp$ such that $\wp'\nabla U(x) < 0$. In the gradient descent method we choose $\wp$ from all the vectors of norm 1 to minimize $\wp'\nabla U(x)$. When we take $\|\wp\|^2 = \wp'\wp$, we obtain $\wp = -\nabla U(x)$. The resulting gradient descent algorithm is an iterative procedure that calculates a series of configurations $x(1), x(2), \ldots$ from an initial configuration $x(0)$. At step n, we want to have $U(x(n) + \epsilon_n\wp_n) \le U(x(n))$, and we must therefore set

$$x(n + 1) = x(n) + \epsilon_n\wp_n,$$

which in the special case $\wp_n = -\nabla U(x(n))$ is written

$$x(n+1) = x(n) - \epsilon_n \nabla U(x(n)). \qquad (7.18)$$

Under certain conditions, this algorithm converges to a configuration $x^\star$ that satisfies $\nabla U(x^\star) = 0$. The choice of ϵ_n is difficult, except in convex cases such as the quadratic case, where we choose $\epsilon_n = \arg\min_\epsilon U(x(n)) - \epsilon \nabla U(x(n))$.

Note that in the gradient descent algorithm, going from $x(n)$ to $x(n+1)$ occurs by the simultaneous changing (known as relaxation) of all x_s, $s \in S$. Other approaches proceed sequentially by relaxing each x_s in turn or by relaxing subsets of $(x_s, s \in S)$. This simply amounts to considering specific directions $\wp$. This is true of the ICM algorithm in the previous subsection, where a single value x_s is relaxed at a time:

$$\begin{aligned} x_s(n+1) &= \arg\max_{x_s \in E_s} P\big[X_s = x_s \mid \check{X}_s = \check{x}_s(n)\big] \\ &= \arg\min_{x_s \in E_s} U\big(x_s, \check{x}_s(n)\big), \qquad (7.19) \\ x_{s'}(n+1) &= x_{s'}(n), \quad \forall s' \neq s. \end{aligned}$$

In this algorithm the direction $\wp$ at step n, written as $\wp_s$, is the direction of the coordinate s, $\wp_s = (0, \ldots, 0, 1, 0, \ldots, 0)'$. We therefore have a coordinate-by-coordinate descent, which will be shown in greater detail in the following example. In other situations it may be advantageous to perform a relaxation by subsets of $(x_s, s \in S)$. For example, when $U(x)$ is written as $U(x^p, x^b)$ where x^p and x^b have a specific meaning for the application concerned, it is natural to relax x^p and x^b alternately. This type of block-by-block relaxation is examined in [160], which gives a proof of convergence toward the overall minimum when U is convex.

EXAMPLE.
Let us look again at the energy of approximation by a spline surface (2.36) described in Section 2.5.3. (page 49). Taking the resolution $\ell = 1$, it is written by setting $s = (i, j)$:

$$\begin{aligned} U(x) = \alpha \sum_{ij} \Big(& [x_{i+1,j} - 2x_{i,j} + x_{i-1,j}]^2 \\ & +2[x_{i+1,j+1} - x_{i,j+1} - x_{i+1,j} + x_{ij}]^2 \\ & +[x_{i,j+1} - 2x_{i,j} + x_{i,j-1}]^2 \Big) \\ +\frac{1}{2} \sum_{(i,j) \in S_0} & (x_{i,j}^2 - 2y_{i,j}x_{i,j}), \end{aligned}$$

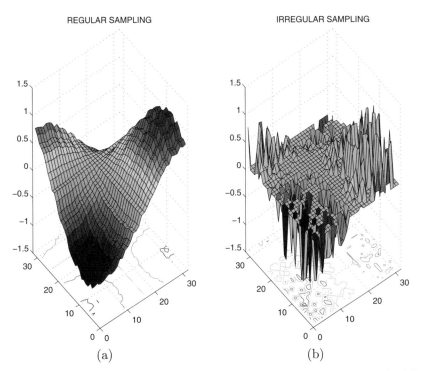

Figure 7.2. Surface sampled regularly and irregularly. (a) $\{y_s, \ s \in S\}$. (b) $\{y_s, \ s \in S_0\}$.

where S_0 is the set of sites carrying the observed values (Figure 7.2(b)). We rewrite this in vector form as follows:

$$U(x) = \frac{1}{2} \langle \underline{x}, \underline{A}\,\underline{x} \rangle - \langle \underline{F}, \underline{x} \rangle.$$

The function U is quadratic. If $\underline{A}$ is positive definite, the overall minimum $\widetilde{x}$ of U is a solution of $\nabla U(\widetilde{x}) = 0$, or $\underline{A}\,\widetilde{x} - \underline{F} = 0$. This linear system is huge, but we can take advantage of the fact that $\underline{A}$ is a block-diagonal sparse matrix. The ICM algorithm amounts to calculating iteratively $\arg\min_{x_s} U(x_s, \breve{x}(n))$, or

$$\frac{\partial}{\partial x_s} U\big(x_s(n+1), \breve{x}(n)\big) = 0.$$

This equation in $x_s(n+1)$ is written as

$$\underline{A}_{s,s} x_s(n+1) + \sum_{s,s':\, s' \neq s} \underline{A}_{s,s'} x_{s'}(n) - \underline{F}_s = 0. \tag{7.20}$$

This iterative diagram is the well-known Gauss–Seidel algorithm, used to solve $\underline{A}\,\widetilde{x} - \underline{F} = 0$ by iteration. We have therefore just proven that the ICM algorithm and the Gauss–Seidel algorithm are equivalent in the quadratic case.

Inside the grid S, the relaxation calculation (7.20) for x_s is

$$
\begin{aligned}
0 = {} & 20\alpha\, x_{i,j}(n+1) - \\
& 8\alpha\, [x_{i-1,j}(n) + x_{i+1,j}(n) + x_{i,j-1}(n) + x_{i,j+1}(n)] + \\
& 2\alpha\, [x_{i-1,j-1}(n) + x_{i+1,j+1}(n) + x_{i-1,j+1}(n) + x_{i+1,j-1}(n)] + \\
& \alpha\, [x_{i-2,j}(n) + x_{i+2,j}(n) + x_{i,j-2}(n) + x_{i,j+2}(n)] + \\
& x_{i,j}(n+1)\mathbf{1}_{(i,j)\in S_0} - y_{i,j}\mathbf{1}_{(i,j)\in S_0},
\end{aligned}
\tag{7.21}
$$

where the calculation at $s = (i,j)$ is very localized: It involves only the values of x that lie within a 5×5 neighborhood around s. The relaxation (7.21) is valid for the sites s that are neither on the circumference of S nor within a distance of 1 from the circumference. On this circumference and in its neighborhood, the relaxation formula must be changed accordingly. There are five special cases.

Figures 7.2 and 7.3 illustrate this optimization using the example from Figure 3.4 (page 59). Then, the noisy observed values used were given on all S (Figure 7.2(a)). Now, we only have observed values on a subset S_0 of the square grid S representing in this case one-third of the sites of S. The sites of S_0 were obtained by random selection in S (Figure 7.2(b)). The regularization parameter is $\alpha = 1/2$, and the initial solution is $x_s(0) = 0$ for all s. The convergence (Figure 7.2) is slower here than in the earlier example given in Section 7.4.1. This is because the initial solution is very far from the final solution, and also because of the reduced numbers of sites of S_0.

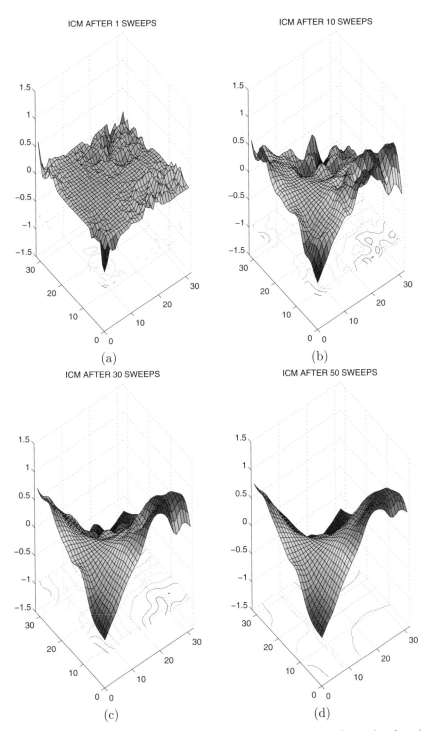

Figure 7.3. Approximation of an irregularly sampled surface. (a, b, c) Intermediate solutions. (d) Final solution.

8
Parameter Estimation

One of the difficulties of using energy models is the need to choose the weighting parameters for each of their terms. In the simple examples given in the previous chapters, the energy expressions contained parameters that were assumed to be known. In real applications, however, these parameters are always unknown. In this chapter we will describe methods that can be used to estimate these parameters. These are general methods covering a wide range of situations. Nonetheless, the estimation of parameters is still a problem and a drawback of energy-based modeling.

There are two levels of difficulty. The first level concerns energies related to observable fields; here we say we are dealing with *complete data*. This estimation problem falls within the classic context of the parametric estimation of spatial stochastic processes. It requires the estimation of the parameters of a model based on direct observations that are assumed to obey the model in an approximate fashion. The second, and far more complicated, level involves parameters associated with hidden fields for which we have no observations. We are then dealing with *incomplete data*. We should emphasize that most real applications fall into this category.

8.1 Complete Data

Setting the problem:
Let us take the family of Gibbs distributions

$$\Pi_\theta(x) = \frac{1}{\mathcal{Z}_\theta} \exp -U(x, \theta),$$

$$U(x, \theta) = \sum_{i=1}^{q} \theta_i U_i(x), \qquad (8.1)$$

$$U_i(x) = \sum_{c \in C_i} V_{(i),c}(x),$$

where $\theta = (\theta_1, \ldots, \theta_q)'$ is a vector of parameters belonging to a domain included in $\mathbb{R}^q$. Let U_i be a neighborhood potential energy whose set of cliques is C_i, where the potential of any clique $c \in C_i$ is $V_{(i),c}$. We have one observed value x^O of a Markov random field X with theoretical distribution

$$P[X = x] = \Pi_{\theta^*}(x).$$

The problem is then to estimate θ^* in accordance with x^O.   △

This being set, from now on we will use the notation $P_\theta(x)$ instead of $\Pi_\theta(x)$. The expression for the energy U is the expression introduced in Chapter 6, but limited to the energy written as U_1. Note that U takes on a special form because it is linear with respect to θ:

$$U(x, \theta) = \theta' \mathbf{U}(x),$$

where we write $\mathbf{U}(x) = (U_1(x), \ldots, U_q(x))'$. A more general form would be

$$U(x, \theta) = \sum_{i=1}^{q} U_i(x, \theta_i),$$

where $U_i(x, \theta_i)$ would not necessarily be linear with respect to θ_i. The most usual form for model building in actual applications is

$$U(x, \theta) = \sum_{i=1}^{q} \theta_i U_i(x, \beta_i),$$

where θ_i and β_i are unknown parameters, and β_i can be a vector. In Part III we will see that the parameters β_i can often be chosen empirically, while the weighting parameters θ_i cannot. That is why we will deal with the form (8.1) first. Below, we present three estimation methods in decreasing order of complexity. The first, the maximum likelihood method, has limited practical utility. It does, however, give a good grasp of the problem of estimating parameters. The second, the maximum pseudolikelihood method is, in our opinion, the definitive method. The third, an empirical method, is less general in scope, but it is simple and it provides an explicit expression for the estimation.

8.1.1 Maximum Likelihood

The maximum likelihood method (Section 1.1.1) is the best known method in statistics for estimating the parameters of a model. This method is important because of the good asymptotic statistical properties of the resulting estimator. The method is based on the following principle.

Definition 8.1. *Let x^O be an observed value from a random field X, and let $\theta^* = (\theta_1^*, \ldots, \theta_q^*)'$ be the unknown vector of parameters. To estimate θ^*, the principle of maximum likelihood consists in choosing the vector $\widehat{\theta}$ that makes the observed value x^O the most probable:*

$$\widehat{\theta} = \arg\max_{\theta} P_\theta \left[X = x^O \right].$$

"Likelihood" is what we call the function $L(\theta) = P_\theta \left[X = x^O \right]$.

This definition is identical to Definition 1.1 (page 7). Because the logarithm is concave, $\widehat{\theta}$ also maximizes $\log L(\theta)$. It would seem impossible to solve this optimization numerically, because the normalization constant $\mathcal{Z}_\theta$ that depends on θ cannot be calculated. Let us, however, look at $\log L(\theta)$ and, first, at its gradient. From

$$\log L(\theta) = -\log\left(\sum_x \exp -\left[\theta' \mathbf{U}(x)\right]\right) - \theta' \mathbf{U}(x^O)$$

we obtain, by vector differentiation,

$$\begin{aligned}
\frac{d}{d\theta} \log L(\theta) &= \frac{\sum_x \mathbf{U}(x) \exp -\theta' \mathbf{U}(x)}{\sum_x \exp -\theta' \mathbf{U}(x)} - \mathbf{U}(x^O) \\
&= \sum_x \mathbf{U}(x) P_\theta(x) - \mathbf{U}(x^O) \\
&= \mathbf{E}_\theta[\mathbf{U}(X)] - \mathbf{U}(x^O).
\end{aligned} \tag{8.2}$$

A similar calculation gives the matrix of the Hessian whose element (i, j) is

$$\begin{aligned}
\frac{\partial^2}{\partial\theta_i \partial\theta_j} \log L(\theta) &= -\sum_x U_i(x) U_j(x) P_\theta(x) \\
&\quad + \left(\sum_x U_i(x) P_\theta(x)\right)\left(\sum_x U_j(x) P_\theta(x)\right).
\end{aligned}$$

Except for its sign, this is the covariance $\mathrm{Cov}(U_i(X), U_j(X))$. This means that, except for its sign, the Hessian is the covariance matrix of the random vector $\mathbf{U}(X)$:

$$\frac{\partial^2}{\partial \theta_i \partial \theta_j} \log L(\theta) \;=\; -\mathrm{Var}(\mathbf{U}(X))_{i,j}.$$

This matrix is positive semidefinite. The log-likelihood is therefore concave. This means that a necessary and sufficient condition for $\widehat{\theta}$ to be the maximum likelihood estimation is for the gradient to vanish:

$$\mathbf{E}_{\widehat{\theta}}[\mathbf{U}(X)] - \mathbf{U}(x^O) \;=\; 0. \tag{8.3}$$

Note. This shows the interesting fact that the maximum likelihood solution is related to the maximum entropy. Referring to Property 5.1 on page 102 and setting $a = \mathbf{U}(x^O)$, we see that the maximum likelihood principle leads us to choose the distribution $P_{\widehat{\theta}}(X = x)$ that maximizes the entropy under the constraint (8.3).

Instead of solving equation (8.3), we can try to maximize $\log L(\theta)$ using a gradient algorithm ((7.18), page 142). Here, in fact, we need to use a stochastic gradient algorithm because the expectation $\mathbf{E}_\theta$ can only be calculated by simulation. This is how Younes [183, 184] implemented this type of approach. At each step k the algorithm simulates a configuration $x(k)$ obtained following a complete sweep of the Gibbs sampler for the transition matrix associated with the distribution $P_{\theta(k)}(x)$, where $\theta(k)$ is the current value of the parameter:

Algorithm 8.1. (Maximum Likelihood for Complete Data).

- *Step 0 :* Select $x(0)$ arbitrarily.
- *Step $(k+1)$:* Transition $k \to (k+1)$.

$$\theta(k+1) \;=\; \theta(k) + \frac{\epsilon}{k+1}\big[\mathbf{U}(x(k+1)) - \mathbf{U}(x^O)\big].$$

At each step, the gradient $\mathbf{E}_{\theta(k)}[\mathbf{U}(X)] - \mathbf{U}(x^O)$ is approximated by $\mathbf{U}(x(k+1)) - \mathbf{U}(x^O)$. We can show that for sufficiently small $\epsilon > 0$, this algorithm converges toward the maximum likelihood solution. The proof of this result is highly technical [183].

This approach is difficult to implement. For practical reasons, we prefer to use more direct methods, even at the cost of losing some of the properties of the maximum likelihood estimator.

8.1.2   *Maximum Pseudolikelihood*

The maximum likelihood principle chooses $\widehat{\theta}$ such that the observed value x^O is the most probable one. In the context of Markov random fields, we would expect x^O to be very strongly spatially consistent with the chosen energy. In other words, we would expect the local specifications $P(x_s^O \mid N_s(x^O))$ to be large. This principle was introduced by Besag [21].

Definition 8.2. *Let x^O be an observed value of a Markov random field X with neighborhood system $\{N_s\}$, and let $\theta^* = (\theta_1^*, \ldots, \theta_q^*)'$ be the unknown vector of parameters. To estimate θ^*, the principle of maximum pseudolikelihood chooses the vector $\widehat{\theta}$ that makes the observed value x^O the most consistent:*

$$\widehat{\theta} = \arg\max_{\theta} \prod_s P_\theta(x_s^O \mid N_s(x^O)).$$

The consistency function $\mathcal{L}(\theta) = \prod_s P_\theta(x_s^O \mid N_s(x^O))$ is called the pseudolikelihood. It is also written as $\mathcal{P}_\theta(x^O)$.

This definition is of interest mainly because it does not involve the normalization constant $\mathcal{Z}_\theta$. The expression (5.9) allows us to write

$$P_\theta(x_s \mid N_s(x)) = \frac{\exp -[\theta' \mathbf{U}(x_s, N_s(x))]}{\sum_\lambda \exp -[\theta' \mathbf{U}(\lambda, N_s(x))]}, \tag{8.4}$$

where $\mathbf{U}(x_s, N_s(x))$ denotes the energy restricted to the support configuration $\{s, N_s\}$. To calculate the maximum pseudolikelihood, we use the logarithm of the pseudolikelihood

$$\log \mathcal{L}(\theta) = \sum_s \log P_\theta(x_s \mid N_s(x)).$$

Like the log-likelihood, this function is concave because its Hessian is the sum of the Hessians

$$\frac{\partial^2}{\partial\theta_i \partial\theta_j} \log P_\theta(x_s \mid N_s(x)) = -\mathrm{Var}[\mathbf{U}(X_s, N_s(X))]_{i,j}, \quad \forall s,$$

where $\mathrm{Var}[\mathbf{U}(X_s, N_s(X))]$ defines a positive semidefinite covariance matrix. For sufficiently large $|S|$, we can show that it is strictly positive, which makes $\log \mathcal{L}(\theta)$ a strictly concave function. This ensures the uniqueness of the maximum pseudolikelihood estimation.

This estimate is also consistent in the following sense. The potential of the distribution (8.1) is spatially translation invariant; X can be seen as the restriction to the finite domain S of a homogeneous Markov random field defined on an infinite domain and having the same specifications as the restricted field. We write $\mathcal{L}_n(\theta)$ to denote the pseudolikelihood and $\widehat{\theta}_n$ to denote the estimate obtained on the $n \times n$ domain S. We then have [86] $\lim_{n \to \infty} \widehat{\theta}_n = \theta^*$. Note that this method is equally applicable to energies that are nonlinear with respect to θ, although there is a risk of losing the uniqueness of the solution.

EXAMPLE 1: *Bernoulli Energy.*
Let us consider the binary field on $E_s = \{0,1\}$ with local specifications
((5.11), page 107):

$$P_\theta[x_s \mid N_s(x)] = \frac{\exp -x_s \delta_s}{1 + \exp -\delta_s} \cdot \qquad (8.5)$$

$$\text{with } \delta_s = \theta_1 + \theta_2 \sum_{t \in N_s} x_t.$$

Its log-pseudolikelihood is

$$\log \mathcal{L}(\theta) = - \sum_s x_s \left(\theta_1 + \theta_2 \sum_{t \in N_s} x_t \right)$$

$$- \sum_s \log \left[1 + \exp - \left(\theta_1 + \theta_2 \sum_{t \in N_s} x_t \right) \right]. \qquad (8.6)$$

The local specifications (8.5) are related to the logistic regression model
which is well known in statistics [1] and image processing [46]. Here this
model attempts to explain x_s by the neighborhood $N_s(x)$. The maximum
of the log-likelihood function $\log \mathcal{L}(\theta) \doteq f(\theta)$ can be calculated numerically
using the Newton–Raphson algorithm. This is a general method for op-
timizing nonlinear concave functions that requires the second derivatives
of $f(\theta)$ [92]. The algorithm starts with an initial solution $\theta(0)$ and generates a sequence $\{\theta(k), \; k = 0, 1, \dots\}$ which typically converges to $\hat{\theta}$. At
each iteration k, the function $f(\theta)$ is approximated in a neighborhood of
$\theta(k)$ by a second-degree polynomial, and the new solution $\theta(k+1)$ is the
location of that polynomial's maximum value. Like the gradient descent
method (Section 7.4.2), it is based on a Taylor approximation. Let G be
the gradient of f, and H its Hessian matrix with elements $(\partial^2 f)/(\partial\theta_i\partial\theta_j)$.
At each iteration k, $f(\theta)$ is approximated near $\theta(k)$ by its Taylor series
expansion of order two: $f(\theta) \approx Q_k(\theta)$, where

$$Q_k(\theta) = f(\theta(k)) + G'_k(\theta - \theta(k)) + \frac{1}{2}(\theta - \theta(k))' H_k(\theta - \theta(k)),$$

where G_k and H_k are G and H, respectively, evaluated at $\theta(k)$. Solving
$\partial Q_k/\partial\theta = G_k + H_k(\theta - \theta(k)) = 0$ gives the new solution

$$\theta(k+1) = \theta(k) - (H_k)^{-1} G_k, \qquad (8.7)$$

when H_k is nonsingular. Now consider the log-likelihood (8.6) and rewrite
the probability (8.5) for $X_s = 1$ as $P_\theta[X_s = 1 \mid N_s(x)] \propto \exp \theta' \bar{x}_s \doteq \pi(\bar{x}_s)$
with $\bar{x}_s = (-1, -\sum_s x_s)$. Let $\vec{x}$ denote the vector of x_s values, $\vec{\pi}$
the vector of $\pi(\bar{x}_s)$ values, $\bar{X}$ the matrix with current line $\bar{x}_s$, and
$\Gamma = \text{Diag}[\dots, \pi(\bar{x}_s)(1 - \pi(\bar{x}_s)), \dots]$. We then obtain $G = \bar{X}'(\vec{x} - \vec{\pi})$,
$H = -\bar{X}'\Gamma\bar{X}$, and after expansion, this gives the kth step (8.7) of the

Newton–Raphson algorithm:

$$\theta(k+1) = \left(\bar{X}'_k \Gamma_k \bar{X}_k\right)^{-1} \bar{X}_k \Gamma_k \, \vec{y}_k, \tag{8.8}$$
$$\text{with} \quad \vec{y}_k = \bar{X}_k \theta(k) + \Gamma_k^{-1}(\vec{x}_k - \vec{\pi}_k).$$

In the second expression in (8.8) we recognize the weighted least-squares estimator (3.11) associated with the model $\vec{Y}_k = \bar{X}_k \theta + \vec{W}$, where $\vec{W}$ is a random vector whose covariance matrix is Γ_k^{-1}. This algorithm for calculating the maximum likelihood is called "iteratively reweighted least squares".

EXAMPLE 2: *Gaussian Energy.*
The local specifications are (Proposition 5.1, page 109)

$$P_\theta[x_s \mid N_s(x)] = \frac{1}{\sqrt{2\pi\sigma^2}} \exp - \left[\frac{1}{2\sigma^2} \left(x_s - \gamma \sum_{t \in N_s} x_t \right)^2 \right],$$

with $\sigma^2 = 1/(2\theta_1)$ and $\gamma = -\theta_2/(2\theta_1)$. These give the log-pseudolikelihood

$$\log \mathcal{L}(\theta) = -\frac{|S|}{2} \log(2\pi\sigma^2) - \frac{1}{2\sigma^2} \sum_s \left(x_s - \gamma \sum_{t \in N_s} x_t \right)^2.$$

This brings us back to a least-squares estimate whose solution in (γ, σ^2) is expressed analytically as follows:

$$\widehat{\gamma} = \frac{\sum_s \left(x_s \sum_{t \in N_s} x_t \right)}{\sum_s \left(\sum_{t \in N_s} x_t \right)^2}, \tag{8.9}$$

$$\widehat{\sigma}^2 = \frac{1}{|S|} \sum_s \left(x_s - \widehat{\gamma} \sum_{t \in N_s} x_t \right)^2.$$

From this we can quite easily deduce $\left(\widehat{\theta}_1, \widehat{\theta}_2\right)$.

8.1.3 Logistic Estimation

This approach was introduced in [67] and [149]. It is related to logistic models. It is very simple, but can only be applied to fields such that $|E_s|$ is small. To keep the explanation simple, let us assume that E_s is the same at every site, so that X is a binary, ternary, or m-ary field.

By taking the ratio of the specifications for two values λ and ν at the site s and the same conditioning, we obtain

$$\log \frac{P_{\theta^*}[X_s = \lambda \mid N_s(x)]}{P_{\theta^*}[X_s = \nu \mid N_s(x)]} = \theta^{*\prime} \left[\mathbf{U}(x_s = \nu, N_s(x)) - \mathbf{U}(x_s = \lambda, N_s(x)) \right].$$

$$\tag{8.10}$$

The method is based on this relationship, which is true for any site s because X is a homogeneous field. The basic concept for the estimation of θ^* is that because x^O comes from the distribution P_{θ^*} and because X is a homogeneous field, it is appropriate to replace the probabilities in the left-hand term with estimates performed on x^O. If we do this for various configurations of the neighborhood $N_s(x)$, we obtain a system of linear equations that might include more equations than unknowns, which can be solved for θ.

• Let us discuss the binary case $E_s = \{0, 1\}$. We consider the Bernoulli energy whose specifications are (8.5). The conditioning is summed up by the value $n_s = \sum_{t \in N_s} x_t$. We therefore write $P_{\theta^*}[X_s = 1 \mid N_s(x)] = P_{\theta^*}[X_s = 1 \mid n_s]$, and (8.5) becomes

$$P[X_s = 1 \mid n_s] \quad \propto \quad \exp -[(1, n_s) \, \theta^*].$$

In this case there are only five different neighborhood configurations, $n_s = 0, \ldots, 4$. We therefore obtain a set of five relationships with the values $\lambda = 1$, $\nu = 0$ and $n_s = 0, \ldots, 4$:

$$\log \frac{P_{\theta^*}[X_s = 1 \mid n_s = j]}{P_{\theta^*}[X_s = 0 \mid n_s = j]} = (-1, -j) \, \theta^*, \quad j = 0, \ldots, 4.$$

The estimate of the specifications is the frequency of occurrence in the observation x^O of the local configuration $\{x_s = 1, n_s\}$ with $n_s = j$, for all $j = 0, \ldots, 4$:

$$\widehat{P}[X_s = 1 \mid n_s = j] = \frac{1}{\sum_{s \in S} \mathbf{1}[n_s = j]} \sum_{s \in S} \mathbf{1}[x_s^O = 1] \, \mathbf{1}[n_s = j].$$

Finally, we have the linear system

$$\log \frac{\widehat{P}[X_s = 1 \mid n_s = j]}{1 - \widehat{P}[X_s = 1 \mid n_s = j]} \approx (-1, -j) \, \theta^*, \quad j = 0, \ldots, 4,$$

written in matrix form as follows:

$$\widehat{G} \approx \mathbf{H}\theta^*, \tag{8.11}$$

where $\mathbf{H}$ has as its row $H^{(j)} = (-1, -j)$. The estimate of θ^* is then obtained by least squares:

$$\widehat{\theta} = \arg\min_{\theta} \, \|G - \mathbf{H}\theta\|^2,$$

for which the solution is $\widehat{\theta} = (\mathbf{H}'\mathbf{H})^{-1}\mathbf{H}G$. In practice, however, it is important to ensure that the system is well defined, meaning that the configurations $(1, n_s = j)$ must be observable and that $\widehat{P} \neq 1 - \widehat{P}$ (if not, we would have to substitute an arbitrarily small value for a zero probability $\widehat{P}$, for example). Note that this method remains applicable in the case of an energy that is nonlinear with respect to θ. It does, however, lose its simplicity, because the system to be solved is not linear.

We can generalize this method to apply it to more complicated Bernoulli energies. To do this, we rewrite (8.10):

$$\log \frac{P_{\theta^*}[X_s = 1 \mid N_s(x)]}{P_{\theta^*}[X_s = 0 \mid N_s(x)]}$$

$$= \sum_{i=1}^{q} \theta_i^* \sum_{c \in C_i \ : \ s \in c} \left[V_{(i),c}(x_s = 0, \check{x}_s) - V_{(i),c}(x_s = 1, \check{x}_s) \right]$$

$$\doteq \sum_{i=1}^{q} \theta_i^* H_i(N_s(x)) \ = \ H(N_s(x)) \ \theta^*.$$

Let $\{E^{(j)}\}$ be the optimum partition (i.e., with lowest cardinal value) of all the neighborhood configurations ($\prod_{t \in N_s} E_t$) for which the specifications are constant on each $E^{(j)}$, which means that they satisfy

$$\forall x \ : \ N_s(x) \in E^{(j)} \ \Rightarrow \ P_{\theta^*}\left[X_s = \lambda \mid N_s(x) \in E^{(j)}\right] \doteq p_{\lambda,j},$$

where $\{p_{\lambda,j}, \lambda \in E_s\}$ is the specification associated with $E^{(j)}$. On $E^{(j)}$, $H(N_s(x)) \doteq H^{(j)}$ is constant. We then obtain

$$\log \frac{P_{\theta^*}\left[X_s = 1 \mid N_s(x) \in E^{(j)}\right]}{P_{\theta^*}\left[X_s = 0 \mid N_s(x) \in E^{(j)}\right]} \ = \ H^{(j)} \theta^*,$$

or, by setting $G_j = \log(p_{1,j}/p_{0,j})$,

$$G_j \ = \ H^{(j)} \theta^*,$$

which leads to the linear system (8.11) after estimating G_j. (Note that $N_s(x) \in E^{(j)}$ corresponds to the old notation $n_s = j$).

We may now move on to the general case of m-ary fields by writing

$$\log \frac{P_{\theta^*}[X_s = \lambda \mid N_s(x)]}{P_{\theta^*}[X_s = \nu \mid N_s(x)]}$$

$$= \sum_{i=1}^{q} \theta_i^* \sum_{c \in C_i \ : \ s \in c} \left[V_{(i),c}(x_s = \nu, \check{x}_s) - V_{(i),c}(x_s = \lambda, \check{x}_s) \right]$$

$$= \sum_{i=1}^{q} \theta_i^* H_i(N_s(x), \lambda, \nu).$$

If we write $\{E^{(j)}\}$ for the optimum partition of ($\prod_{t \in N_s} E_t$) associated with the local specifications $\{p_{\lambda,j}\}$, we have

$$\log \frac{P_{\theta^*}\left[X_s = \lambda \mid N_s(x) \in E^{(j)}\right]}{P_{\theta^*}\left[X_s = \nu \mid N_s(x) \in E^{(j)}\right]} \ = \ H^{(\lambda,\nu,j)} \theta^*, \tag{8.12}$$

which gives us the linear system

$$G_{\lambda,\nu,j} = H^{(\lambda,\nu,j)}\theta^*, \quad \lambda > \nu.$$

This can be written as $G = \mathbf{H}\theta^*$. With this new generalization, estimating G very soon becomes difficult because the observed values of the configurations $\{x_s = \lambda, E^{(j)}\}$ in x^O are increasingly sparse as the model becomes more complicated. This increases the risk of having no observed values at all for certain configurations. Finally, applying certain assumptions, we can show that the estimator obtained is almost certainly convergent with respect to P_{θ^*}: $\lim_{n\to\infty}\hat{\theta}_n = \theta^*$ [104].

This method remains valid if E_s is dependent in a regular fashion on the site s, provided that the spatial invariance is preserved. The remaining problem is the practical matter of managing the partitions of the neighborhood configurations $E^{(j)}$.

8.2   Incomplete Data

In this situation we have a dual aim because the unknowns are both the vector of parameters θ^* and the hidden configuration x^*. When the parameters are assumed to be known, the estimation of the hidden field is covered in Chapter 6. When they are not, there are two possible approaches. The first involves estimating the parameters before estimating x^*. The second seeks to estimate the parameters and the hidden field at the same time.

Setting the problem:
Let there be the following family of Gibbs distributions:

$$\Pi_\theta(x,y) = \frac{1}{\mathcal{Z}_\theta}\exp -U(x \mid y, \theta),$$

$$U(x \mid y, \theta) = \sum_{i=1}^{q_1}\theta_{1,i}U_{1,i}(x) + \sum_{i=1}^{q_2}\theta_{2,i}U_{2,i}(y \mid x), \qquad (8.13)$$

$$U_{\ell,i}(x) = \sum_{c\in C_{\ell,i}}V_{(\ell,i),c}(x),$$

where $\theta' = (\theta_1', \theta_2')$ with $\theta_1' = (\theta_{1,1}, \ldots, \theta_{1,q_1})$ and $\theta_2' = (\theta_{2,1}, \ldots, \theta_{2,q_2})$. The vector of parameters θ belongs to a domain included in $\mathbb{R}^q$ with $q = q_1 + q_2$. The neighborhood potential energy $U_{\ell,i}$ has the set of cliques $C_{\ell,i}$, where the potential of any clique $c \in C_{\ell,i}$ is $V_{(\ell,i),c}$. We have an incomplete observation y^O of a Markov random field (X, Y) whose theoretical distribution is

$$P[X = x, Y = y] = \Pi_{\theta^*}(x,y).$$

From here, the problem is to estimate θ^* given y^O, and above all to estimate an optimum configuration x^* with respect to a Bayesian cost function. Bear in mind that the energy $U(x \mid y)$ is that of both (X, Y) and $(X \mid Y)$ (refer to (6.3) and (6.4), page 116).   △

We set $P_\theta(x)$ for $\Pi_\theta(x)$. We write U as follows:

$$U(x \mid y, \theta) = \theta' \, \mathbf{U}(x, y)$$
$$= \theta_1' \, \mathbf{U}_1(x) + \theta_2' \, \mathbf{U}_2(y, x), \qquad (8.14)$$

where we write $\mathbf{U}_1(x) = (U_{1,1}(x), \dots, U_{1,q_1}(x))'$ and $\mathbf{U}_2(y, x) = (U_{2,1}(y \mid x), \dots, U_{2,q_2}(y \mid x))'$. This is a far more difficult problem than the case of complete data. We will present our arguments along the same lines as in the previous section. This first method, maximum likelihood, is chiefly of educational interest. The second method, pseudolikelihood, is an adaptation of the EM algorithm to the case of Gibbs distributions. It is important because of the results it obtains and the methodological framework it provides. The third method is based on a pragmatic approach that allows effective modeling for low-level imaging applications. Like the logistic method for complete data, it will be mainly based on local specifications. The first two methods give a simultaneous estimate of the parameters and the hidden field, which the third method does not.

8.2.1 Maximum Likelihood

Because X is not observable, we consider the marginal distribution

$$P_\theta[Y = y] = \sum_x P_\theta[X = x, Y = y],$$

and the associated likelihood

$$L(\theta) = P_\theta[Y = y^O].$$

The maximum likelihood estimate is therefore $\widehat{\theta} = \arg\max_\theta L(\theta)$. In contrast to the case of complete data, $L(\theta)$ is not concave, which means that several local maxima can exist. A necessary condition for a maximum is for the gradient of the log-likelihood to be zero. The following lemma is crucial for both the maximum likelihood and the EM algorithm in the next paragraph.

Lemma 8.1. *The gradient of the log-likelihood is written as*

$$\frac{d}{d\theta} \log L(\theta) = \mathbf{E}_\theta \left[\frac{d}{d\theta} \log P_\theta(X, y^O) \,\middle|\, Y = y^O \right], \qquad (8.15)$$

where the expectation is relative to the distribution $P_\theta(x \mid y^O)$.

Proof:

Using the notation $L(\theta) = P_\theta(y^O)$ and the symbol $'$ for differentiation of multivariate function, we can write

$$\frac{\mathrm{d}}{\mathrm{d}\theta} \log L(\theta) = \frac{P'_\theta(y^O)}{P_\theta(y^O)} = \frac{1}{P_\theta(y^O)} \sum_x P'_\theta(x, y^O)$$

$$= \sum_x \frac{P'_\theta(x, y^O)}{P_\theta(x, y^O)} \frac{P_\theta(x, y^O)}{P_\theta(y^O)}$$

$$= \sum_x \left[\frac{\mathrm{d}}{\mathrm{d}\theta} \log P_\theta(x, y^O) \right] P_\theta(x \mid y^O)$$

$$= \mathbf{E}_\theta \left[\frac{\mathrm{d}}{\mathrm{d}\theta} \log P_\theta(X, y^O) \mid Y = y^O \right]. \qquad \square$$

This expression can be interpreted as follows. If X were observable, then the *joint* likelihood

$$L_J(\theta) = P_\theta(x^O, y^O) \qquad (8.16)$$

would be accessible, and the gradient of its logarithm $\frac{\mathrm{d}}{\mathrm{d}\theta} \log P_\theta(x^O, y^O)$ would be used to optimize it. But because this is not true, we use the mean of the gradients $\frac{\mathrm{d}}{\mathrm{d}\theta} \log P_\theta(x, y^O)$ with respect to the probabilities $P_\theta(x \mid y^O)$. This expression, then, can be seen as the *posterior mean gradient* of the joint likelihood. This trick is based on Bayesian principles; we eliminate the need for the unknown field X by averaging with respect to a posterior distribution.

From this result we deduce an expression that allows us to calculate the maximum likelihood.

Lemma 8.2. *The gradient of the log-likelihood is written as*

$$\frac{\mathrm{d}}{\mathrm{d}\theta} \log L(\theta) = \mathbf{E}_\theta[\mathbf{U}(X, Y)] - \mathbf{E}_\theta\left[\mathbf{U}(X, y^O) \mid Y = y^O\right], \quad (8.17)$$

where the first expectation is with respect to the distribution $P_\theta(x, y)$, and the second is with respect to $P_\theta(x \mid y^O)$.

Proof:

We use the expression for the gradient (8.2) obtained for the case of a complete observation:

$$\frac{\mathrm{d}}{\mathrm{d}\theta} \log P_\theta(x, y) = \mathbf{E}_\theta[\mathbf{U}(X, Y)] - \mathbf{U}(x, y).$$

By substituting this into (8.15), we obtain

$$\mathbf{E}_\theta \left[\frac{d}{d\theta} \log P_\theta(X, y^O) \middle| Y = y^O \right]$$

$$= \sum_x \left[\frac{d}{d\theta} P_\theta(x, y^O) \right] P_\theta(x \mid y^O)$$

$$= \sum_x \left[\mathbf{E}_\theta[\mathbf{U}(X,Y)] - \mathbf{U}(x, y^O) \right] \log P_\theta(x, y^O)$$

$$= \mathbf{E}_\theta[\mathbf{U}(X,Y)] - \mathbf{E}_\theta \left[\mathbf{U}(X, y^O) \mid Y = y^O \right].$$

$\square$

A standard gradient algorithm to maximize the log-likelihood $L(\theta)$ would look something like this (7.18):

$$\theta(k+1) = \theta(k) - \epsilon \left[\mathbf{E}_{\theta(k)}[\mathbf{U}(X,Y)] - \mathbf{E}_{\theta(k)} \left[\mathbf{U}(X, y^O) \mid y^O \right] \right].$$

This would, however, lead to a very complicated simulation to estimate the expectations at each stage. That is why Younes [183, 185] uses a stochastic gradient algorithm that makes it unnecessary to pause at each step to wait for the expectation estimation procedure to stabilize. This technique uses two Gibbs samplers simultaneously at each step k. The first sampler is associated with the distribution $P_{\theta(k)}(x, y)$. After a complete sweep, it generates a pair of configurations $x(k+1), y(k+1)$. The second sampler is associated with $P_{\theta(k)}(x \mid y^O)$, and generates a configuration written as $x^{|y}(k+1)$.

Algorithm 8.2. (Maximum Likelihood for Incomplete Data).

- *Step 0 :* Choose $x(0)$ and $\theta(0)$ as a "best guess".
- *Step $(k+1)$:* Transition $k \to (k+1)$.

$$\theta(k+1) = \theta(k)$$
$$+ \frac{\epsilon}{k+1} \left[\mathbf{U}(x(k+1), y(k+1)) - \mathbf{U}(x^{|y}(k+1), y^O) \right].$$

At each step, the gradient $\mathbf{E}_{\theta(k)}[\mathbf{U}(X,Y)] - \mathbf{E}_{\theta(k)} \left[\mathbf{U}(X, y^O) \mid Y = y^O \right]$ is approximated by $\mathbf{U}(x(k+1), y(k+1)) - \mathbf{U}(x^{|y}(k+1), y^O)$. Although [183, 185] gives theoretical results for the convergence of the algorithm and comments about its use, this is still a difficult algorithm to apply. Finally, we note that the posterior distribution $P_{\theta(k)}(x \mid y^O)$ allows us to estimate the hidden field at the same time as the parameters at each step of the algorithm.

- EM ALGORITHM

This algorithm was proposed by Baum et al. [18] to deal with the problem of one-dimensional hidden Markov chains. It was later taken up by

Dempster et al. [66] for use in various incomplete observation situations. Although it only calculates a local maximum and its convergence is hard to predict, it has come into widespread use because it provides an estimation scheme that can be adapted to various situations. We will describe it briefly, once again using our own notation, because the concept remains the same.

As before, we wish to maximize the likelihood $L(\theta) = P_\theta(Y = y^O)$. A necessary condition to obtain a maximum is that the gradient of the logarithm of L must vanish, or equivalently, according to (8.15), the posterior mean gradient of $\log L_J(\theta)$ must vanish (8.16):

$$\mathbf{E}_\theta \left[\frac{\mathrm{d}}{\mathrm{d}\theta} \log P_\theta(X, y^O) \,\middle|\, Y = y^O \right] = 0. \tag{8.18}$$

In this expression the unknown parameter appears in both the gradient and the expectation. The idea is therefore to obtain a running estimate $\theta(k)$ relating to the expectation in order to simplify the solving of (8.18). The algorithm can be stated as follows: Given an initial solution $\theta(0)$, calculate at step $(k+1)$ the running estimate $\theta(k+1)$ that is a solution of

$$\mathbf{E}_{\theta(k)} \left[\frac{\mathrm{d}}{\mathrm{d}\theta} \log P_\theta(X, y^O) \,\middle|\, Y = y^O \right] = 0. \tag{8.19}$$

At any step k, the left-hand side of this equation can be rewritten as

$$\frac{\mathrm{d}}{\mathrm{d}\theta} \mathbf{E}_{\theta(k)} \left[\log P_\theta(X, y^O) \,\middle|\, Y = y^O \right].$$

(The expression $\frac{\mathrm{d}}{\mathrm{d}\theta} \mathbf{E}_\theta \left[\log P_\theta(X, y^O) \mid Y = y^O \right]$, however, does not give (8.15)). The algorithm can therefore be expressed in terms of maximizing $\mathbf{E}_{\theta(k)} \left[\log P_\theta(X, y^O) \middle| Y = y^O \right]$ instead of solving the equation for the vanishing gradient. This is the classic expression of the EM algorithm.

Algorithm 8.3. (EM algorithm).

- *Step 0 :* Choose $\theta(0)$ as a "best guess".
- *Step $(k+1)$:* Transition $k \to (k+1)$.
  Calculate $\theta(k+1)$, a solution of:

$$\max_\theta \mathbf{E}_{\theta(k)} \left[\log P_\theta(X, y^O) \middle| Y = y^O \right]. \tag{8.20}$$

When $U(x \mid y, \theta)$ is linear with respect to θ, this expectation tends to be strictly concave with respect to θ, and the maximum is well-defined. The letters EM in the name of the algorithm stand for the two tasks that are performed at each step: calculating the expectation (Task E) and maximization (Task M). The following result gives a justification for this algorithm.

Property 8.1. *If* $\{\theta(k)\}_{k\geq 0}$ *is the series of running estimates of the EM algorithm, then the series of marginal likelihoods* $\{L(\theta(k))\}_{k\geq 0}$ *is an increasing series.*

Proof:
The proof uses Jensen's inequality: If f is a concave function, and Z is a random variable, then $f(\mathbf{E}[Z]) \geq \mathbf{E}[f(Z)]$. We set $\theta(k+1) = \theta'$, $\theta(k) = \theta$, and $y^O = y$. We wish to show that $P_{\theta'}(y) \geq P_{\theta}(y)$. We obtain

$$
\begin{aligned}
\log \frac{P_{\theta'}(y)}{P_{\theta}(y)} &= \log \left[\frac{1}{P_{\theta}(y)} \sum_x P_{\theta'}(x,y) \right] \\
&= \log \left[\sum_x \frac{P_{\theta}(x \mid y)}{P_{\theta}(x,y)} P_{\theta'}(x,y) \right] \\
&= \log \mathbf{E}_{\theta} \left[\frac{P_{\theta'}(X,Y)}{P_{\theta}(X,Y)} \middle| Y = y \right] \\
&\geq \mathbf{E}_{\theta} \left[\log \frac{P_{\theta'}(X,Y)}{P_{\theta}(X,Y)} \middle| Y = y \right] \quad \text{(according to Jensen)} \\
&= \mathbf{E}_{\theta} \left[\log P_{\theta'}(X,Y) \middle| Y = y \right] - \mathbf{E}_{\theta} \left[\log P_{\theta}(X,Y) \middle| Y = y \right],
\end{aligned}
$$

which is positive at the output of Step $(k+1)$ of the algorithm. $\qquad\square$

It is not feasible to solve (8.19) or (8.20) for the case of Markov random fields. To do so would require the use of a stochastic algorithm just as complicated as the one suggested for calculating the maximum likelihood in the case of incomplete data.

8.2.2 Gibbsian EM Algorithm

To work around this problem, a modification to this algorithm that can help in certain situations is suggested in [40]. It involves replacing the joint likelihood $L_J(\theta) = P_{\theta}(x,y)$ given in (8.20) with the joint pseudolikelihood $\mathcal{L}_J(\theta)$. The likelihood is written as

$$
L_J(\theta) = P_{\theta}(x,y) = P_{\theta_1}(x)P_{\theta_2}(y \mid x),
$$

where P_{θ_1} and P_{θ_2} are the likelihoods of the fields X and $(Y \mid X)$, respectively. The associated parameters are $\theta_1 = (\theta_{1,1}, \ldots, \theta_{1,q_1})$ and $\theta_2 = (\theta_{2,1}, \ldots, \theta_{2,q_2})$ (8.13). Similarly, the pseudolikelihood is written as

$$
\mathcal{L}_J(\theta) = P_{\theta}(x,y) = \mathcal{P}_{\theta_1}(x) \, \mathcal{P}_{\theta_2}(y \mid x),
$$

where $\mathcal{P}_{\theta_1}$ and $\mathcal{P}_{\theta_2}$ are the pseudolikelihoods of the fields X and $(Y \mid X)$ respectively. If we assume that $(Y \mid X = x)$ is governed by the distribution

(6.2),

$$P_{\theta_2}(y \mid x) = \prod_s P_{\theta_2}(y_s \mid x), \tag{8.21}$$

it turns out that $P_{\theta_2} = \mathcal{P}_{\theta_2}$. With the introduction of $\mathcal{P}$, Step $(k+1)$ of the algorithm must calculate

$$\max_\theta \mathbf{E}_{\theta(k)}\left[\log \mathcal{P}_\theta(X, y^O) \mid Y = y^O\right], \tag{8.22}$$

or equivalently, it must solve

$$\mathbf{E}_{\theta(k)}\left[\frac{\mathrm{d}}{\mathrm{d}\theta} \log \mathcal{P}_\theta(X, y^O) \bigg| Y = y^O\right] = 0. \tag{8.23}$$

This equation breaks down into two equations, one in θ_1 and the other in θ_2:

$$\mathbf{E}_{\theta(k)}\left[\frac{\mathrm{d}}{\mathrm{d}\theta_1} \log \mathcal{P}_{\theta_1}(X) \bigg| Y = y^O\right] = 0, \tag{8.24}$$

$$\mathbf{E}_{\theta(k)}\left[\frac{\mathrm{d}}{\mathrm{d}\theta_2} \log \mathcal{P}_{\theta_2}(y^O \mid X) \bigg| Y = y^O\right] = 0. \tag{8.25}$$

It is useful to introduce the pseudolikelihood when $\frac{\mathrm{d}}{\mathrm{d}\theta} \log \mathcal{P}_\theta(X, y^O)$ has an expression that allows us to solve (8.23). Let us look at a very simple example. We assume $\frac{\mathrm{d}}{\mathrm{d}\theta} \log \mathcal{P}_\theta(X, y^O) = A(X, y^O)\theta + b(X, y^O)$. The equation can then be written as $\mathbf{E}_{\theta(k)}\left[A(X, y^O) \mid Y = y^O\right]\theta + \mathbf{E}_{\theta(k)}\left[b(X, y^O) \mid Y = y^O\right] = 0$, in which the expectations are calculated by a Monte Carlo simulation using a Gibbs sampler associated with $P_{\theta(k)}(x \mid y^O)$. This is called a Gibbsian EM algorithm after the Gibbs–Markov prior model and because the Gibbs sampler is used. Below we look at a few situations in which this approach can be used.

The expression $\frac{\mathrm{d}}{\mathrm{d}\theta} \log \mathcal{P}_\theta(X, y^O)$ must be expressed in analytical form as soon as the energy U is fixed. This can be done directly by differentiation. For an energy that is linear with respect to θ, a new expression for (8.23) in terms of $\mathbf{U}$ can be established. Let us apply this to θ_1 and the equation (8.24), which we write as

$$\sum_s \mathbf{E}_{\theta(k)}\left[\frac{\mathrm{d}}{\mathrm{d}\theta_1} \log P_{\theta_1}(X_s \mid N_s(X)) \bigg| Y = y^O\right] = 0,$$

where $P_{\theta_1}(X_s \mid N_s(X)) \propto \exp -[\theta' \ \mathbf{U}(X_s, N_s(X))]$ according to (8.4). Using (8.2), we obtain

$$\frac{\mathrm{d}}{\mathrm{d}\theta_1} \log P_{\theta_1}(x_s \mid N_s(x)) = \mathbf{E}_{\theta_1}^s[\mathbf{U}_1(X_s, N_s(x))] - \mathbf{U}_1(x_s, N_s(x))$$

$$\doteq \overline{\mathbf{U}}_{\theta_1}(., N_s(x)) - \mathbf{U}_1(x_s, N_s(x)),$$

where $\mathbf{E}_{\theta_1}^s$ is the expectation with respect to $\{P_{\theta_1}(x_s \mid N_s(x)),\ x_s \in E_s\}$. Finally, the equivalent of equation (8.24) is

$$\sum_s \mathbf{E}_{\theta(k)}\left[\overline{\mathbf{U}}_{\theta_1}(., N_s(X)) - \mathbf{U}_1(X_s, N_s(X))\Big| Y = y^O\right] = 0. \quad (8.26)$$

A similar expression can be obtained for (8.25), but this is less useful, because in any case the fidelity energy often has a simple expression, like the one in (8.21).

The Gibbsian EM algorithm was first proposed for use on hidden m-ary Markov random fields. We now discuss its use for various prior energies.

- BINARY HIDDEN FIELD

Let us consider the energy seen in (6.7):

$$U(x \mid y) = \theta_1 U_1(x) + \theta_2 U_2(y \mid x) = \theta_1 \sum_{<s,t>} x_s x_t + \theta_2 \sum_s y_s x_s.$$

It defines the posterior distribution of a binary Markov random field encoded on $\{-1, 1\}$ and degraded by a multiplicative binary noise. The local specifications of X are

$$P_{\theta_1}(x_s \mid N_s(x)) = \frac{\exp{-(\theta_1 x_s n_s)}}{\exp(\theta_1 n_s) + \exp{-(\theta_1 n_s)}},$$

where $n_s = \sum_{t \in N} x_t$, and the local specifications of $(Y \mid X = x)$ are

$$P_{\theta_2}(y_s \mid \breve{y}, x) = P(y_s \mid x_s) = \frac{\exp{-(\theta_2 x_s y_s)}}{\exp{\theta_2} + \exp{-\theta_2}}.$$

Let us rewrite equation (8.24) using (8.26). After expansion,

$$\mathbf{E}_{\theta_1}^s[\mathbf{U}_1(X_s, N_s(x))] - \mathbf{U}_1(x_s, N_s(x))$$

$$= \sum_{x_s = -1, 1} x_s n_s P[X_s = x_s \mid n_s] - x_s n_s$$

$$= -\frac{e^{\theta_1 n_s} - e^{-\theta_1 n_s}}{e^{\theta_1 n_s} + e^{-\theta_1 n_s}} n_s - x_s n_s,$$

equation (8.26) is written

$$\sum_s \mathbf{E}_{\theta(k)}\left[n_s(X)\left(\tanh(\theta_1 n_s(X)) + X_s\right) \mid Y = y^O\right] = 0$$

$$\sum_s \sum_x n_s(x)\left(\tanh(\theta_1 n_s(x)) + x_s\right) P_{\theta(k)}(x \mid y^O) = 0,$$

in which we wrote $n_s(X)$ to emphasize the fact that n_s is a random variable that is a function of X. The hyperbolic tangent is written as tanh. Let us set $f_s(x) = n_s(x)[\tanh(\theta_1 n_s(x)) + x_s]$, where $f_s(x)$ is dependent on the type of neighborhood configuration $N_s(x)$ and the value x_s. There are five types of neighborhood configuration given by the values of n_s: $\{-4, -2, 0, 2, 4\}$. We are thus in a situation similar to that described in Section 8.1.3, with the difference that here we have local marginal distributions on (s, N_s), not local specifications as in Section 8.1.3. We partition the set $E_s \times \left(\prod_{t \in N_s} E_t\right)$ of local image configurations so that f_s is constant on each part. We write this partition as $\{E^{(j)}, j = 1, \ldots, 5\}$. For example $E^{(1)} = \{(x_s, N_s(x)) \text{ such that } (x_s, n_s) = (-1, 4) \text{ or } (x_s, n_s) = (1, -4)\}$ for which $f_s(x) = 4(\tanh 4\theta_1 - 1) \doteq f^{(1)}(\theta_1)$. Our equation is therefore written as

$$\sum_{j=1}^{5} f^{(j)}(\theta_1) \sum_{x} \sum_{s} \mathbf{1}\left[(x_s, N_s(x)) \in E^{(j)}\right] P_{\theta(k)}(x \mid y^O) = 0,$$

and we then write

$$\sum_{j=1}^{5} f^{(j)}(\theta_1) \, P_{\theta(k)}(E^{(j)} \mid y^O) = 0.$$

Finally, the five probabilities $P_{\theta(k)}(E^{(j)} \mid y^O)$ are estimated according to a Gibbs sampler associated with the posterior distribution $P_{\theta(k)}(x \mid y^O)$. Let us write these estimates as $\widehat{P}^{(j)}$. After calculating the four other $f^{(j)}(\theta_1)$, the estimate $\theta_1(k+1)$ is then the solution of the nonlinear equation

$$4(\widehat{P}^{(1)} + \widehat{P}^{(5)}) \tanh(4\theta_1) + 2(\widehat{P}^{(2)} + \widehat{P}^{(4)}) \tanh(2\theta_1)$$
$$+ 4(\widehat{P}^{(5)} - \widehat{P}^{(1)}) + 2(\widehat{P}^{(4)} - \widehat{P}^{(2)}) = 0. \quad (8.27)$$

To estimate θ_2, we use a direct differentiation that gives

$$\sum_{s} \mathbf{E}_{\theta(k)}\left[y_s X_s + \tanh(\theta_2) \middle| Y = y^O\right] = 0,$$

and is then handled like the previous estimation. We can also calculate an estimate $x(k)$ of the hidden field. At each iteration, the Gibbs sampler generates a sequence of configurations that makes this estimation possible. This is explained in greater detail in the following example.

• m-ARY HIDDEN FIELD

First Formulation.
This is an extension of the binary case: For all s, $E_s = \{0, \ldots, m-1\}$. This time, we do not set a special energy expression. As before, (8.26) is

reformulated in the following steps:

$$\sum_s \mathbf{E}_{\theta(k)}\big[f_s(X)|Y = y^O\big] = 0,$$

$$\sum_s \sum_x f_s(x)\, P_{\theta(k)}\big(x\,|\,y^O\big) = 0,$$

$$\sum_j f^{(j)}(\theta_1)\, P_{\theta(k)}\big(E^{(j)}\,|\,y^O\big) = 0,$$

where $\{E^{(j)}\}$ denotes the partition of the local configurations $E_s \times (\prod_{t\in N_s} E_t)$ such that $f_s(x)$ is constant for all part $E^{(j)}$. Its value is written as $f^{(j)}$. As in Section 8.1.3, this approach is valid for situations where there is a sufficiently small number of parts $E^{(j)}$ that it is feasible to estimate them using the Gibbs sampler.

Second Formulation.
In this context we will introduce another way of using a Gibbs algorithm that does not require the solving of nonlinear equations as in (8.27). It hinges on the fact that the estimation of θ_1 has no direct utility for the estimation of the hidden field x^*. (The Gibbs sampler used to perform the Bayesian estimation of x^* uses the local specifications $P(x_s \mid N_s(x))$ of the hidden field X, but not their parametric expression with respect to θ.) The idea, then, is to overparametrize the problem with a view to estimating these specifications rather than their parameters.

Looking again at the distribution (6.2) for $(Y \mid X = x)$,

$$P_{\theta_2}(y \mid x) = \prod_s P_{\theta_2}(y_s \mid x) = \mathcal{P}_{\theta_2}(y \mid x),$$

the joint pseudolikelihood is written as

$$\begin{aligned}
\mathcal{P}_\theta(x,y) &= \mathcal{P}_{\theta_1}(x)\, \mathcal{P}_{\theta_2}(y \mid x) \\
&= \prod_s P_{\theta_1}(x_s \mid N_s(x)) \prod_s P_{\theta_2}(y_s \mid x). \qquad (8.28)
\end{aligned}$$

Let $\{E^{(ij)} = \{i, E^{(j)}\}\}$ be the partition of the local configurations $E_s \times (\prod_{t\in N_s} E_t)$ such that $P(x_s \mid N_s(x))$ is constant over any part $E^{(ij)}$, and its value is written as p_{ij}. This implies that

$$p_{ij} = P\big[X_s = i \,\big|\, N_s(x) \in E^{(j)}\big], \quad i = 0,\ldots, m-1.$$

Overparametrizing replaces the parameters θ_1 by

$$\Theta_1 = (p_{ij}), \quad i = 0,\ldots, m-1.$$

In the binary case above, we have $\Theta_1 = (p_{1j})_{j=1}^5$, where j denotes the five neighborhood configurations characterized by the value of n_s.

To simplify our arguments, we consider the following situation:

$$P(y_s \mid x) = P(y_s \mid x_s)$$
$$\text{whith} \quad q_{\ell i} \doteq P[Y_s = \ell \mid X_s = i].$$

We then set

$$\Theta_2 = (q_{\ell i}), \quad i = 0, \dots, m - 1,$$

with $\ell = 0, \dots, L-1$ if y_s is discretized on L levels. More generally, i could be associated with a partition of the local configurations of x. The pseudolikelihood (8.28) is written by setting $\Theta = (\Theta_1, \Theta_2)$,

$$\mathcal{P}_\Theta(x, y) = \prod_{ij} (p_{ij})^{n_{ij}(x)} \prod_{\ell i} (q_{\ell i})^{m_{\ell i}(x,y)},$$

with the following definitions:

$$n_{ij}(x) = \sum_s \mathbf{1}\big[x_s = i, \ N_s(x) \in E^{(j)}\big],$$
$$m_{\ell i}(x, y) = \sum_s \mathbf{1}[y_s = \ell, \ x_s = i].$$

Step k of the Gibbsian EM algorithm cannot directly use equation (8.23). It must be expanded from the maximization (8.22) with the constraints that (p_{ij}) at fixed j and $(q_{\ell i})$ at fixed i are probability vectors. We must then use the method of Lagrange multipliers, which seeks to maximize the Lagrangian

$$\mathbf{E}_{\Theta(k)}\big[\log \mathcal{P}_\Theta(X, y^O) \mid Y = y^O\big] + \sum_j \lambda_j \left(\sum_i p_{ij} - 1\right)$$
$$+ \sum_i \nu_i \left(\sum_\ell q_{\ell i} - 1\right),$$

where λ_j and ν_i are the Lagrange multipliers. We can then see that

$$\mathbf{E}_{\Theta(k)}\big[\log \mathcal{P}_\Theta(X, y^O) \mid Y = y^O\big]$$
$$= \sum_x \mathcal{P}_{\Theta(k)}(x \mid y^O)\left[\sum_{ij} n_{ij}(x) \log p_{ij} + \sum_{\ell i} m_{\ell i}(x, y^O) \log q_{\ell i}\right],$$

and we then obtain

$$\mathbf{E}_{\Theta(k)}\left[\log \mathcal{P}_{\Theta}\left(X, y^{O}\right) \mid Y = y^{O}\right]$$

$$= \sum_{ij}\left[\sum_{x} n_{ij}(x)P_{\Theta(k)}\left(x \mid y^{O}\right)\right] \log p_{ij}$$

$$+ \sum_{\ell i}\left[\sum_{x} m_{\ell i}\left(x, y^{O}\right)P_{\Theta(k)}\left(x \mid y^{O}\right)\right] \log q_{\ell i}$$

$$= \sum_{ij} \mathbf{E}_{\Theta(k)}\left[n_{ij}(X) \mid Y = y^{O}\right] \log p_{ij}$$

$$+ \sum_{\ell i} \mathbf{E}_{\Theta(k)}\left[m_{\ell i}\left(X, y^{O}\right) \mid Y = y^{O}\right] \log q_{\ell i}.$$

When the gradient of the Lagrangian vanishes, we obtain $\Theta(k + 1)$:

$$p_{ij}(k + 1) = \frac{\mathbf{E}_{\Theta(k)}\left[n_{ij}(X) \mid Y = y^{O}\right]}{\sum_{i} \mathbf{E}_{\Theta(k)}\left[n_{ij}(X) \mid Y = y^{O}\right]} \tag{8.29}$$

$$q_{\ell i}(k + 1) = \frac{\mathbf{E}_{\Theta(k)}\left[m_{\ell i}(X, y^{O}) \mid Y = y^{O}\right]}{\sum_{i} \mathbf{E}_{\Theta(k)}\left[m_{\ell i}(X, y^{O}) \mid Y = y^{O}\right]}.$$

• m-ARY HIDDEN FIELDS AND GAUSSIAN NOISE

When we take a specific distribution for $P_{\theta_2}(y_s \mid x_s)$, we can try to find the expression for the estimate of θ_2 directly. To do this, we can easily rewrite

$$\mathbf{E}_{\Theta(k)}\left[\log \mathcal{P}_{\theta_2}(y \mid X) \mid Y = y^{O}\right]$$

$$= \sum_{i}\sum_{s} P_{\Theta(k)}\left[X_s = i \mid y^{O}\right] \log P_{\theta_2}[y_s \mid X_s = i]. \tag{8.30}$$

For example, in the Gaussian case, $(Y_s \mid X_s = i) \sim LG(\mu_i, \sigma^2)$ and $\theta_2 = (\mu_0, \ldots, \mu_{m-1}, \sigma^2)'$, we have

$$\log P_{\theta_2}[y_s \mid X_s = i] = -\log \sqrt{2\pi}\sigma - \frac{(y_s - \mu_i)^2}{2\sigma^2}.$$

We substitute this into (8.30) and, after the gradient of (8.30) vanishes, we obtain

$$\mu_i(k + 1) = \frac{\sum_{s} y_s\, P_{\Theta(k)}\left[X_s = i \mid y^{O}\right]}{\sum_{s} P_{\Theta(k)}\left[X_s = i \mid y^{O}\right]}, \tag{8.31}$$

$$\sigma^2(k + 1) = \frac{\sum_{i}\sum_{s} (y_s - \mu_i(k + 1))^2\, P_{\Theta(k)}\left[X_s = i \mid y^{O}\right]}{\sum_{i}\sum_{s} P_{\Theta(k)}\left[X_s = i \mid y^{O}\right]}. \tag{8.32}$$

We will now give details of the calculation of the expectations $\mathbf{E}_{\Theta(k)}$ and the probabilities $P_{\Theta(k)}\left(X_s = i \mid y^{O}\right)$ required for the final calculation of the estimates (8.29), (8.31), and (8.32) of Step $k + 1$. To do this, we consider the Gibbs sampler whose transition matrix is based on $P_{\Theta(k)}(x\,|\,y^O)$. We write $(x(0), x(1), \ldots, x(N))$ to denote the configurations obtained after N complete sweeps of the sampler (these configurations should also be indexed by k). We then have

$$\mathrm{E}_{\Theta(k)}\big[n_{ij}(X)\,\big|\,Y = y^O\big] \approx \frac{1}{N - n_0} \sum_{n=n_0+1}^{N} n_{ij}(x(n)), \qquad (8.33)$$

$$P_{\Theta(k)}\big[X_s = i\,\big|\,Y = y^O\big] \approx \frac{1}{N - n_0} \sum_{n=n_0+1}^{N} \mathbf{1}[x_s(n) = i], \quad (8.34)$$

where n_0 is an instant from which we consider that the sampler's Markov chain has converged. By substituting these last two expressions into (8.31) and (8.32), we obtain

$$\mu_i(k+1) = \frac{\sum_s \sum_n y_s\,\mathbf{1}[x_s(n) = i]}{\sum_s \sum_n \mathbf{1}[x_s(n) = i]},$$

$$\sigma^2(k+1) = \frac{\sum_i \sum_s \sum_n (y_s - \mu_i(k+1))^2\,\mathbf{1}[x_s(n) = i]}{\sum_i \sum_s \sum_n \mathbf{1}[x_s(n) = i]}.$$

This shows in particular that $\mu_i(k+1)$ is the mean of the gray scale of the sites labeled i in the images $x(n_0 + 1), \ldots, x(N)$.

Finally, at each step $k + 1$, an estimate of the hidden field can be calculated according to the Bayesian MPM rule,

$$\widehat{x}_s(k+1) = \max_{\lambda} P_{\Theta(k)}\big(X_s = \lambda\,\big|\,y^O\big), \qquad (8.35)$$

where the posterior probability is approximated by (8.34). It is not necessary to perform this calculation for each step; it is necessary only at the last step. Using $\widehat{x}(k)$ gives the optimum initialization of the Gibbs sampler for Step $k + 1$. As we watch the progress of the algorithm, it is interesting to visualize the sequence of the reconstructions $\widehat{x}(k)$.

We can summarize this method by giving the Gibbsian EM algorithm for this case (m-ary/Gauss). For other situations, the algorithm would be similar but would depend on the formal expression of the estimates. As in the case of the standard EM algorithm, the convergence of this algorithm is not well known. We can show that it diverges when the estimation formulas are repeated while taking into account the fact that the variance depends on the region. Experience confirms this observation. A basic mathematical analysis of this topic is given in [187].

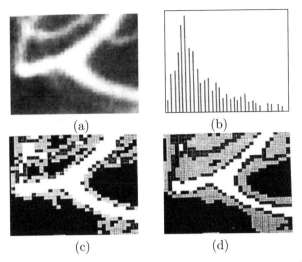

Figure 8.1. Segmentation of a radiographic image into four regions. (a) Original image. (b) Image grayscale histogram. (c) First iteration. (d) Fifth iteration.

Algorithm 8.4. (Gibbsian EM algorithm, m-ary/Gauss).

- *Step 0* : Select $\widehat{x}(0)$ arbitrarily,
  $(p_{ij})_{i=0}^{m-1}(0)$ is the uniform distribution,
  $\theta_2(0)$ is a rough empirical estimate computed from y^O.

- *Step $(k+1)$* : Transition $k \to (k+1)$.
    - *Task E* : Calculate expectations (8.33) and probabilities (8.34) from

        $$(x(0), \ldots, x(n), \ldots, x(N))$$

        obtained via Gibbs sampler based on $P_{\Theta(k)}(x \mid y^O)$.
    - *Task M* : Calculate $\Theta_1(k+1)$ and $\theta_2(k+1)$ according to (8.31) and (8.29).
    - *Reconstruction* : Estimate hidden field by calculating $\widehat{x}(k+1)$ according to (8.35) and (8.34).

We should give the following additional information about the above algorithm. The initialization $\widehat{x}(0)$ has no effect on the result. This means that it is possible to start out with a constant-value image. We calculate $\theta_2(0)$ on the image grayscale histogram. For example, $\mu_i(0)$ is the $\frac{i+1}{m+1}$th quantile of this histogram, and $\sigma(0) \propto \frac{\mu_0(0)-\mu_{m-1}(0)}{m-1}$. Experimentally, we observe that a poor initialization of θ_2 slows down convergence but does not affect the quality of the result. Figure 8.1 illustrates the use of this algorithm for the purpose of segmenting y into four regions. This is a case of transforming y into a piecewise constant image, where the level of each piece corresponds to the expectation of Y_s on the piece. The algorithm was initialized by $\mu_2(0) < \mu_1(0) < \mu_3(0)$. Note that the histogram is uni-modal and does not give any clues about the four regions.

- GAUSSIAN HIDDEN FIELD

We will now give the analytical expressions for the estimates of the parameters of a hidden Markov random field of Gaussian energy. At step $k+1$, equation (8.24) must be solved,

$$\mathbf{E}_{\theta(k)} \left[\frac{\mathrm{d}}{\mathrm{d}\theta_1} \log \mathcal{P}_{\theta_1}(X) \middle| Y = y^O \right] = 0,$$

where the log-pseudolikelihood with respect to $\theta_1 = (\gamma, \sigma_X^2)$ is (Proposition 5.1, page 109)

$$\log \mathcal{P}_{\theta_1}(x) = -\frac{|S|}{2}\log(2\pi\sigma_X^2) - \frac{1}{2\sigma_X^2}\sum_s \left(x_s - \gamma\sum_{t\in N_s}x_t\right)^2.$$

This equation leads directly to

$$\gamma(k+1) = \frac{\mathbf{E}_{\theta(k)}\left[\sum_s(X_s\sum_{t\in N_s}X_t)\mid Y = y^O\right]}{\mathbf{E}_{\theta(k)}\left[\sum_s(\sum_{t\in N_s}X_t)^2\mid Y = y^O\right]},$$

$$\sigma_X^2(k+1) = \frac{1}{|S|}\mathbf{E}_{\theta(k)}\left[\sum_s\left(X_s - \gamma(k+1)\sum_{t\in N_s}X_t\right)^2\middle| Y = y^O\right].$$

These are the expressions for the estimators of the maximum pseudo likelihood (8.9), weighted according to the posterior distribution. As before, these expressions are calculated approximately based on a sample $(x(n_0+1),\ldots,x(N))$ obtained using the Gibbs sampler associated with $P_{\theta(k)}(x \mid y^O)$:

$$\gamma(k+1) \approx \frac{\sum_{n=n_0+1}^N\sum_s[x_s(n)\sum_{t\in N_s}x_t(n)]}{\sum_{n=n_0+1}^N\sum_s[\sum_{t\in N_s}x_t(n)]^2},$$

$$\sigma_X^2(k+1) \approx \frac{1}{(N-n_0)|S|}\sum_{n=n_0+1}^N\sum_s\left[x_s(n) - \gamma(k+1)\sum_{t\in N_s}x_t(n)\right]^2.$$

To be able to write out the whole algorithm, we would also need a distribution for $(Y \mid X = x)$. This is done in Chapter 12, in the context of tomography.

8.2.3 Bayesian Calibration

This method is not an estimation in the statistical sense, but rather a parameter calibration technique. It cannot be used in all circumstances, but it is the preferred method of those who use Markov modeling. Its use is best demonstrated through examples. Now we present two very similar approaches.

• Looking again at the logistic estimation in Section 8.1.3, we start by considering the estimation of the parameters θ_1 of the prior energy. Once again referring to (8.12), we have the following expression for the hidden field X:

$$\log \frac{P_{\theta_1}\left[X_s = \lambda \mid N_s(x) \in E^{(j)}\right]}{P_{\theta_1}\left[X_s = \nu \mid N_s(x) \in E^{(j)}\right]} = H^{(\lambda,\nu,j)} \, \theta_1. \tag{8.36}$$

In the context of logistic estimation, the local specifications $p_{\lambda,j} = P_{\theta_1}\left[X_s = \lambda \mid N_s(x) \in E^{(j)}\right]$ that appear in this expression are estimated via the frequency of occurrence of the local configurations $\{\lambda, E^{(j)}\}$ on the observed image x^O. In the case of a hidden field, however, we do not have this type of observed value. The idea then is to *follow Bayesian logic through to its conclusion* by allowing the user to select some of the values $p_{\lambda,j}$ [8]. A small set of triplets $\{(\lambda_l, \nu_l, E^{(j_l)}); l = 1, \ldots, L\}$ is therefore chosen, and for these we can also select values p_{λ_l,j_l} and p_{ν_l,j_l} given the prior information that allowed us to define $U_{\theta_1}(x)$. We therefore obtain the linear system

$$\log \frac{p_{\lambda_l,j_l}}{p_{\nu_l,j_l}} = H^{(\lambda_l,\nu_l,j_l)}\theta, \quad l = 1, \ldots, L,$$

which is solved by the method of least squares. This approach is based on the fact that the optimization algorithms essentially use the local specifications $p_{\lambda,j}$ in the Bayesian estimation of x. With this calibration approach, we are seeking to ensure that the specifications are consistent with our prior information. When the application allows, this approach can be extended to calibrating all the parameters (θ_1, θ_2) of the posterior energy. It should be sufficient to use the specifications $P(x_s \mid N_s(x), y)$ of the posterior distribution. We must then select a set of quadruplets $\{(\lambda_l, \nu_l, E^{(j_l)}, E_l(y)); l = 1, \ldots, L\}$ for which we choose the values of the specifications. We write $E_l(y)$ to denote a local configuration or state that is a function of y with respect to $p(y \mid x)$. An example showing a calibration of (θ_1, θ_2) is given in Chapter 9.

• A similar and more often used approach is known as the method of *qualitative boxes* [6, 9, 55]. Instead of trying to calibrate the parameters using equations as in (8.36), calibration is performed using inequalities, such as

$$a_1^{(l)} \leq H^{(l)}\theta \leq a_2^{(l)}, \quad l = 1, \ldots, L ,$$

where a_1 and a_2 are scalar values. This defines a polyhedron (which might be empty) in R^q, known as the qualitative box, in which θ is chosen.

EXAMPLE: *Fitting of discrete curves to a scalar field.*
This involves fitting regular curves to spatial data. It is similar to spline fitting, but in this case the situation is more complicated. The energy used here is based on the energy that will be built in Chapter 11 (page 235). We have a field written as $\{d_s, \ s \in S\}$ that is seen as the degradation of a set of discrete planar curves (see Figure 11.4 page 237). The field d is actually y after application of an edge detector. The hidden field is a binary field $\{x_s, \ s \in S\}$ such that $x_s = 1$ if s belongs to a curve and $x_s = 0$ otherwise. The field d is quantitative; $d_s \in [-b, b]$, and d is a scrambled version of a certain configuration x. When d_s is close to b, it is very likely that s will be on a curve, unlike the case where d_s is close to $-b$, since the value $d_s \approx 0$ is a source of uncertainty. The energy is

$$U(d, x) = U_1(x) + U_2(d, x) = \sum_s x_s H_{\vartheta_s}(x) - \sum_s x_s d_s. \quad (8.37)$$

The prior energy shows how regular the curves are,

$$H_{\vartheta_s}(x) = \theta_1 \mathbf{1}[n_s = 0] + \theta_2 \mathbf{1}[n_s = 1]$$
$$+ \mathbf{1}[n_s = 2]\left(\theta_{3,1}\mathbf{1}[a_s \leq \pi/2] - \theta_{3,2}\mathbf{1}[a_s > \pi/2]\right)$$
$$+ \mathbf{1}[n_s \geq 3]\,\theta_4(n_s)\,,$$

where the following notation is used

- ϑ_s is the 3×3 window centered on the site s (excluding s),
- $n_s = \mathrm{card}\{t \in \vartheta_s \mid x_t = 1\}$,
- a_s, defined only when $n_s = 2$, is the absolute value of the angle between $s - s'$ and $s - s''$, where s', s'' are the two sites of ϑ_s such that $x_{s'} = 1$ and $x_{s''} = 1$.

All θ are positive. The value θ_1 penalizes curves that are reduced to a point, θ_2 and $\theta_{3,2}$ control the length of curves, and $\theta_{3,1}$ penalizes strong curvatures. To obtain curves that are one pixel wide, θ_4 strongly penalizes site clusters such that $x_s = 1$, and its value depends on the local configuration of the clusters.

The calibration for this example was performed by Coldefy [55]. Its main elements are as follows. Let us write x^n for a configuration containing only one curve g_n, which has length n. In this case $g_{n+1} = g_n \cup \{s'\}$, meaning that s' extends g_n and therefore

$$\frac{P[X = x^{n+1} \mid d]}{P[X = x^n \mid d]} \propto P[X_{s'} = 1 \mid N_{s'}(x^n), d], \quad (8.38)$$

where $\{N_s\}$ is the 5×5 neighborhood system associated with the cliques $\{\vartheta_s\}$. The Bayesian calibration consists in choosing specific intervals containing the specifications (8.38) for some configurations. In particular, we have $U(x^0 \mid d) = 0$ and $U(x^1 \mid d) = \theta_1 - d_s$. Because isolated sites are undesirable, we impose the condition

$$\frac{P(x^1 \mid d)}{P(x^0 \mid d)} \leq 0.1.$$

This leads to $\theta_1 \geq d_s - \log(0.1)$, which is written as follows in the most unfavorable case:

$$\theta_1 \geq b - \log(0.1). \tag{8.39}$$

Likewise, x^2, which contains $g_2 = \{s, s'\}$, is undesirable, and so we prefer x^0. When estimating the hidden field using a Gibbs sampler that modifies one site at a time, we must therefore be able to pass through x^1 to reach x^0. We therefore set

$$\frac{P(x^2 \mid d)}{P(x^1 \mid d)} \leq 1.$$

With $U(x^2 \mid d) = 2\theta_2 - d_s - d_{s'}$, we therefore obtain $2\theta_2 \geq d_{s'} + \theta_1$, which is written as follows in the most unfavorable case:

$$\theta_2 \geq \frac{1}{2}(b + \theta_1). \tag{8.40}$$

For $n > 2$, if g_n is regular, its energy is $U(x^n \mid d) = 2\theta_2 - (n - 2)\theta_{3,2} - \sum_{s \in g_n} d_s$. If g_{n+1} is also regular, we have

$$\frac{P(x^{n+1} \mid d)}{P(x^n \mid d)} = \exp -(d_{s'} + \theta_{3,2}).$$

To favor the estimation of long, regular curves, we impose the condition

$$0.75 \leq \frac{P(x^n \mid d)}{P(x^{n+1} \mid d)} \leq 1,$$

leading to $0 \leq d_{s'} + \theta_{3,2} \leq -\log(0.75)$, which is written as follows in the case of indecision

$$0 \leq \theta_{3,2} \leq -\log(0.75). \tag{8.41}$$

The value $\theta_{3,1}$ penalizes irregular, strongly curved curves. Let g_n be a regular curve and $\widehat{g}_n$ an irregular curve that is different from g_n at only one site $\widehat{s}$, a site with strong curvature compared to its counterpart s on g_n. We therefore have $U(\widehat{x}^n \mid d) = 2\theta_2 - (n - 3)\theta_{3,2} + \theta_{3,1} - \sum_{s \in \widehat{g}_n} d_s$. We set

$$\frac{P(\widehat{x}^n \mid d)}{P(x^n \mid d)} \leq 0.1,$$

which leads to $\theta_{3,1}-d_{\widehat{s}} \geq \log(0.1)-\theta_{3,2}-d_s$. If we assume that $d_{\widehat{s}} \approx d_s$, this can be simplified to $\theta_{3,1} \geq \log(0.1)-\theta_{3,2}$. Note that in this case, the ratio of probabilities is equivalent to the ratio of the two local specifications, as in (8.36). In fact, the frame found is insufficient, because here again we must ensure that we can go from $\widehat{x}^n$ to x^n when using the Gibbs sampler. This requires the use of the θ_4 terms, as described in [55]. Finally, the frames (8.39), (8.40), and (8.41), as well as those yet to be found, constitute a qualitative box from which the user selects the parameters according to considerations like those mentioned above.

Part III

Modeling in Action

9
Model-Building

There is a goal, but no path;
that which we call a path is hesitation.
Franz Kafka.

We are interested in discrete energy models only. The methodology was given in Chapter 7. We must now go into more depth to obtain a method that is practical for use in real applications. In this chapter, we introduce model-building using a concrete example that will illustrate the methodology. This example is relevant to both Part I and Part II. It involves a representation using discretized regular curves, but in a more complicated situation than the classic case of smoothing with nonparametric spline functions. The purpose is to estimate a family of regular curves of unknown number and length that will link sparse variably shaped "spots" distributed in a sequence of images.

9.1 Multiple Spline Approximation

Let us introduce this example gradually, starting with the problem of one-dimensional spline approximation. Remember that in this case, the data are n points in the plane $\{(y_i, t_i), i = 1, \ldots, n\}$ with known abscissas t_i, and ordinates y_i that are observed values seen as the result of a random variable $LG(g_i, \sigma^2)$ with unknown parameters, where g_i is the point of the curve discretized with respect to t_i. Each abscissa is associated with one

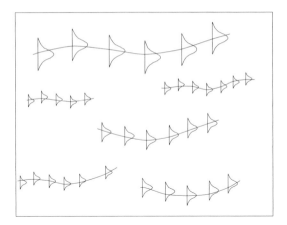

Figure 9.1. Multiple spline approximation.

and only one observation, and we approximate the entire set of data with a unique curve g.

In a similar situation we might have Gaussian probability distributions $LG(\mu_i, \sigma_i^2)$ with given parameters instead of the scalar values y_i. At each abscissa t_i, these distributions give a confidence interval inside which the curve is required to pass. These confidence intervals are located on straight lines that are perpendicular to the t-axis. We may therefore consider that the datum at t_i is the vector $y_i = (\mu_i, \sigma_i^2)$. In this case, the fidelity energy is

$$U_2(y \mid g) = \sum_i \frac{1}{\sigma_i^2}(\mu_i - g_i)^2. \tag{9.1}$$

When the covariances σ_i^2 are constant, this expression is formally equivalent to the usual spline energy expression.

We now complicate the situation. Let us imagine that we have n_i observed values at each t_i. The data are $\{(\mathbf{y}_i, t_i), i = 1, \ldots, n\}$ with $\mathbf{y}_i = (y_{i,1}, \ldots, y_{i,n_i})$ and $y_{ij} = (\mu_{ij}, \sigma_{ij}^2)$. Our goal is to determine a family of regular curves $\{g^\ell, \ell = 1, .., \kappa\}$ that fit these data as well as possible. Figure 9.1 illustrates our point.

The problem is that we have no partition of the data that would take us back to the previous case. In other words, the data assigned to each of the curves g^ℓ and the number of curves κ are unknown. The curves are also of unknown length. This is an ill-posed problem. Only initial information can eliminate any ambiguity, as Figure 9.1 illustrates. Let us now discuss our example. Figure 9.1 can be interpreted as a search for two-dimensional paths across confidence regions. A three-dimensional version of this problem is encountered in the field of biology (3-D electrophoresis analysis) as well as in geophysics (3-D focal analysis for 2-D seismic migration), where *confidence regions* and associated *paths* are specific to the field of use. Instead of explaining these abstract entities, we refer the reader to the specialist literature [53]. The extension to the three-dimensional case involves two-dimensional confidence regions located in planes perpendicular to the t-axis. The problem stated above can be transposed as follows.

9.1.1 *Choice of Data and Image Characteristics*

Let us assume that the values t_i are equidistant abscissas written as $(t = 1, \ldots, n) \doteq T$. In fact, in real situations the confidence regions are not directly obtainable; they must be *extracted* from the observed images. These images form a series $\mathcal{I}_t$ of parallel images indexed by t, which thereby form a "cube" of data with coordinates (t, u_1, u_2). Typically, $\mathcal{I}_t$ consists of variably shaped spots on a more or less uniform background (see Figure 9.2(a)). Figure 9.2(b) represents the side of the cube that is perpendicular to the images $\mathcal{I}_t$ (the coordinates of $\mathcal{I}_t$ and the side are (u_1, u_2) and (t, u_1), respectively; (see Figure 9.4)). The vertical line shown gives the position with respect to t of the image $\mathcal{I}_t$ represented in Figure 9.2(a).   The confidence regions are taken from these spots. We must emphasize that the preliminary phase of defining and extracting image characteristics is a crucial preliminary step for all modeling. We briefly illustrate this point on the image $\mathcal{I}_t$ of Figure 9.2. Each spot is likened to a degraded and partially observed Gaussian surface. Following an estimation procedure that we will not describe in detail, we obtain the parameters $\{\Lambda_t^k, k = 1, \ldots, n_t\}$ of these surfaces. The parameters of the kth spot are represented by $\Lambda_t^k = (M_t^k, \mu_t^k, V_t^k)$, where M_t^k is the mass of the nonnormalized Gaussian surface, μ_t^k is its average position, and V_t^k is its covariance matrix, in which the variances are written as $(\sigma_1^2)_t^k$ and $(\sigma_2^2)_t^k$. We write $G_t^k(u)$ to denote the Gaussian surface, and we consider the following two-dimensional interval:

$$\mathcal{W}_t^k = \left[(\mu_1)_t^k \pm 2(\sigma_1)_t^k \right] \times \left[(\mu_2)_t^k \pm 2(\sigma_2)_t^k \right].$$

To begin with, let us assume that the windows $\mathcal{W}_t^k$ do not intersect. If we define the confidence regions as the truncated Gaussian surfaces:

$$R_t^k(u) = \begin{cases} G_t^k(u) & \text{if } u \in \mathcal{W}_t^k, \\ 0 & \text{otherwise}, \end{cases}$$

then, because the windows do not intersect, we define the image R_t of the confidence regions as

$$R_t(u) = \sum_{k=1}^{n_t} R_t^k(u).$$

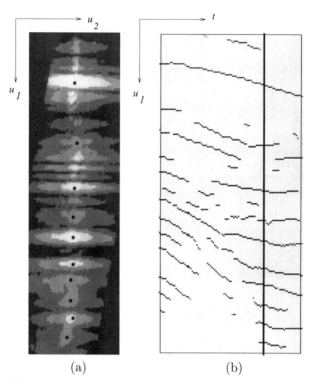

(a)   (b)

Figure 9.2. (a) An image $\mathcal{I}_t$. (b) Projection of 3-D paths regularized on a plane perpendicular to the images $\mathcal{I}_t$.

In real images, the windows can intersect, and we therefore define

$$R_t(u) = \max_{k=1}^{n_t} R_t^k(u),$$

$$\tau_t(u) = \arg\max_{k=1}^{n_t} R_t^k(u),$$

where $\tau_t(u)$ is the number of the confidence region associated with the site u. We use R_t to denote the image $\mathcal{I}_t$ where the spots have been smoothed and truncated. Figure 9.3 illustrates the results of this kind of preprocessing.

Finally, the data to be used in our problem are

$$y_{t,u} = \left(R_t(u), \Lambda_t^{\tau_t(u)} \right).$$

Because the images $\mathcal{I}_t$ are observed on a common grid $S^I \subset \mathbf{Z}^2$, the data y are indexed by $T \times S^I \doteq S$, belonging to $\mathbf{Z}^3$. In image analysis, the sites $s = (u, t)$ of this domain are called voxels. We now write the data in the

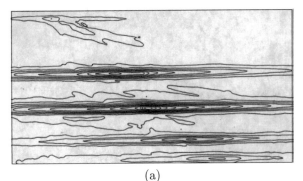

(a)

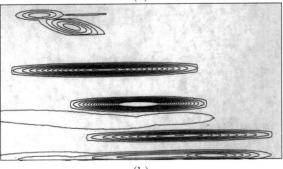

(b)

Figure 9.3. (a) Image $\mathcal{I}_t$. (b) Image $R_t(u)$.

following simplified form:

$$y_s = (R_s, \Lambda_s), \quad \forall s \in S.$$

If s is in a confidence region, in other words, if $R_s \neq 0$, then Λ_s denotes the vector of parameters of this confidence region. Otherwise, Λ_s is the "empty" vector. We write R_s to denote the "degree" of confidence in s for this region.

9.1.2 Definition of the Hidden Field

Here, we aim to determine a family of discretized regular curves that best fit the regions of confidence. In other words, we wish to find the three-dimensional paths that best connect these regions. This type of path is a continuous directed polygonal line whose vertices belong to some of the regions of confidence and whose every edge connects two regions in distinct images R_t. We call this type of directed edge a *pointer*, and we write the pointer as $x_s = (a_s, b_s, c_s)$, where a_s and b_s are the coordinates on the S^I grid, and c_s is the coordinate on the t-axis (Figure 9.4). A family of paths

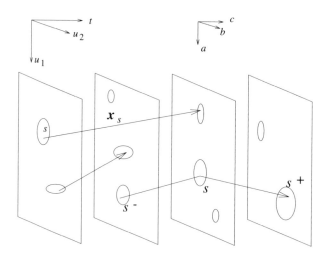

Figure 9.4. Configuration of pointers.

is represented by a configuration of pointers x_s indexed over all S,

$$x = \{x_s = (a_s, b_s, c_s) \in \mathsf{Z}^3, \, s \in S\},$$

such that:

- $c_s \geq 0$, where c_s is the range of the pointer x_s.

- If $c_s > 0$, then s and $s^+ \dot{=} s + x_s$ are the nodes adjacent to the pointer x_s.

- If $c_s = 0$, then s is either the end of a path or a site not located on a path.

We have now clearly extended the definition of pointers to the entire grid S. We use x to denote a multivariate vector field, where each x_s is defined by three variables. We write 0_s for the pointer whose range is zero.

We wish to impose the condition that not more than one path may pass through any given confidence region. This is achieved by applying the following constraint:

(C1) $\forall \, s = (t, u)$ and $s' = (t, u')$ such that $x_s \neq 0_s$ and $x_{s'} \neq 0_{s'}$, we have $\|s - s'\| > \delta_1$.

Finally, we impose an additional condition that limits the range of the pointers:

(C2) $\forall\, s : c_s \leq \delta_2$.

With these constraints, the set E of pointer configurations is re-duced to a subset E_C that constitutes our working set $E_C = \{x \in E$ such that x satisfies (C1) and (C2)$\}$.

9.1.3   Building an Energy

Remember that our aim is to find a family of paths that have certain qualities and that provide the best possible fit with the data y. The terms "qualities" and "best possible" are used with respect to a set of initial information that the user sets and then expresses in the form of an energy on E_C. A family of paths that satisfies this information will have a low energy, and one that does not will have a high energy. The energy has the following form:

$$U(x \mid y, \theta) = U_1(x, \theta_1) + U_2(y \mid x, \theta_2)$$
$$= \sum_{i=1}^{2} \theta_{1,i} U_{1,i}(x) + \sum_{i=1}^{2} \theta_{2,i} U_{2,i}(y \mid x).$$

There are four pieces of initial information: Two of them are so-called prior information, referring to the qualitative aspects of the paths, and the other two concern the data–path dependency. To describe this energy, we begin by setting $\mathbf{1}[c_s \neq 0] \dot{=} \mathbf{1}_s$.

Figure 9.5 shows the configuration of pointers at one stage of their regu-larization. As in the case of spline approximation, we wish to have regular paths. Thus, the first term $U_{1,1}$ quantifies the changes of slope along the paths:

$$U_{1,1}(x) = \sum_{s} \mathbf{1}_s \mathbf{1}_{s+} \left(\left| \frac{a_s}{c_s} - \frac{a_{s+}}{c_{s+}} \right| + \left| \frac{b_s}{c_s} - \frac{b_{s+}}{c_{s+}} \right| \right).$$

The second term $U_{1,2}$ must favor families that have few, but very long, paths. It is written as follows:

$$U_{1,2}(x) = \sum_{s} \mathbf{1}_s (1 - \mathbf{1}_{s+}) - \beta \sum_{s} \mathbf{1}_s,$$

where β is a positive parameter set empirically. The term $U_{1,2}$ is therefore composed of two terms. The first gives the number of paths in x by counting the number of path ends. The second is the total path length or, more specifically, the number of pointers making up the paths.

The paths that are chosen must pass through sites with a high degree of confidence. The third term $U_{2,1}$ therefore gives the sum of the degrees of confidence along the vertices of the paths:

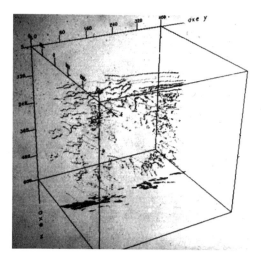

Figure 9.5. Pointers undergoing regularization.

$$U_{2,1}(y \mid x) = -\sum_s \mathbf{1}_s R_s.$$

The last term, $U_{2,2}$, which is more closely related to the application, quantifies the homogeneity of the confidence regions along each of the paths:

$$U_{2,2}(y \mid x) = \sum_s \mathbf{1}_s \, d(\Lambda_s, \Lambda_{s+}).$$

The distance $d(\Lambda_s, \Lambda_{s+})$ is defined on the basis of Kullback information $\mathcal{K}$ between the two Gaussian distributions on R^2 with parameters (μ_s, V_s) and (μ_{s+}, V_{s+}), respectively [124, 20]. The information $\mathcal{K}$ was used earlier ((3.16), page 69). Using G and G_+ to denote these distributions, d is defined by

$$d(\Lambda_s, \Lambda_{s+}) = \mathcal{K}(G, G_+) + \mathcal{K}(G_+, G),$$
$$\mathcal{K}(G, G_+) = -\int G(u) \log\left(\frac{G_+(u)}{G(u)}\right) du.$$

The distance d is known as the Kullback divergence. In our example, it is written

$$d(\Lambda, \Lambda_+) = \frac{1}{2}(\mu - \mu_+)'(V^{-1} - V_+^{-1})(\mu - \mu_+) + \frac{1}{2}\mathrm{trace}(V^{-1}V_+ + V_+^{-1}V).$$

With this energy defined, we may wonder how its final expression was obtained and how the parameters θ can be determined. This is discussed in the next section.

9.2   Markov Modeling Methodology

In this text the terms *modeling* and *model-building* are synonymous. Note that the energy mentioned above was given without any reference to the formalism of Markov random fields, just as the classic spline approximation was in the first place (see Part I, Chapter 2). This energy is derived from a neighborhood potential, however, which means that any configuration of pointers (or family of paths) can be seen as the result of a conditional Markov random field $(X \mid Y = y)$:

$$P_\theta[X = x \mid Y = y] \propto \exp -U(x \mid y, \theta).$$

At any site $s \in S$, the neighborhood N_s is a discrete parallelepiped centered on $s = (t, u)$, with dimension $2\delta_2 + 1$ for the t-coordinate and $(2\delta_1 + 1) \times (2\delta_1 + 1)$ for the $u = (u_1, u_2)$-coordinates. This is a consequence of the constraints (C1) and (C2). The Markov property is stated as follows:

$$P_\theta(x_s \mid \breve{x}_s, y) = P_\theta(x_s \mid N_s(x), y).$$

This Bayesian–Markovian interpretation also allows us to set out a methodology for effective model-building, in terms of estimation in particular. The above energy expression is the result of an interactive procedure that can be summarized as follows.

Method 9.1.

(i) *Choose an expression for U,*

(ii) *Estimate the parameters θ,*

(iii) *Calculate the estimate $\widehat{x}$,*

(iv) *If $\widehat{x}$ satisfies the initial information, end of procedure; otherwise, go to (i).*

This is a general method that uses the classic threesome for any modeling procedure: choice of model, estimation, model diagnostic checking. For points (ii) and (iii), however, its implementation is not general, as seen in Part II. We will now illustrate this procedure, giving modeling details, for the energy under consideration.

9.2.1   Details for Implementation

To determine the parameters, we return to the Bayesian calibration approach described in Section 8.2.3 . This approach is based on the local specifications of the conditional Markov random field $(X \mid Y = y)$, which are written as $P_\theta(x_s \mid N_s(x), y)$. Looking again at (8.10) on page 154,

$$\log \frac{P_\theta(x_s \mid N_s(x), y)}{P_\theta(x_s = 0_s \mid N_s(x), y)} = \theta'[\mathbf{U}(x_s = 0_s, N_s(x), y) - \mathbf{U}(x_s, N_s(x), y)]$$

$$\doteq \sum_{i=1}^{2} \theta_{1,i} H_{1,i}(N_s(x), x_s) +$$

$$\sum_{i=1}^{2} \theta_{2,i} H_{2,i}(N_s(x), x_s, y)$$

$$= \theta' H(N_s(x), x_s, y),$$

where y is limited to the neighborhood $\{s, N_s\}$. Locally, the functions $H_{j,i}$ are related to the regularity of the path, to the presence or absence of a path end, to the the degree of confidence, and to the homogeneity of the confidence regions, respectively. When $\mathbf{1}_s \neq 0_s$, we have

$$H_{1,1}(N_s(x), x_s) = -\mathbf{1}_{s-} \left(\left| \frac{a_s}{c_s} - \frac{a_{s-}}{c_{s-}} \right| + \left| \frac{b_s}{c_s} - \frac{b_{s-}}{c_{s-}} \right| \right)$$

$$-\mathbf{1}_{s+} \left(\left| \frac{a_s}{c_s} - \frac{a_{s+}}{c_{s+}} \right| + \left| \frac{b_s}{c_s} - \frac{b_{s+}}{c_{s+}} \right| \right),$$

$$H_{1,2}(N_s(x), x_s) = -(1 - \mathbf{1}_{s+}) + \mathbf{1}_{s-} + \beta,$$

$$H_{2,1}(N_s(x), x_s, y) = R_s,$$

$$H_{2,2}(N_s(x), x_s, y) = -d(\Lambda_s, \Lambda_{s-}) - d(\Lambda_s, \Lambda_{s+}), \tag{9.2}$$

where s^- denotes the possible position of the pointer that precedes s on the path under consideration: $s = s^- + x_{s-}$. The specifications $P_\theta(x_s \mid N_s(x), y)$ and $P_\theta(x_s = 0_s \mid N_s(x), y)$ are given by the neighborhood configuration, which depends on both pointers, x_{s-} and x_{s+}, and on the data y_{s-}, y_s, and y_{s+}. Given these, the $H_{j,i}$ can be calculated according to the expressions (9.2). To determine θ, we take a set of *extreme local configurations* $\mathcal{E}$. A local configuration $e \in \mathcal{E}$ is said to be extreme if $P_\theta(x_s \mid N_s(x), y)$ and $P_\theta(x_s = 0_s \mid N_s(x), y)$ are close to 0 or 1. The Bayesian calibration approach consists in choosing these local specifications arbitrarily, without attempting to estimate them from the data y. We write $\{(p^e, p_0^e)\ e \in \mathcal{E}\}$ to denote the chosen values, and H^e to denote the corresponding vectors. Figure 9.6 represents, in two-dimensional form, two extreme local configurations, with specifications $p^e = 0.99$ and $p^e = 0.01$. In Figure 9.6(a), the path has no change in slope, confidence is high (represented by the symbol $\circ$ placed at site s), and the regions of confidence are highly homogeneous (represented by ellipses of the same size at s and s^+). For Figure 9.6(b), the inverse situation applies.

We thus have a system that is linear with respect to θ:

$$\log \frac{p^e}{p_0^e} = \theta' H^e, \quad e \in \mathcal{E},$$

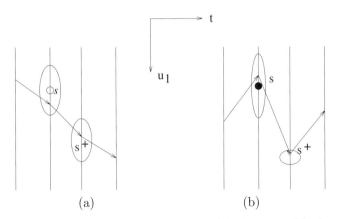

Figure 9.6. Two extreme local configurations. (a) $p^e = 0.99$. (b) $p^e = 0.01$.

containing more equations than unknowns. We can find θ via a least-squares solution. Let $\widehat{\theta}$ be the result.

An initial validation of this result can be obtained before reaching Step (iv) of the general procedure. Although the user's choice of energy expression is dictated by the initial information, it can give rise to local specification values that do not match the characteristic specifications of this same information. This is a serious problem because the specifications control the hidden field estimation algorithms. One way to check whether the calculated specifications are a good match for the desired specifications is to calculate the residual errors, as follows:

$$r^e = \log \frac{p^e}{p_0^e} - \widehat{\theta}' H^e, \quad \forall \, e \in \mathcal{E},$$

For the model to be suitable, a necessary condition is that these residuals must be small. If they are not, the components $\widehat{\theta}' H^e$ must be reviewed to determine qualitative changes to H^e, i.e., to the expression for the po-tentials. This is analogous to the multiple regression methodology [73]. In our context, we can supplement this check with an additional step. When all the residuals are small, we take a second set of local configurations on which the residuals are calculated using the $\widehat{\theta}$ obtained beforehand.

The final validation, performed in Step (iv), consists of the expert's judgement. This is often the only check that counts. Figures 9.5 and 9.7(a) give an outline of an initial configuration. After the hidden field estimation, regularization gives paths that are similar to those in Figures 9.2(b) and 9.7(b). In this case, minimization was performed using a deterministic relaxation algorithm from the ICM family (refer to Section 7.4.1 and [8]), where the initial configuration $x(0)$ was defined as follows. Let $S_0 \subset S$ be

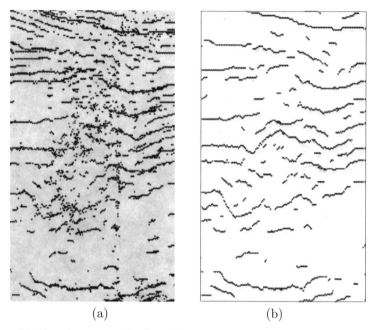

(a)   (b)

Figure 9.7. Regularization of 3-D paths: projection in a plane perpendicular to the images $\mathcal{I}_t$. (a) Initial paths $x(0)$. (b) Paths after regularization.

the set of sites that are the centers of confidence regions. We then set

$$s^+ = \arg\min_{s'}\{\|s' - s\|,\ s' \in N_s \text{ and } s' \in S_0,\ t' > t\},\ \forall s \in S_0,$$

where $s' = (t', u')$. The initial configuration is:

$$x_s(0) = \begin{cases} s^+ - s & \text{if } s \in S_0, \\ 0 & \text{otherwise.} \end{cases}$$

Figure 9.7 illustrates this regularization as a projection in the (t, u_1)-plane.

10
Degradation in Imaging

This chapter deals with the processing of degradation on radiographic images. In many cases, the methods discussed can also be applied to fields other than radiography. The radiographic image acquisition process includes a physical phase in which the image is formed by radiation and is disturbed by blurring or scattering effects [17]. The second phase is the detection of the radiation by a detector characterized by a response function. In the third phase, the measuring instruments deliver a final image to which a measurement noise is added. We will present models for denoising, deblurring, scatter analysis, and sensitivity function estimation.

The denoising model with explicit discontinuities (Section 10.1.1) provides an important example. It involves a pair of hidden fields $X = (X^p, X^b)$, and can be applied to other situations (Chapter 11, Section 12.2). Furthermore, in this example the prior energy is easy to analyze via its local specifications, providing an insight into how one energy can make various tasks cooperate with each other. In Section 10.2, which deals with blurring, our main aim is to illustrate the concept of an ill-posed problem. Section 10.3 discusses scatter in radiographic images. This problem requires us to draw upon the laws of physics in order to establish the degradation model. Finally, Section 10.4 concerns the restoration of an image from a fusion of two images. Although restoration, i.e., denoising, was dealt with in Section 10.1, here we must estimate the transfer function between the two images. This estimation will use tricubic regression splines (Section 3.2.1).

10.1   Denoising

We assume that the degradation comes from an additive Gaussian noise. We recall the degradation model from Chapter 5:

$$Y = \mathcal{O}(X^p, W) = X^p + W. \tag{10.1}$$

We write X^p to denote the field before degradation. The noise is represented by the random variables W_s, which are independent and Gaussian $LG(0, \sigma_W^2)$. Note that X^p and W are independent. Despite its simplicity, this model has great practical utility [19]. It will be made more complicated at a later stage, in particular when we consider a variance that is dependent on the theoretical signal intensity X_s. This assumption is characteristic of radiography. The restoration problem can be stated as follows: We wish to denoise the image while preserving the edge discontinuities of the original image. Classically, denoising is performed by smoothing, for example using spline surfaces whose second-order energy is

$$U(g \mid y) = \theta \int \left[(D^{2,0}g)^2 + 2(D^{1,1}g)^2 + (D^{0,2}g)^2 \right] dt\, du + \sum_s \left(y_s - g(t_s, u_s) \right)^2.$$

This smoothing is appropriate only if the edge regions and the regions outside the edges can be approximated by a unique spline surface of constant order. However, with this type of smoothing, we observe an attenuation of edge contrast. The smoothing operation must therefore take the edges into account, for example by taking care not to straddle an edge. The edges are unknown, however, and the presence of noise makes them difficult to detect. The solution is therefore to perform denoising and edge detection together. We emphasize that in this case, edge detection is a secondary task whose only purpose is to improve the denoising. This means that we are not necessarily seeking to achieve the same quality of edge detection that we would desire if edge detection were our primary aim.

10.1.1   Models with Explicit Discontinuities

• METHODOLOGY

Geman and Geman [84] have given a clear introduction to this type of simultaneous processing. We now refer to the methodology of inverse problems discussed in Section 6.1. To choose the configurations of the hidden field X, it is natural to consider the pairs $x = (x^p, x^b)$ where x^p denotes an image configuration before the degradation (10.1), and x^b denotes a configuration of edges. The random field X^p, like the field Y, is defined on the grid of pixels S^p, while x^b is defined on the grid of edge elements S^b, a dual grid of S^p as shown in Figure 10.1 for a 4×4 grid S^p. The symbol $\circ$ represents the *pixel sites*, and the symbol $+$ represents the *edge sites*. We write $S = S^p \cup S^b$ to denote the grid of X. The field X_s^p has values in E^p,

Figure 10.1. Mixed grid $S = S^p \cup S^b$ (S^p : ○, S^b : +, ↔: edge).

for example $E^p = \{0, \ldots, 255\}$, and X^b_s has values in $E^b = \{0, 1\}$:

$$x^b_s = \begin{cases} 1 & \text{if } s \text{ belongs to an edge,} \\ 0 & \text{otherwise.} \end{cases}$$

This encoding does not impose much on the edges. We do, however, distinguish between the edge sites located between two horizontal sites and those located between two vertical sites, which we call "vertical edge sites" and "horizontal edge sites" respectively. These names imply that when $x^b_s = 1$, there is at s an edge element whose direction is either vertical (↕) or horizontal (↔) according to the type of edge site s, as shown in the lower right-hand corner of Figure 10.1.

If X is modeled by a Markov random field with neighborhood system $\{N_s, s \in S\}$ and prior energy $U_1(x)$, then, given the independence of the W_s, we know from Chapter 6 that $(X \mid Y = y)$ is a Markov random field with the same neighborhoods as X and with energy

$$U(x \mid y) = U_1(x) + U_2(y \mid x),$$

with $U_1(x) = \sum_{i=1}^{q_1} \theta_i U_{1,i}(x)$ and $U_2(y \mid x) = \frac{1}{2\sigma^2_W} \sum_{s \in S^p} (y_s - x^p_s)^2$,

where the last expression comes from the Gaussian distribution of the W_s. Using the general notation from Chapter 6, we rewrite

$$U(x \mid y, \Theta) = U_1(x, \Theta_1) + U_2(y \mid x, \sigma^2_W),$$

with $\Theta = (\theta_1, \ldots, \theta_{q_1}, \theta_{q_1+1})$ and $\theta_{q_1+1} = \sigma^2_W.$

To solve the restoration problem, we must first build an energy U_1, and then determine an estimation $\widehat{\Theta}$ of the parameters Θ. The restored image will then be the result of the optimization with respect to x of $U(x \mid y, \widehat{\Theta})$.

● DEFINITION OF PRIOR ENERGY

We must define a prior energy $U_1(x, \Theta_1)$ that will allow us to smooth the noise while preserving the edges. To do this, we take the neighborhood

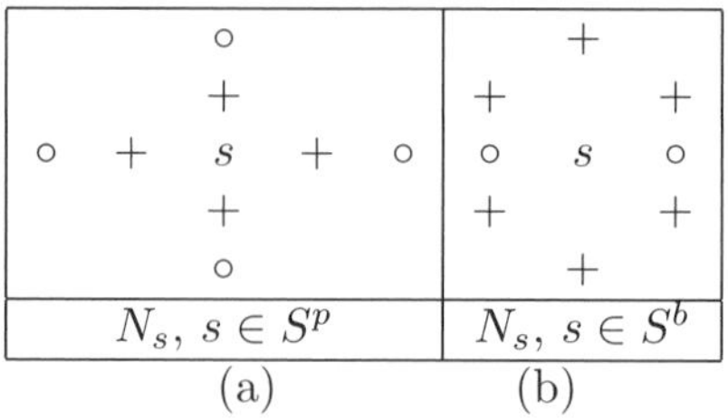

$\begin{array}{c} \circ \\ + \\ \circ \;\; + \;\; s \;\; + \;\; \circ \\ + \\ \circ \end{array}$	$\begin{array}{c} + \\ + \qquad + \\ \circ \;\; + \quad s \quad \circ \\ + \qquad + \\ + \end{array}$
$N_s,\; s \in S^p$	$N_s,\; s \in S^b$
(a)	(b)

Figure 10.2. Mixed neighborhood system. (a) On S^p. (b) On S^b.

system on S illustrated in Figure 10.2. Except for the sites located on the boundary of the grid, the neighborhood of any pixel site consists of the four nearest pixel sites and the four nearest edge sites. Figure 10.2(b) shows the neighborhood of a vertical edge site. The neighborhood of a horizontal edge site can be deduced from this; it would be obtained by a rotation through $\pi/2$. This neighborhood consists of the two nearest pixel sites and the six nearest edge sites that are not separated from s by a pixel site.

MODEL 1.
The prior energy is expected to contain three pieces of information [38]. The first is that, except for the edge regions, the restored image x^p must be smooth. The second is that the edges are associated with highly discontinuous regions in the image x^p. The third is that the edges are organized.

The first piece of information can be expressed as follows:

$$\mathrm{E}[X_s^p \mid N_s(x^p)] = \gamma \sum_{t \in N_s,\; t \in S^p} x_t^p, \qquad (10.2)$$

$$\mathrm{Var}[X_s^p \mid N_s(X^p)] = \sigma_X^2(0).$$

Its purpose is to smooth the noise. This energy means that for the low-energy configurations of x^p, we have: $x_s^p \approx \gamma \sum_{t \in N_s,\; t \in S^p} x_t^p$. In Section 5.3.3 we saw that these assumptions define a Gaussian Markov random field whose energy is the sum of two terms:

$$\theta_1 U_{1,1}(x^p) + \theta_2 U_{1,2}(x^p) = \theta_1 \sum_{s \in S^p} (x_s^p)^2 + \theta_2 \sum_{<s,t>} x_s^p x_t^p.$$

Note that $\gamma = -\frac{\theta_2}{2\theta_1}$ and $\sigma_X^2(0) = \frac{1}{2\theta_1}$, and therefore θ_1 must be positive.
    The second piece of information must give rise to an energy term that expresses an interaction between x^p and x^b, and whose purpose is edge detection:

$$\theta_3 U_{1,3}(x^p, x^b) = -\theta_3 \sum_{<s,t>} \left(x_s^p - x_t^p\right)^2 x_{k<s,t>}^b.$$

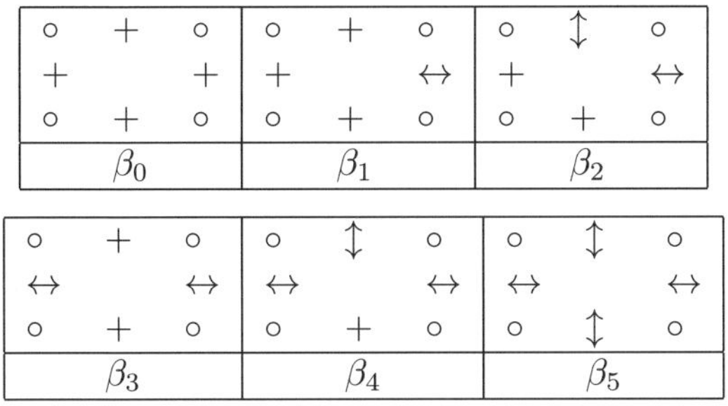

o	+	o	o	+	o	o	$\updownarrow$	o
+		+	+		$\leftrightarrow$	+		$\leftrightarrow$
o	+	o	o	+	o	o	+	o
	β_0			β_1			β_2	

o	+	o	o	$\updownarrow$	o	o	$\updownarrow$	o
$\leftrightarrow$		$\leftrightarrow$	$\leftrightarrow$		$\leftrightarrow$	$\leftrightarrow$		$\leftrightarrow$
o	+	o	o	+	o	o	$\updownarrow$	o
	β_3			β_4			β_5	

Figure 10.3. Edge configurations on fourth-order cliques.

The index $k{<}s,t{>}$ denotes the edge site between the two neighboring pixel sites s and t of S^p, and θ_3 is a positive parameter.

The third piece of information concerns the spatial organization of the edge elements. It leads to an energy term that favors the local continuity of edge elements, to the detriment of interruptions and intersections. Of the cliques from the neighborhood system $\{N_s, \ s \in S^b\}$, we keep only the fourth-order cliques, i.e., those consisting of four neighboring edge sites. On these cliques, the edge element configurations (plus or minus rotations through $\pi/2$) are illustrated in Figure 10.3. The β_i denote the potentials of these configurations.

The associated energy term is therefore

$$\theta_4 U_{1,4}\left(x^b\right) \ = \ \theta_4 \sum_{i=0}^{5} \beta_i n_i\left(x^b\right) ,$$

where $n_i\left(x^b\right)$ is the number of cliques with potential β_i in x^b. The potentials β_i are modeling parameters that are chosen according to our prior knowledge of how the edges are organized. Note that the potential of this energy is not canonical. In canonical form, the expression for the general model contains all the cliques from the neighborhood system $\{N_s, \ s \in S^b\}$:

$$U_{1,4}(x^b) = \beta_1' \sum_{C_1} x_s^b + \beta_2' \sum_{C_2} x_{s_1}^b x_{s_2}^b + \beta_3' \sum_{C_2'} x_{s_1}^b x_{s_2}^b$$

$$+ \ \beta_4' \sum_{C_3} x_{s_1}^b x_{s_2}^b x_{s_3}^b + \beta_5' \sum_{C_4} x_{s_1}^b x_{s_2}^b x_{s_3}^b x_{s_4}^b.$$

Here, C_4 is the set of fourth-order cliques that were kept (as described above), C_3 is the set of cliques consisting of three neighbors, C_1 is the set of singletons $s \in S^b$, C_2 are the neighbors at a distance of $\sqrt{2}$, and C_2' are the neighbors at a distance of 1. With $\beta_0 = 0$ and disregarding the conditions on the boundary of S^b, we can show [38] that the β_i and the β_i'

are linearly related:

$$\beta_1 = \frac{1}{2}\beta_1', \qquad \beta_2 = \beta_1' + \beta_2', \qquad \beta_3 = \beta_1' + \beta_3',$$

$$\beta_4 = \frac{3}{2}\beta_1' + 2\beta_2' + \beta_3' + \beta_4', \qquad \beta_5 = 2\beta_1' + 4\beta_2' + 2\beta_3' + 4\beta_4' + \beta_5'.$$

This clearly shows why the canonical form and its parameters are difficult to interpret in the context of modeling.

Finally, the prior energy is

$$U_1(x, \Theta_1) = \sum_{i=1}^{2} \theta_i U_{1,i}(x^p) + \theta_3 U_{1,3}(x^p, x^b) + \theta_4 U_{1,4}(x^b).$$

MODEL 2.

Another formulation, first suggested in [84], does not consider the explicit smoothing term:

$$U_1(x, \Theta_1) = \theta_1 U_{1,1}(x^p, x^b) + \theta_2 U_{1,2}(x^b) \qquad (10.3)$$

$$= \theta_1 \sum_{<s,t>} \phi\left(\frac{x_s^p - x_t^p}{\beta}\right)(1 - x_{k<s,t>}^b) + \theta_2 \sum_{i=0}^{5} \beta_i n_i(x^b).$$

This expression contains the same edge organization term as before. Note that β is a scaling parameter. In [90], the choice of smoothing function ϕ is

$$\phi(u) = \frac{-1}{1 + u^2}. \qquad (10.4)$$

For this function, which was already seen in Section 2.2 in (2.17), the low-energy configurations are those that have few local variations but might contain high-amplitude edges, since these are penalized at the same order of magnitude when the amplitude is such that $\phi(u)$ is close to 0. This can be seen as a gradual thresholding of the high values of u, allowing easier calibration of θ_1. Note that here, the term $U_{1,1}$ handles both detection and smoothing. Because the explicit smoothing term is absent, a quadratic interaction term $\phi(u) = u^2$ would not be satisfactory in this second model. It would place too high a penalty on high-amplitude edges.

• ANALYSIS OF MODEL 1

To explain exactly how the interaction term prevents the smoothing from attenuating the discontinuities in x^p, we start by writing the local specifications of the field X^p. For any configuration x, we consider the partition of S^p into five segments $\{S^p(q), q = 0, \ldots, 4\}$, where $S^p(q)$ is the set of sites s such that the neighborhood N_s contains exactly q edge elements, i.e., q edge sites of value 1.

Proposition 10.1. *On each of the segments $S^p(q)$, $P(x_s^p \mid N_s(x))$ is a Gaussian distribution $LG(\mu_X(q), \sigma_X^2(q))$ with*

$$\mu_X(q) = \gamma_1(q) \sum_{t \in N_s} x_t^p + \gamma_2(q) \sum_{t \in N_s} x_t^p x_{k<s,t>}^b,$$

$$\sigma_X^2(q) = \frac{1}{2(\theta_1 - \theta_3 q)}, \quad \gamma_1(q) = -\theta_2 \sigma_X^2(q), \quad \gamma_2(q) = -2\theta_3 \sigma_X^2(q).$$

Proof:
This is a standard proof. With $0_s = 0$, we obtain

$$U_1\left(x^p, x^b\right) - U_1\left(x_{/s}^p, x^b\right) = \left(x_s^p\right)^2 \left(\theta_1 - \theta_3 \sum_{t \in N_s} x_{k<s,t>}^b\right)$$

$$+ x_s^p \left(\theta_2 \sum_{t \in N_s} x_t^p + 2\theta_3 \sum_{t \in N_s} x_t^p x_{k<s,t>}^b\right),$$

$$\doteq \left(x_s^p\right)^2 c_1 + x_s^p c_2 = c_1\left(x_s^p + c_2'\right)^2 - c_1 c_2'^2,$$

where $c_2' = c_2/(2c_1)$. The local specification is $P(x_s^p \mid N_s(x)) \propto \exp -\left[U_1\left(x^p, x^b\right) - U_1\left(x_{/s}^p, x^b\right)\right]$. If we consider the continuous version of the distribution, the normalization constant is $\sqrt{\pi/c_1} \exp\left(c_1 c_2'^2\right)$. We therefore have $P\left(x_s^p \mid N_s(x)\right) = \sqrt{\pi/c_1}^{-1} \exp -\left[c_1\left(x_s^p + c_2'\right)^2\right]$. This means that $P(x_s^p \mid N_s(x))$ is a Gaussian distribution with variance $\sigma_X^2(q) = 1/(2c_1)$ and expectation $\mu_X(q) = -c_2'$. $\qquad\square$

On $S^p(0)$, the distribution is the usual Gaussian distribution (10.2). Near an edge, however, the distribution changes. In particular, the value of the pixels located on the other side of the border is subtracted from the usual average. If this operation were not performed, the amplitude of the edges would be attenuated during smoothing.

The local specifications of X^b also help us to understand the role of the interaction term in edge detection. The expression $P(x_k^b \mid N_k(x)) \propto \exp -\left[U_1(x^p, x^b) - U_1(x^p, x_{/k}^b)\right]$ gives us

$$P[X_{k<s,t>}^b = 1 \mid N_{k<s,t>}(x)] = \frac{\exp\left[\theta_3\left(x_s^p - x_t^p\right)^2 - \theta_4 f_{k<s,t>}\left(x^b\right)\right]}{1 + \exp[\theta_3\left(x_s^p - x_t^p\right)^2 - \theta_4 f_{k<s,t>}\left(x^b\right)]},$$
(10.5)

where $f_k\left(x^b\right) = U_{1,4}\left(x^b\right) - U_{1,4}\left(x_{/k}^b\right)$ is dependent only on $\{x_t^b, t \in N_s\}$. We can therefore see that, in the absence of discontinuities in the x_s^p surface, i.e., $x_s^p - x_t^p = 0$, the local specifications are those of a Markov random field X^b independent of X^p. When $x_s^p - x_t^p \neq 0$, however, the conditional probability of having $X_s^b = 1$ increases and approaches 1 for strong discontinuities. In this case, the role of $f_k\left(x^b\right)$ becomes insignificant, θ_3 is

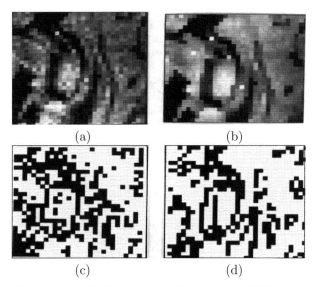

Figure 10.4. Denoising an NMR image. (a) Noisy image. (b) Restored image. (c) Edges before restoration. (d) Edges after restoration.

a scaling parameter, and θ_4 is a parameter representing the competition between detection and organization, entities given by $\theta_3\left(x_s^p - x_t^p\right)^2$ and $\theta_4 f_{k<s,t>}\left(x^b\right)$, respectively.

● RESTORATION EXAMPLE

The two models described in this paragraph are suitable for piecewise constant images. In the next paragraph we will see that one way to go farther is to define this model not directly on the basis of the gray levels, but on the differences between them to the first or second order. Note also that these models are spatially invariant because their parameters are independent of the position in the image. As a result, the above estimation must sometimes be performed locally and restricted to subimages so that the parameter estimation follows the spatial progress of the image. Figure 10.4(a) was restored using Model 1. It shows the restoration of an NMR image at 0.1 tesla. This image was degraded by a thermal noise for which the white noise model is particularly appropriate. Restoration is performed on a 32×32 subimage. For this size, it is reasonable to assume spatial invariance. Although the piecewise constant image model is not really valid, we can see by looking at the organization of the edges that the model is fairly well suited to this restoration job. We now describe how the parameters of this model were estimated.

Parameter Estimation.[1] The maximum pseudolikelihood principle is used as follows.

Proposition 10.2. *On each of the segments $S^p(q)$, $P(y_s \mid N_s(y, x^b))$ is a Gaussian distribution $LG(\mu_Y(q), \sigma_Y^2(q))$ with*

$$\mu_Y(q) = c\gamma_1(q) \sum_{t \in N_s} y_t + c\gamma_2(q) \sum_{t \in N_s} y_t \, x^b_{k<s,t>}.$$

On $S^p(0)$, the variance is

$$\sigma_Y^2(0) = \sigma_X^2(0) + \sigma_W^2 \left[1 + 4c\gamma_1^2(0)\right], \quad \text{with} \quad c = \frac{\sigma_X^2}{\sigma_W^2 + \sigma_X^2}.$$

This property is specific to the Gaussian case, and allows us to construct an ad hoc maximum pseudolikelihood estimation. Let us look at the estimation of θ_1 and θ_2. Let $\tilde{x}^b$ be an edge configuration obtained using any edge detector. We must calculate the following:

$$\max_{(\theta_1, \theta_2)} \prod_{s \in S^p(0)} P(y_s \mid N_s(y, \tilde{x}^b)). \tag{10.6}$$

Strictly speaking, this is not the maximum pseudolikelihood, because $Y = X + W$ is not a Gaussian Markov random field on $S^p(0)$ even though X is. (The spectral density of a Gaussian Markov random field is $f_X(\omega) = \sigma_X^2/Q(e^{i\omega})$, where Q is a polynomial [104]. Because the spectral density of W is constant, we can see that the spectral density of $X + W$ is not the inverse of a polynomial as is $f_X(\omega)$). In any case, if we set

$$\rho = -c\frac{\theta_2}{2\theta_1}, \tag{10.7}$$

$$\sigma_Y^2(0) = \frac{1}{2\theta_1} + \sigma_W^2 \left(1 + 4\frac{\rho^2}{c}\right), \tag{10.8}$$

then expression (10.6), an "extended pseudolikelihood", is written as

$$\prod_{s \in S^p(0)} \frac{1}{\sqrt{2\pi\sigma_Y^2(0)}} \exp - \left(\frac{1}{2\sigma_Y^2(0)} \left(y_s - \rho \sum_{t \in N_s} y_t\right)^2\right).$$

Its maximum with respect to $(\rho, \sigma_Y^2(0))$ is the least-squares solution

$$\hat{\rho} = \frac{\sum_{s \in S^p(0)} \left(y_s \sum_{t \in N_s} y_t\right)}{\sum_{s \in S^p(0)} \left(\sum_{t \in N_s} y_t\right)^2},$$

$$\hat{\sigma}_Y^2(0) = |S^p(0)|^{-1} \sum_{s \in S^p(0)} \left(y_s - \hat{\rho} \sum_{t \in N_s} y_t\right)^2.$$

---

[1]This paragraph can be omitted at the first reading.

We substitute these estimates into (10.7) and (10.8), and then we must solve a system of two equations in four unknowns: $\theta_1, \theta_2, \sigma_W^2, c$. Two other equations are chosen. In real images, the spatial correlation $\gamma_1(0) = -\frac{\theta_2}{2\theta_1}$ is usually close to the upper limit, which for our model, is $\frac{1}{4}$ assuming stationarity. We then set $\frac{1}{4} = -\frac{\theta_2}{2\theta_1}$. We can eliminate the unknown σ_W^2 using a signal-to-noise ratio R, which appears when we express θ_1 according to (10.7):

$$\theta_1 = \left[2\sigma_Y^2(0)(1-R)\right]^{-1}.$$

The ratio R is a value between 0 and 1 and is chosen empirically.

The estimation of θ_3 would be performed in a similar way. The estimation of θ_4 is obtained via a Bayesian calibration (Section 8.2.3) based on the local specifications of the edges (10.5), in which the estimation $\widehat{\theta}_3$ replaces θ_3:

$$P\left[X_{k<s,t>}^b = 1 \mid N_{k<s,t>}(x)\right] = \frac{\exp\left[\widehat{\theta}_3\left(x_s^p - x_t^p\right)^2 - \theta_4 f_{k<s,t>}\left(x^b\right)\right]}{1 + \exp\left[\widehat{\theta}_3\left(x_s^p - x_t^p\right)^2 - \theta_4 f_{k<s,t>}\left(x^b\right)\right]}.$$

10.1.2   Models with Implicit Discontinuities

• EQUIVALENCE WITH THE SIMPLIFIED EXPLICIT APPROACH

Let us simplify model 2 given in (10.3) by omitting the edge organization term and by considering a new function ϕ written as ρ :

$$U_1(x^p, x^b) = \theta_1 \sum_{<s,t>} \rho\left(\frac{x_s^p - x_t^p}{\beta}\right)\left(1 - x_{k<s,t>}^b\right),$$

$$\rho(u) = u^2 - 1,$$

with the following associated posterior energy:

$$U\left(x^p, x^b \mid y\right) = U_1\left(x^p, x^b\right) + \frac{1}{2\sigma_W^2}\sum_s\left(y_s - x_s^p\right)^2. \qquad (10.9)$$

In this particular case, we note that

$$\min_{x^b} U\left(x^p, x^b\right) = \sum_{<s,t>} \rho^-\left(\frac{x_s^p - x_t^p}{\beta}\right),$$

$$\rho^-(u) = \left(u^2 - 1\right)^-,$$

where ρ^- is the truncated quadratic function $\rho^-(u) = 0$ if $u^2 - 1 > 0$. Because the values of x_k^b are independent of each other, the minimization with respect to x^b is obtained by minimizing at each site $k < s, t >$. Therefore, when $u^2 - 1 > 0$, this minimization gives $x_{k<s,t>}^b = 1$, and otherwise it gives $x_{k<s,t>}^b = 0$. It follows that the minimization of (10.9) is equivalent to minimizing the energy without taking x^b into account, and defined by

$$U^-(x^p \mid y) = \theta_1 \sum_{<s,t>} \rho^- \left(\frac{x_s^p - x_t^p}{\beta} \right) + \frac{1}{2\sigma_W^2} \sum_s (y_s - x_s^p)^2. \quad (10.10)$$

We emphasize that this equivalence holds true only when the edge organization term is absent. In certain situations, however, this term can be essential.

This approach can be generalized to other functions ρ^-, which we now write as ϕ, as follows. The equation (10.10) can be written in more general form as

$$U^-(x^p \mid y) = \theta_1 \sum_c \phi \left(\frac{Q_c(x^p)}{\beta} \right) + \frac{1}{2\sigma_W^2} \sum_s (y_s - x_s^p)^2, \quad (10.11)$$

where c denotes a clique and Q_c denotes a clique potential. On the other hand, if we set $\widetilde{x}_k^b = 1 - x_k^b$, (10.9) can be rewritten as follows:

$$U(x^p, \widetilde{x}^b \mid y) = \theta_1 \sum_{c=k<s,t>} \left[\left(\frac{x_s^p - x_t^p}{\beta} \right)^2 \widetilde{x}_c^b - \widetilde{x}_c^b \right] + \frac{1}{2\sigma_W^2} \sum_s (y_s - x_s^p)^2,$$

where this expression corresponds to the function $\rho(u) = u^2 - 1$. A more general form of this function is

$$U(x^p, \widetilde{x}^b \mid y) = \theta_1 \sum_c \left[\left(\frac{Q_c(x^p)}{\beta} \right)^2 \widetilde{x}_c^b + \zeta(\widetilde{x}_c^b) \right] + \frac{1}{2\sigma_W^2} \sum_s (y_s - x_s^p)^2.$$

$$(10.12)$$

We might then wish to know under what conditions on ϕ in (10.11) there exists a dual energy with explicit discontinuities of the form (10.12). Because the $\widetilde{x}_c^b$ are independent, this question amounts to finding the conditions on ϕ for the existence of a function ζ such that $\phi(u) = \inf_{0 \le a} (u^2 a + \zeta(a))$. The solution is given in [90]:

Theorem 10.1. *If the function $\phi(u)$ defined on $[0, \infty)$ has the properties $\phi(0) = -1$, $\phi(\sqrt{u})$ concave, and $\lim_{u \to \infty} \phi(u) = 0$, then there exists a function $\zeta(a)$ defined on $[0, M]$ such that*

$$\phi(u) = \inf_{0 \le a} (u^2 a + \zeta(a)),$$

when ζ is strictly decreasing from $\zeta(0) = 0$ to $\zeta(M) = -1$.

Geometrically, $\phi(u)$ appears as the lower limit of a quadratic family. The function (10.4) satisfies the assumptions of the theorem. Note that with this duality, we end up in a paradoxical situation: By trying to eliminate the hidden field x^b for the sake of simplicity, we have made optimization more difficult. With the explicit formulation, when $Q_c(x^p)$ is linear with respect to x^p, the term $(Q_c(x^p)/\beta)^2$ is quadratic, and therefore $(X^p \mid x^b)$ is a Gaussian field. Furthermore, the $\left(\tilde{X}_c^b \mid x^p\right)$ are independent. As a result, the optimization is better with $U(x^p, \tilde{x}^b \mid y)$ than with $U^-(x^p \mid y)$.

• HIGHER-ORDER SMOOTHING

As we have already stated, the energies we have considered have favored configurations consisting of regions with a constant gray level. Using the energy with implicit discontinuities to restore an image which is not piecewise constant, results in an image with a "snakeskin" appearance. After optimization, false discontinuities are placed on the regions with a low slope. As for spline surfaces, we then tend to work with higher orders to adapt to the piecewise constant images, in terms of either gradient or curvature. For the case of the formulation with implicit discontinuities $(x \equiv x^p)$, we consider the function (10.4):

$$\phi(u) = \frac{-1}{1 + |u/\beta|^\gamma}, \tag{10.13}$$

where $1 \le \gamma \le 2$. Small values of β favor the configurations with close gray levels (see Figure 2.1, page 32). This type of function that penalizes big differences is used in robust approximation methods (Section 2.3.2). The first-order prior energy is written

$$U_1(x) = \theta_1 \sum_{c=<s,t>} \phi(x_s - x_t) \doteq \theta_1 \sum_{c \in C_1} \phi(Q_c^1(x)), \tag{10.14}$$

where C_1 is the set of cliques $< s, t >$ of the first-order model, and $Q_c^1(x) = x_s - x_t$. By setting $s = (s_1, s_2)$, we obtain

$$U_1(x) = \theta_1 \sum_s [\phi(x_{s_1+1,s_2} - x_{s_1,s_2}) + \phi(x_{s_1,s_2+1} - x_{s_1,s_2})].$$

At any s, the differences are the discretized versions of the partial derivatives of a continuous surface, let us say $g(s_1, s_2)$. The low-energy configurations are those with zero derivatives: $D^{1,0}g = D^{0,1}g = 0$. For the second-order case, the second-order derivatives vanish, $D^{2,0}g = D^{0,2}g = D^{1,1}g = 0$, giving the following prior energy:

$$U_1(x) = \theta_1 \sum_s \big[\phi(x_{s_1+2,s_2} - 2x_{s_1+1,s_2} + x_{s_1,s_2})$$

$$+ \phi(x_{s_1,s_2+2} - 2x_{s_1,s_2+1} + x_{s_1,s_2})$$

$$+ 2\, \phi(x_{s_1+1,s_2+1} - x_{s_1+1,s_2} - x_{s_1,s_2+1} + x_{s_1,s_2}) \big].$$

This can be rewritten in terms of its cliques C_2. There are two types of clique: those consisting of three aligned neighboring sites, $\{(s_1+2, s_2), (s_1+$

$1, s_2), (s_1, s_2)\}$ or $\{(s_1, s_2 + 2), (s_1, s_2 + 1), (s_1, s_2)\}$, and those consisting of four neighboring sites arranged in a square, $\{(s_1 + 1, s_2 + 1), (s_1 + 1, s_2), (s_1, s_2 + 1), (s_1, s_2)\}$. We then write

$$U_1(x) \doteq \theta_1 \sum_{c \in C_2} \phi(Q_c^2(x)). \tag{10.15}$$

We handle the third-order case in the same way:

$$U_1(x) \doteq \theta_1 \sum_{c \in C_3} \phi(Q_c^3(x)), \tag{10.16}$$

but we omit the details of C_3 and Q^3. The low-energy configurations of orders 1, 2, and 3 are therefore those that are locally constant, planar, and quadratic, respectively.

10.2   Deblurring

Here, we seek to restore images that have been corrupted by both convolution blurring and a noise. The formation of the blurring is assumed to be spatially invariant. The observed degraded image y is the result of the addition of a noise to the blurred image $f = \mathcal{H} * x$:

$$Y = \mathcal{O}(X, W) = \mathcal{H} * X + W. \tag{10.17}$$

We write $W = \{W_s\}$ to denote a Gaussian white noise with variance σ^2, and $\mathcal{H}$ is the blurring filter

$$(\mathcal{H} * X)_s = \sum_{u \in F} h_u X_{s-u}. \tag{10.18}$$

The coefficients $\{h_u, u \in F\}$ define the *point spread function* $\mathcal{H}$. Its support is $F = \{u = (k, l); \ -k_0 \le k \le k_0; \ -l_0 \le l \le l_0\}$. We set

$$\sum_{u \in F} h_u = 1, \tag{10.19}$$

to preserve the mean value of the initial image after its corruption by blurring. We write (k_0, l_0) to represent the extent of the blurring. The point spread function is often Gaussian, as follows:

$$h_u = \frac{1}{\kappa} \exp -\frac{\|u\|^2}{2\sigma^2},$$

where κ is the constant of normalization to 1, to satisfy (10.19). In the case of a defocusing blur, $\mathcal{H}$ is written as

$$h_u = \frac{1}{\kappa r^2},$$

where r is the radius of the focal spot. In this paragraph, we assume that $\mathcal{H}$ is known. Deblurring is a difficult problem, and the results are often less than perfect. We will now seek to explain some of the difficulties involved.

10.2.1   A Particularly Ill-Posed Problem

We begin with a continuous representation of the problem. Taking a bounded function $x(s)$ that is measurable on R^2, we define its blurred version as

$$f(s) = \int_{R^2} h(u)\, x(s-u)\, du, \qquad (10.20)$$

and we assume its point spread function to be Gaussian:

$$h(u) = \frac{1}{2\pi\sigma^2} \exp -\frac{\|u\|^2}{2\sigma^2}. \qquad (10.21)$$

We must estimate x given f and h. This is an ill-posed problem, and as we will see, the restoration of x becomes much more complicated in the presence of a noise. We can easily see why this problem is ill-posed if we look at the analogy between the effect of blurring and diffusion processes such as the one that comes out of the heat equation. This idea is explained by Carasso et al. [30] (see also [113]). Given an image $x(s)$, the heat equation in the plane $s = (s_1, s_2)$ is written as follows for any instant $t > 0$:

$$\frac{\partial f}{\partial t}(s, t) = \Delta f(s, t) \quad \text{with} \quad f(s, 0) = x(s). \qquad (10.22)$$

The symbol Δ represents the Laplacian $\Delta f = \frac{\partial^2 f}{\partial s_1^2} + \frac{\partial^2 f}{\partial s_2^2}$. The unique solution to this equation at time t is

$$f(s, t) = \frac{1}{4\pi t} \int_{R^2} x(s-u) \exp -\frac{\|u\|^2}{4t} du, \qquad (10.23)$$

and this is the same as (10.20) with $\sigma^2 = 2t$. The blurred image f can therefore be interpreted as the spatial temperature distribution at the instant t, given the initial distribution $f(s, 0) = x(s)$. The image restoration problem is therefore a case of solving the diffusion equation by reversing time: a classic example of an ill-posed problem in mathematical physics.

To examine the effect of a noise on this problem, let us establish the expression for f in the Fourier domain. Here, $x(s)$ is defined on $S = \{(s_1, s_2); \ 0 \le s_1 \le 2\pi; \ 0 \le s_2 \le 2\pi\}$, and x can be extended to x^p, a 2π-periodic function of s_1 and s_2, said to be S-periodic. If f^p is an S-periodic function that is a solution of the diffusion equation, then f^p

defined on S and written f is a solution of

$$\frac{\partial f}{\partial t}(s,t) = \Delta f(x,t) \quad \text{with} \quad f(s,0) = x(s),$$

$$f(0, s_2, t) = f(2\pi, s_2, t), \qquad \frac{\partial f}{\partial s_1}(0, s_2, t) = \frac{\partial f}{\partial s_1}(2\pi, s_2, t),$$

$$f(s_1, 0, t) = f(s_1, 2\pi, t), \qquad \frac{\partial f}{\partial s_2}(s_1, 0, t) = \frac{\partial f}{\partial s_2}(s_1, 2\pi, t).$$

The S-periodicity allows us to represent f as a Fourier series with the functions $\phi_{mn}(s) = \exp(ims_1 + ins_2)$ indexed by $\mathbb{Z}^2$:

$$f(s,t) = \sum_{m,n \in \mathbb{Z}} a_{mn}(t)\phi_{mn}(s)$$

$$a_{mn}(t) = \langle f(.,t), \phi_{mn} \rangle = \frac{1}{4\pi^2} \int_S f(s,t)\phi_{mn}(-s)ds.$$

This expression is written for a fixed t. The scalar product is associated with the norm L^2. We then obtain

$$\Delta f(s,t) = -\sum_{m,n}(m^2 + n^2)a_{mn}(t)\phi_{mn}(s). \tag{10.24}$$

For S-periodic functions, Δ is a symmetric operator $\langle \Delta f, g \rangle = \langle f, \Delta g \rangle$ resulting from the orthogonality of complex exponential functions. The eigenvectors of Δ are the ϕ_{mn} with associated eigenvalues λ_{mn} as indicated in (10.24):

$$\Delta\phi_{mn} = \lambda_{mn}\phi_{mn}, \quad \lambda_{mn} = -(m^2 + n^2), \quad \|\phi_{mn}\|^2 = 1.$$

The coefficients $a_{mn}(t)$ of the solution f of the diffusion equation are expressed in terms of the coefficients b_{mn} of the initial image x. If

$$x(s) = \sum_{m,n} b_{mn}\phi_{mn}(s),$$

then at time t, the function

$$f(s,t) = \sum_{m,n} b_{mn}e^{\lambda_{mn}t}\phi_{mn}(s), \tag{10.25}$$

satisfies (10.22), and therefore $a_{mn}(t) = b_{mn}\exp(\lambda_{mn}t)$.

Let us now discuss the effects of a noise. Assume that x has norm 1, and that f is degraded by a noise concentrated on the characteristic function $\phi_{10,10}$. This noise is chosen to have norm 10^{-40}. The noisy image is therefore

$$\widetilde{f}(s,t) = f(s,t) + 10^{-40}\phi_{10,10}(s).$$

Let $\widetilde{x}$ be the initial field corresponding to $\widetilde{f}$ via the diffusion equation. We will examine the situation where the noise is not taken into account, and

$\widetilde{f}$ is restored instead of f. Using (10.25), which takes us from the $a_{mn}(t)$ to the b_{mn}, we obtain

$$\widetilde{x}(s) = x(s) + e^{200t} 10^{-40} \phi_{10,10}(s),$$

since $\lambda_{10,10} = -200$. Taking $t = 1$, for example, we obtain

$$\|\widetilde{x}(s) - x(s)\| = e^{200} 10^{-40} > 10^{46}.$$

We can therefore see that two very similar images f and $\widetilde{f}$ can correspond to initial images x and $\widetilde{x}$ that are very different. In other words, even a small error in the evaluation of the noise can lead to a very large restoration error. This is an ill-posed problem. It is therefore essential to take the noise into account.

10.2.2 Model with Implicit Discontinuities

As we have said, deblurring is a difficult problem, and the results tend to be disappointing. Occasionally, however, restorations of remarkably good quality can be achieved. This occurs in particular with methods based on the function ϕ given in (10.13). Let us look again at the degradation model (10.17) and (10.18), for which we assume an independent Gaussian noise $LG(0, \sigma^2)$. Following on from Section 10.1, we consider the following posterior energy:

$$U(x \mid y) = \frac{1}{2\sigma^2} \sum_s \left(y_s - \sum_{u \in F} h_u x_{s-u} \right)^2 + \theta \sum_{c \in C_k} \phi(Q_c^k(x)),$$

where $Q_s^k(x)$ is defined to order $k = 1, 2, 3$ in (10.14), (10.15), and (10.16). Referring to (5.9), we obtain the local specifications of the Markov random field $(X \mid y)$:

$$P[x_s \mid N_s(s), y]$$

$$\propto \exp - \left[\frac{1}{2\sigma^2} \sum_{v \in F} \left(y_{s-v} - \sum_{u \in F} h_u x_{s-v-u} \right)^2 + \theta \sum_{c:\, s \in c} \phi(Q_c^k(x)) \right].$$

The prior energy U_1, introduced for the purpose of denoising, will now be used for deblurring. The difference between these two situations comes from the fidelity term in $U(x \mid y)$.

A few practical comments about the use of this energy would now be appropriate. Firstly, the calculation costs involved are very high. The fidelity term of $P(x_s \mid N_s(x), y)$ requires $|F|^2$ contributions to be calculated. This means that for a 5×5 blur, the sum $\sum_{v \in F}$ requires the calculation of 625 contributions. The calculation is shortened by approximating this sum according to $\sum_{v \in \widetilde{F}}$ where $\widetilde{F}$ is a subset of F. Furthermore, the minimization of U that is performed by one of the relaxation techniques described in Chapter 7 can very easily be made simpler to calculate. We recall that relaxation is performed by successive updates of the value of x at each site in accordance with local specifications. To do this, instead of using the local distribution $\{P[X_s = \lambda \mid N_s(x), y], \lambda \in E_s\}$ on all E_s, we restrict it to a subset of E_s calculated according to the values of x on $\{s, N_s\}$ obtained in the previous updates. This trick can be used in many other applications.

The convergence of the optimization algorithm is another problem, particularly in the case of a model of order $k = 3$. The algorithm that starts out with the blurred image y as its initial solution tends to remain trapped near y because of the complicated nature of the configurations that are allowed by the third-order model. One way to alleviate this severe disadvantage is to take a hierarchical approach which involves starting the restoration with $k = 1$, and when it converges, continuing with $k = 2$ until that converges, and finally using $k = 3$. The experimental results of [90] demonstrate the advantages of this hierarchical approach, whose utility extends beyond the context discussed in our example.

10.3   Scatter

[2] Scatter causes the degradation of image quality in radiographic images. It occurs when, instead of moving in a straight line, the photons from the radiation source collide with the atoms of the matter through which they are passing (Figure 10.5). The photons are then either absorbed (photoelectric effect) or scattered (Compton effect or Rayleigh effect). When they emerge from the matter, the scattered photons constitute a radiant image x that can be acquired using radiographic film, for example. This collision process is stochastic, and is described by *transport equations*. The radiant image can be simulated using Monte Carlo methods. To do this, we must set the parameters of the radiographic system and the parameters that describe the structure of the matter to be examined. This type of simulation provides images without the need to perform actual experiments. This is an important consideration in the field of nondestructive testing, where some experiments are very complicated and expensive. This method is also useful in calibrating the system parameters, resulting in improved image quality. The simulation of radiant images therefore requires the direct problem to be dealt with, and this is far more difficult than the situations described by the degradation models encountered so far. This direct problem is discussed in the next subsection. The inverse problem, which seeks

---

[2]This section can be omitted at the first reading.

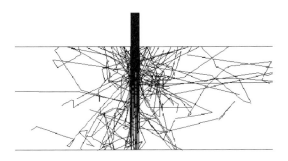

Figure 10.5. Examples of photon paths in the material.

to eliminate the scattering effect from a real image y, is discussed in a later subsection.

10.3.1 Direct Problem

This paragraph concerns the simulation of photon transport in a material that may contain pockets that are empty or filled with a different material. The photons are emitted by a high-power gammaradiographic source (power up to 1.33 MeV). When the object being examined is very thick (several centimeters), the photons are subjected to several collisions that the simulation process must take into account for the resulting scatter to be represented. Here, the radiographic chain is limited to the transport of photons in the material (this excludes the detector). This means that we will look at the results of the simulations in terms of the radiant images made up by the distribution of photons as they exit the object. A Monte Carlo simulation is used. This simulation estimates the probability that a photon emitted by the source will reach a surface element $\mathcal{V}$ on the radiant image detector. According to the law of large numbers, the estimation will be given by the proportion $N_{\mathcal{V}}/N$ of the number of photons $N_{\mathcal{V}}$ that have reached $\mathcal{V}$ out of the N photons emitted by the source.

Physicists have been working on the problem of simulating photon (or neutron) transport since the early 1950s [181]. They formalized the problem using a "natural" Markov chain obtained directly from physical principles. Later, a more advanced mathematical formalism was developed. This established the connection between the transport equation and the Kolmogorov equations for a Markov chain [125]. For our purposes, we will use an approach somewhere in between the two.

● Natural Markov Chain

The successive collisions of a photon in the material $\mathcal{M}$ are described by a Markov chain $\{Z_n, n \geq 0\}$. The state variable Z_0 is that of the source.

The variables Z_n, $n \geq 1$, are those of any collisions in the material. In practice the chain has finite length, and the last variable Z_n relates either to the premature halting of the photon in the material or to its arrival at the image detector. The variables Z_n are random vectors

$$Z_n = (S_n, \Lambda_n),$$

with state (s_n, λ_n) in $E = \mathcal{M} \times \mathcal{E} \subset \mathbb{R}^3 \times \mathbb{R}^+$. For $n \geq 1$, s_n denotes the position in $\mathcal{M} \subset \mathbb{R}^3$ of the nth collision, and λ_n denotes the energy of the photon emerging from s_n.[3] Clearly, s_1 must belong either to the material or to the detector. Unlike what we saw in Chapter 7, $\{Z_n\}$ is a Markov chain with continuous states. Despite this difference, a Monte Carlo simulation method can be used. Instead of the transition probability matrix, we define the kernel K of the probability of transition. For any measurable function, this allows us to write the expectation of the conditional random variable $[f(Z_n) \mid Z_{n-1} = z]$ with respect to the probability distribution $K(z,.)$:

$$\mathbf{E}[f(Z_n) \mid Z_{n-1} = z] = \int f(z_n) \, K(z, dz_n) \doteq \int f(z_n) dK(z, z_n).$$

This is a general definition. For our application, it is helpful to consider the special case $f(z_n) = \mathbf{1}_{z_n \in B}$ for which we have

$$\mathbf{E}[\mathbf{1}_{Z_n \in B} \mid Z_{n-1} = z] = P[Z_n \in B \mid Z_{n-1} = z]. \tag{10.26}$$

For this case, we write the following for any measurable interval B,

$$K(z; B) \doteq P[Z_n \in B \mid Z_{n-1} = z],$$

and we write $K^{(n)}$ to denote the n-stage transition kernel

$$K^{(n)}(z; B) \doteq P[Z_n \in B \mid Z_0 = z].$$

The kernels are defined according to

$$K^{(1)}(z; B) = K(z; B),$$

$$K^{(2)}(z; B) = \int_{u \in E} dK(z; u) K(u; B),$$

$$K^{(n)}(z; B) = \int_{u \in E} dK(z; u) K^{(n-1)}(u; B), \quad n > 1.$$

Note that this last expression can be written as

$$K^{(n)}(z_0; B) = \int_{(z_1, \ldots, z_{n-1}) \in E^{n-1}} dK(z_0; z_1) \ldots dK(z_{n-2}; z_{n-1}) K(z_{n-1}; B)$$

$$= \int_{(z_1, \ldots, z_{n-1}) \in E^{n-1}} \left[\prod_{k=1}^{n-1} dK(z_{k-1}; z_k) \right] K(z_{n-1}; B)$$

$$\doteq \mathbf{E}_{K^{(n-1)}(z_0;.)}[K(Z_{n-1}, B)]. \tag{10.27}$$

---

[3] Do not confuse the present notation $\mathcal{S}_n$ with the notation for a grid S.

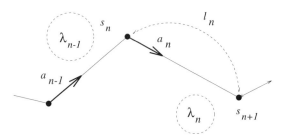

Figure 10.6. Collision parameters of the probabilistic transport model.

The natural Markov chain reflects the physics of the collisions. Let us consider the nth collision $Z_n = (\mathcal{S}_n, \Lambda_n)$, as indicated on Figure 10.6. Three types of random event can occur: photoelectric absorption ($phot$), Compton scattering ($Comp$), or Rayleigh scattering ($Rayl$). For each type of scattering, the direction a_n of the photon after collision is governed by a known probability distribution. The same applies to the distance ℓ_n covered between s_n and s_{n+1}. We set

$$a_n = \frac{s_{n+1} - s_n}{\|s_{n+1} - s_n\|}, \quad \ell_n = \|s_{n+1} - s_n\|.$$

Instead of using $z_n = (s_n, \lambda_n)$, we can also work with

$$z_n = (a_n, \ell_n, \lambda_n).$$

Based on the known position of the source and its emission law, the (a_n, ℓ_n) define the s_n. We will therefore keep the notation z_n for the state (a_n, ℓ_n, λ_n). At the instant of the nth collision, we write C_n for the discrete random variable whose occurrences c are $\{Phot, Comp, Rayl\}$. The transition kernel is then written as

$$dK(z_{n-1}; z_n)$$
$$= \sum_c \pi(z_n \mid z_{n-1}, C_n = c)\, P[C_n = c \mid \lambda_{n-1}] \qquad (10.28)$$
$$= \sum_c \pi_1(a_n | z_{n-1}, C_n = c)\, \pi_2(\ell_n, \lambda_n | a_n, z_{n-1}, C_n = c)\, P[C_n = c | \lambda_{n-1}],$$

where the second expansion follows the Bayes formula. Special distributions for π_1 and π_2 are derived from physical principles [181]. This means that in a situation where the Compton effect occurs, π_1 is given by the Klein–Nishina law, which is here written as $\mathcal{K}$:

$$\pi_1(a_n \mid z_{n-1}, C_n = Comp) = \mathcal{K}((a_n - a_{n-1}), \lambda_{n-1}).$$

The expression of $\mathcal{K}(\phi, \lambda)$ for the angular deviation ϕ of the photon, subject to the incident energy λ, is

$$[1 + \cos^2 \phi]\left[\frac{1}{1 + \frac{\lambda}{\lambda_e}(1 - \cos \phi)}\right]^2\left[1 + \frac{(\frac{\lambda}{\lambda_e})^2(1 - \cos \phi)^2}{(1 + \frac{\lambda}{\lambda_e}(1 - \cos \phi))(1 + \cos^2 \phi)}\right],$$

where λ_e is a known physical constant. In a situation where the Rayleigh effect occurs, π_1 is given by this last expression, in which $\lambda = 0$. The π_2 distribution, for $c \in \{Comp, Rayl\}$, is exponential:

$$\pi_2(\ell_n, \lambda_n \mid a_n, z_{n-1}, C_n = c) \propto$$
$$\mu(\lambda_n)\, e^{-\mu(\lambda_n)\ell_n} \, \mathbf{1}[\lambda_n = \hbar(\lambda_{n-1}, a_n - a_{n-1})], \quad (10.29)$$

where the attenuation function μ is a characteristic of the material, and

$$\hbar(\lambda, \phi) = \frac{\lambda}{1 + \frac{\lambda}{\lambda_e}(1 - \cos \phi)}.$$

● QUANTITY TO BE ESTIMATED

At the output of $\mathcal{M}$, the photons of the radiant image collide with a detector whose purpose is to count the photons on a regular grid. Let us assume that each element of the grid is an element of volume v and that these volumes together constitute a partition of the detector. Similarly, the set of energies is partitioned into intervals written as $\bar{\lambda}$ with center λ.

   Our aim is to use a Monte Carlo method to estimate the probability $Q(z_0; \mathcal{V})$ that a photon emitted by the radiographic source Υ will reach $\mathcal{V} = (v, \bar{\lambda})$ in a manner such that $\sum_{\mathcal{V}} Q(z_0; \mathcal{V}) = 1$. Similarly, we write $Q(z_n; \mathcal{V})$ for the probability that a photon will reach $\mathcal{V}$ from its collision z_n in $\mathcal{M}$. Reaching $\mathcal{V}$ means that if z_n is the last collision in $\mathcal{M}$, then in the case of nonabsorption it must be true that $z_{n+1} \in \mathcal{V}$. Note that the detector at v does not deliver a value equivalent to the probability $Q(z_0; \mathcal{V})$, but an average

$$\xi(v) = \sum_{\lambda} Q(z_0; (v, \bar{\lambda}))\, r(\lambda), \quad (10.30)$$

where $r(\lambda)$, which is a probability distribution, denotes the sensitivity function (or response function) of the detector at the different energies.

   It is clear that $Q(z_0; \mathcal{V})$ is the sum of the probability $K(z_0; \mathcal{V})$ of directly reaching $\mathcal{V}$ and the probability of reaching $\mathcal{V}$ after undergoing at least one collision:

$$Q(z_0; \mathcal{V}) = K(z_0; \mathcal{V}) + \int_{z_1 \in E} dK(z_0; z_1)\, Q(z_1; \mathcal{V}). \quad (10.31)$$

This equation is known as the Fredholm integral equation in Q. If we write j for the number of collisions between z_0 and $\mathcal{V}$, and consider (10.27), it can be expressed as a function of K only:

$$Q(z_0; \mathcal{V}) = K(z_0; \mathcal{V}) + \sum_{j=1}^{\infty} K^{(j+1)}(z_0; \mathcal{V})$$

$$= K(z_0; \mathcal{V}) + \sum_{j=1}^{\infty} \mathbf{E}_{K^{(j)}(z_0;.)}[K(Z_j; \mathcal{V})]. \qquad (10.32)$$

The radiant image (10.30) is then

$$\xi(v) = \sum_{\lambda} K(z_0; \mathcal{V}) \, r(\lambda) + \sum_{j=1}^{\infty} \sum_{\lambda} K^{(j+1)}(z_0; \mathcal{V}) \, r(\lambda)$$

$$\doteq \varphi_0(v) + \sum_{j=1}^{\infty} \varphi_j(v). \qquad (10.33)$$

For $j > 0$, φ_j is the scatter of order j, which refers to photons that have had precisely j collisions.

● MONTE CARLO CALCULATION

Our aim is to estimate the probability $Q(z_0; \mathcal{V}) = \sum_{j=0}^{\infty} K^{(j+1)}(z_0; \mathcal{V})$. In practice, there is a finite number of collisions (in our experiments, the maximum number of collisions is 10). It is therefore appropriate to truncate this series to

$$Q(z_0; \mathcal{V}) = \sum_{j=0}^{J} K^{(j+1)}(z_0; \mathcal{V}).$$

Its estimation will therefore be of the following form: $\widehat{Q}(z_0; \mathcal{V}) = \sum_{j=0}^{J} \widehat{K}^{(j+1)}(z_0; \mathcal{V})$. Let there be m independent copies $\{Z_n^{(i)}, 0 \leq n \leq J+1\}_{i=1,\ldots,m}$ of the Markov chain $\{Z_n, 0 \leq n \leq J+1\}$ of kernel K. In practice, the process $\{Z_n^{(i)}, 0 \leq n \leq J+1\}$ is the simulation by sequential random selection of the collisions according to the distributions that constitute (10.28). The process stops when the photon is absorbed via the photoelectric effect or when it reaches the detector. In (10.32), because $K(Z_j; \mathcal{V})$ is not observable, we write $Q(z_0; \mathcal{V})$ as in (10.26):

$$Q(z_0; \mathcal{V}) = \sum_{j=0}^{J} \mathbf{E}_{K^{(j+1)}(z_0;.)}[\mathbf{1}_{Z_{j+1} \in \mathcal{V}}].$$

The estimator $Q(z_0; \mathcal{V})$ comes from this, and each of its terms has the following estimator:

$$\widehat{K}^{(j+1)}(z_0; \mathcal{V}) = \frac{1}{m} \sum_{i=1}^{m} \mathbf{1}_{Z_{j+1}^{(i)} \in \mathcal{V}}.$$

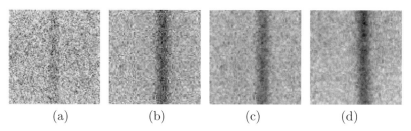

(a)   (b)   (c)   (d)

Figure 10.7. Monte Carlo simulation of a radiographic system. (a) Simulated radiant image. (b) Simulated image with complete radiographic system. (c) Image b with development blur added. (d) Actual radiographic image.

The law of large numbers justifies this estimator, which is unbiased:

$$\mathbf{E}_{K^{(j+1)}(z_0;.)}\left[\widehat{K}^{(j+1)}(z_0;\mathcal{V})\right]$$

$$= \int_{(z_1,\ldots,z_{j+1})\in E^{j+1}}\left[\prod_{k=1}^{j+1}dK(z_{k-1};z_k)\right]\mathbf{1}_{z_{j+1}\in\mathcal{V}}$$

$$= K^{(j+1)}(z_0;\mathcal{V}). \tag{10.34}$$

Figure 10.7(a) is a simulated radiant image of size 5 mm^2 on a 100×100 grid. The object being examined is an iron cube, 7 cm thick, containing a parallelepiped-shaped vertical notch positioned vertically with respect to the source. The iridium 192 source has a radius of 3 mm. This image was obtained by simulating the transport of 400 million photons. The simulator used also allows the Monte Carlo simulation of the effect of a detector consisting of silver-oxide film, and a front screen and a rear screen made of lead [24]. For information, Figure 10.7(b) shows the resulting simulated image. The intensifying effects of the screens can here be observed. In Figure 10.7(c), a blurring effect was added via a simple convolution. Figure 10.7(d) is the actual radiographic image with the parameters chosen for the simulations.

10.3.2 Inverse Problem

Given a radiographic image y, one must decide how to process it in order to remove the undesirable effects of scatter. If we write $\widetilde{x}$ to denote the ideal image that would be obtained in the complete absence of collisions, a general model would be

$$Y = (\widetilde{x} + x) * \mathcal{H} + W, \tag{10.35}$$

where $\widetilde{x}$ and x correspond to φ_0 and $\sum_{j>0}\varphi_j$, respectively, in equation (10.33). We write W to represent an acquisition noise, and $\mathcal{H}$ is a convolution kernel that is characteristic of the components of the radiographic system [156]. We will see that under certain conditions, x is expressed as a function of $\widetilde{x}$: $x = F(\widetilde{x})$. The image to be restored is then $\widetilde{x}$.

We have just looked at a radiographic image simulation method that takes the photon scatter process into account. The resulting image $x(v)$ is the sum of the images corresponding to the scatters of different orders. Although this simulation provides a good representation of the physical process, it is not usually suitable for handling the inverse problem because of the large number of calculations required. In certain situations, however, the inverse problem can be handled thanks to a simplified formulation for photon transport. The first simplification involves the assumption that the scatter is essentially of first order: $x = \varphi_1$. The result of the transport given by (10.30) is then expressed analytically by a triple integral in s. To keep the expressions simple, we will consider the Compton effect only.

● ANALYTICAL FORMULATION OF FIRST SCATTER

Classically, the analytical expression for $x(u)$ is derived directly from physical considerations, and precedes the probabilistic formulation that leads to the Monte Carlo calculation. Here, we will do the reverse. To establish the analytical formula, we will gradually abandon the probabilistic point of view. This requires us to change our notation slightly. Note that $E = \mathcal{M} \times \mathcal{E}$, where $\mathcal{M} = D_1 \times D_2$, and D_2 is the working region of the detection plane. We now have a v that continuously travels on D_2. Similarly to (10.30), we have

$$
x(v) = \int_{\mathcal{E}} x(v, \lambda')\, r(\lambda')\, d\lambda'
$$
$$
= \int_{\mathcal{E}} \left[\int_{\mathcal{M} \times \mathcal{E}} p_I(\Upsilon; (s, \lambda))\, P_0[(s, \lambda); (v, \lambda')]\, ds\, d\lambda \right] r(\lambda')\, d\lambda'.
$$

In physical terms, the function $p_I(\Upsilon; (s, \lambda))$ represents the "density" of incident photons of energy λ at s, and $x(v)$ is a flux. If we permute the two integrals, we can see that for fixed (λ, s, v), only one λ' is accessible: $\lambda' = \hbar(\lambda, s, v)$, where s and v define the direction a of the function $\hbar$ in equation (10.29). Therefore, with $\lambda' = \hbar(\lambda, s, v)$, $x(v)$ is simplified as follows:

$$
x(v) = \int_{\mathcal{M} \times \mathcal{E}} p_I(\Upsilon; (s, \lambda))\, P_0\big[(s, \lambda); (v, \lambda')\big]\, r(\lambda')\, ds\, d\lambda.
$$

Let $\pi(\lambda)$ be the "density" of the photons of energy λ emitted by the source. We define

$$
p_I(\Upsilon; (s, \lambda)) = \pi(\lambda)\, e^{-\mu(\lambda, \Upsilon, s)\, \ell(\Upsilon, s)},
$$
$$
P_0[(s, \lambda); (v, \lambda')] = \kappa(a(s, v), \lambda)\, e^{-\mu(\lambda', s, v)\, \ell'(s, v)},
$$

where the second of these equations is equivalent to (10.28), in which only the Compton effect is considered. We write ℓ to denote the length traveled

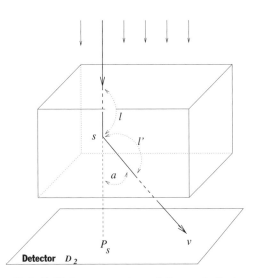

Figure 10.8. Collision parameters of the analytical model.

in the matter between the source and $s \in \mathcal{M}$, and ℓ' to denote the distance between s and v (see Figure 10.8). The direction a is equivalent to the angle induced by $\overline{sv}$ and $\overline{sP_s}$, where P_s is the projection of s on the detector plane D_2:

$$a(s, v) = \left(\widehat{\overline{sv}, \overline{sP_s}}\right) \quad \text{with} \quad P_s = \text{projection of } s \text{ on } D_2.$$

The function $\mu(\lambda', s, v)$ is the attenuation between s and v for the energy λ'. By substituting the expressions for p_I and P_0 into $x(v)$, we obtain the desired analytical expression

$$x(v) = \int_{\mathcal{M} \times \mathcal{E}} \pi(\lambda) \, e^{-\mu(\lambda, \Upsilon, s) \, \ell(\Upsilon, s) - \mu(\lambda', s, v) \, \ell'(s, v)} \kappa(a(s, v), \lambda) \, r(\lambda') \, ds \, d\lambda.$$

$$(10.36)$$

The image x is called the scattered flux image.

- INVERSE PROBLEM FOR AN APPROXIMATE FORMULA

If there were no collisions, and therefore no scatter, the expression (10.36) would be simplified as follows:

$$x_0(v) = \int_{\mathcal{E}} \pi(\lambda) \, e^{-\mu(\lambda, \Upsilon, v) \, \ell(\Upsilon, v)} \, r(\lambda) \, d\lambda. \qquad (10.37)$$

From this point onward, we will consider that the source is located at infinity. This means that we can assume the direction of the incident photons to be perpendicular to the plane D_2. The image x_0 is called the direct flux image.

The inverse problem raises the following question: Can we determine the direct flux image x_0 from the scattered flux image x? The monograph [28] shows that the answer to this question can be yes, if certain fairly strong assumptions are applied to simplify the problem.

Result. *Let us take a radiographic system with a medium-powered monochromatic source, and an object made of a single material with attenuation μ placed at a great distance from the detector. The image x of first-order Compton scattered flux is the two-dimensional convolution by a kernel $\mathbb{H}$ of a functional of the direct flux $\widetilde{x} = x_0 / x_\emptyset$:*

$$x = -x_\emptyset \left(\widetilde{x} \, \frac{\log \widetilde{x}}{\mu} \right) \star \mathbb{H}. \tag{10.38}$$

We write $x_\emptyset$ for the constant value of the image in the absence of any material.

Proof:
We start with the general formula (10.36) and introduce the required simplifications one by one. The distance between the material being examined and D_2 must be large for the following simplifications to be valid:

- The scattering angle a is constant for all points $s \in \mathcal{M}$ such that $\mathbf{P}_s = w$, for all $w \in D_2$.

- The attenuation $\exp -[\mu(\lambda, .)\ell + \mu(\lambda', .)\ell']$ is constant for all the points $s \in \mathcal{M}$ such that $\mathbf{P}_s = w$, and identical to that of the undiverted photons arriving at w.

The triple integral $\int_{\mathcal{M}} ds$ in (10.36) is therefore reduced to a double integral $\int_{D_2} dw$. If we set

$$\mathbb{H}(\overline{v \mathbf{P}_s}, \lambda) \doteq \kappa(a(s,v), \lambda) \ r(\lambda')/r(\lambda) \,,$$

(10.36) can be written

$$x(v) = \int_{D_2 \times \mathcal{E}} \pi(\lambda) \, e^{-\mu(\lambda, \Upsilon, w) \, \ell(\Upsilon, w)} \, \mathbb{H}(\overline{vw}, \lambda) \, r(\lambda) \, \ell(\Upsilon, w) \, dw \, d\lambda.$$

Now, for the case of a medium-powered (150 keV) x-ray source, the angles a vary little with λ, which means that $\mathbb{H}$ can be expressed independently of λ. Restricting ourselves to this case, we obtain

$$x(v) = \int_{D_2} \left[\int_{\mathcal{E}} \pi(\lambda) \, e^{-\mu(\lambda, \Upsilon, w)\ell(\Upsilon, w)} \, r(\lambda) \, d\lambda \right] \mathbb{H}(\overline{vw}) \, \ell(\Upsilon, w) \, dw$$

$$= \int_{D_2} x_0(w) \, \mathbb{H}(\overline{vw}) \, \ell(\Upsilon, w) \, dw.$$

The image x therefore appears as the convolution of x_0 by the convolution kernel $\mathbb{H}\ell$. The function $\ell(\Upsilon, w)$ is unknown, however, as is $\mu(\lambda, \Upsilon, w)$, which depends on the composition of the material on the path of the photon between Υ and w. This means that in many situations where we are trying to reconstruct the thickness of an object made of a single material, ℓ is the main unknown (see Section 12.3). Let us look again at equation (10.37). According to the theorem of the mean, there exists a $\widetilde{\lambda} \in \mathcal{E}$ such that

$$x_0(v) \;=\; e^{-\mu(\widetilde{\lambda}, \Upsilon, v)\ell(\Upsilon, v)} \int_{\mathcal{E}} \pi(\lambda) \; r(\lambda) \; d\lambda \doteq \widetilde{x}(v) \; x_\emptyset,$$

where $x_\emptyset = \int \pi(\lambda) r(\lambda) d\lambda$ is the image obtained in the absence of any object between the source and the detector. Let us consider the following situation: a monochromatic source and a single material. In this case, only one frequency is present, and

$$\widetilde{x}(v) \;=\; e^{-\mu(\widetilde{\lambda}, \Upsilon, v)\ell(\Upsilon, v)}$$

allows us to calculate $\ell(\Upsilon, v) = -\big(\log \widetilde{x}(v)\big)/\mu\big(\widetilde{\lambda}, \Upsilon, v\big)$. We then obtain the desired convolution (10.38):

$$x(v) \;=\; -x_\emptyset \int_{D_2} \mathbb{H}(\overline{vw}) \; \widetilde{x}(w) \; \frac{\log \widetilde{x}(w)}{\mu(\widetilde{\lambda}, \Upsilon, w)} \; dw. \qquad\qquad \square$$

• RESTORATION-DECONVOLUTION PROCEDURE.

Now that this relationship has been explained, let us discuss the restoration of $\widetilde{x}$. We rewrite this model as $x = F(\widetilde{x})$. In the absence of noise and of convolution by $\mathcal{H}$, the general model (10.35) has the result that the direct flux image $\widetilde{x}$ satisfies

$$y \;=\; \widetilde{x} + F(\widetilde{x}).$$

The solution is the result of an iterative fixed-point algorithm. With the initial solution $\widetilde{x}^{(0)} = x_\emptyset$, the nth iteration gives

$$\widetilde{x}^{(n)} \;=\; y - F\left(\widetilde{x}^{(n-1)}\right).$$

The following proposition justifies this algorithm.

Proposition 10.3. *There exists a sufficiently large object–detector distance for the function F to be a contracting function:*

$$\exists\, 0 < \tau < 1 \;:\; \forall x, x', \;\; \|F(x) - F(x')\| \le \tau \|x - x'\|,$$

where τ is expressed explicitly according to the experimental parameters.

The proof, the expression for τ, and a study of the convergence of the algorithm are given in [28].

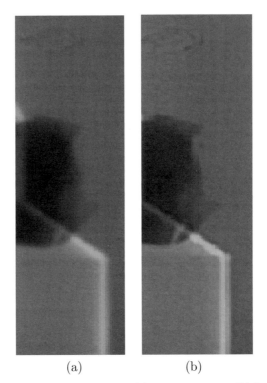

(a)   (b)

Figure 10.9. Restoration-deconvolution. (a) Original image. (b) Processed image.

In the presence of noise and of a convolution by $\mathcal{H}$, we consider the general model (10.35) $Y = \xi \star \mathcal{H} + W$, where $\xi = \tilde{x} + F(\tilde{x})$. Processing occurs in two stages. Firstly, we restore ξ according to a Markov regularization method (Section 10.2.2). Secondly, we solve $\hat{\xi} = \tilde{x} + F(\tilde{x})$, where $\hat{\xi}$ is the result of the first stage. Figure 10.9 illustrates this process for the nondestructive testing using x-rays of a metallic assembly [156]. A 150-kV source is used, and the walls concerned are thin. As a result, the scatter is relatively weak compared to that observed on thicker parts examined with stronger sources. Even so, on the processed image, Figure 10.9(b), we can see increased contrast and the removal of blurring that was present on the original image Figure 10.9(a).

10.4   Sensitivity Functions and Image Fusion

In nondestructive testing (x-ray or γ-ray), situations arise in which several radiographic films are stacked and exposed at the same time during acquisition. This can occur for several reasons; the main reason is to compensate for the effects of defects and artifacts caused by the acquisition conditions. We will discuss two types of situation: In this section restoration is our goal, and in Chapter 11 we wish to perform a detection.

The fusion-restoration problem discussed in this section was approached by Wang et al. [178, 13]. To illustrate the problem, let us consider the pair of images shown in Figures 10.13(a) and 10.13(b), page 224. These images were obtained at the same time from a pair of stacked films $\mathcal{F}$ and $\mathcal{F}'$ of different sensitivities. Each of the films has specific properties, some that are useful, and some that are detrimental to the radiographic examination. The film $\mathcal{F}$ gives images with a low level of noise, but its response function is nonlinear. The film $\mathcal{F}'$, on the other hand, has a linear response function, but produces very noisy images when used under extreme conditions. These noisy images can also be degraded by large stains, which are artifacts caused by the experimental conditions. Finally, we assume the sensitivity functions to be unknown.

Our aim is to combine the two images provided by this pair of films to obtain a noise-free, stain-free image expressed in the grayscale of $\mathcal{F}'$, i.e., with a linear sensitivity function. To express these degradations in more detail, we consider an example whose image pair is shown in Figures 10.10(a) and 10.10(f). This pair was created artificially according to the sequence of transformations shown in Figure 10.10. The image in Figure 10.10(a) is assumed to be noise-free. The image in Figure 10.10(c) was obtained by modifying the gray levels of the initial image (Figure 10.10(a)) according to an increasing nonlinear transformation H, known as a transfer function and represented in Figure 10.10(b), and then by adding Gaussian stains (Figures 10.10(d) and 10.10(e)), and finally by adding white noise. Note that H is defined without taking the stains into account. As a result, in the absence of degradation by stains, the problem is simply to change the grayscale. On the other hand, the presence of stains leads to local modifications in the grayscale, and the problem becomes difficult. In nondestructive testing situations, of course, we have only the pair, and therefore to combine the images optimally we must estimate the transfer function and the stains. Because the fusion process eliminates noise and stains, we call this a restoration. The result of this processing is given in Figure 10.11.

10.4.1 A Restoration Problem

• DEGRADATION MODEL

We have two radiographic images of an object, obtained using the films $\mathcal{F}$ and $\mathcal{F}'$. These images are sampled on a common grid S and discretized on a scale L of gray levels ℓ. Let these two discretized images be called (y, y'). They represent the same view of the object, without any displacement between them.

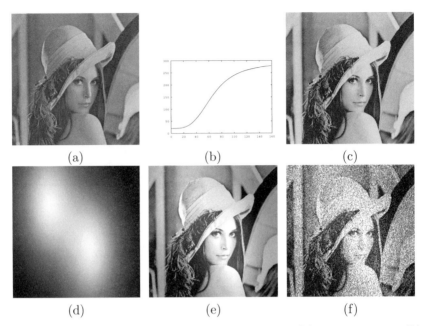

Figure 10.10. Simulation of an artificial degradation. (a) Image $X = x$. (b) Function H and its estimated version are nearly the same. (c) Theoretical image $H(x)$. (d) Image T of stains. (e) Stained image $H(x) + T$. (f) Stained and noisy image $Y' = H(x) + T + \epsilon'$.

The image y is assumed to be a noisy version of a theoretical image x according to the degradation model

$$Y_s = X_s + \epsilon_s, \quad \forall s \in S, \tag{10.39}$$

where ϵ_s is a white noise of variance σ^2. The model for Y' is more compli-cated because of the possible presence of stains, which are represented by a deterministic field T with index S:

$$Y'_s = X'_s + T_s + \epsilon'_s, \quad \forall s \in S, \tag{10.40}$$

where $T_s \neq 0$ if s belongs to a stain, and $T_s = 0$ otherwise. Because the stains are large and smooth, we assume that their local variations are small in the following sense:

Hypothesis 10.1. *There exists a positive number κ such that for all sites $s, u \in S$, we have*

$$\|s - u\| < \kappa \implies T_s \approx T_u.$$

The approximation $\approx$ should be understood in the sense of Taylor's theorem, where T_s is the discretization of a continuously differentiable surface $T(s)$, meaning that as $h \to 0$, there exists $\eta \in [0, 1]$ such that

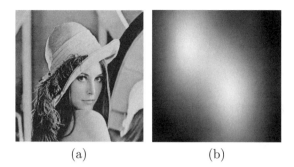

(a)   (b)

Figure 10.11. Fusion-restoration. (a) Image $\widehat{H}(X)$. (b) Estimated stains $\widehat{T}$.

$T(s+h) - T(s) = h'\nabla T(s+\eta h)$. The slow variations in T mean that $h'\nabla T$ is negligible for sufficiently small h. In the images shown in Figure 10.10, of size 512×512, small h corresponds to about twenty pixels, i.e., $\kappa = 20$.

The noises ϵ_s and ϵ'_s are white noises with variance σ^2 and σ'^2, respectively. The specifications of the films $\mathcal{F}$ and $\mathcal{F}'$ give

$$\sigma'^2 \gg \sigma^2,$$

as seen in Figure 10.10. This is not essential, and this assumption could be omitted, although at the cost of certain complications in the mathematical formulation. In x-ray radiography, a classic but rarely used assumption stipulates that the variance of the noise depends on signal intensity. We will look at this assumption concerning ϵ'_s at the end of this subsection. For the time being, we assume that the variance σ'^2 is constant.

- TRANSFER FUNCTION

To combine the images y and y', we need to estimate the transfer function H between the grayscales of the films $\mathcal{F}$ and $\mathcal{F}'$ on the basis of the data y and y'. The transfer function is defined as follows.

Definition 10.1. *In the absence of noise and stains, when the film $\mathcal{F}$ has a response of value ℓ at site s, then the film $\mathcal{F}'$ has a response of value $H(\ell)$ at the same site. The function H is called the transfer function between the sensitivities of the two films.*

Using (10.39) and (10.40), this means that

$$X'_s = H(X_s).$$

We can therefore rewrite (10.40) according to

$$Y'_s = H(X_s) + T_s + \epsilon'_s. \tag{10.41}$$

• RESTORATION

In the models (10.39) and (10.40) we find two theoretical versions x and x' of a single entity called the "object" in the context of nondestructive testing. These versions are not faithful reproductions of the object because of the unknown sensitivity function (grayscale) of each film. Similarly, the restoration operation that eliminates noise and stains can only provide an image that is known to within a sensitivity function. Between x and x', it is preferable to restore x', which is defined via a linear sensitivity function.

This type of restoration takes advantage of the qualities of the two films while leaving out their faults. If we tried to restore the object with y' only, the result would be given on a linear scale, but its shape would be very inaccurate because y' is degraded by a strong noise and stains. If we tried to restore the object with y only, the resulting shape would have better detail but would be given on a nonlinear scale, which is undesirable. By considering y and y' together, we can restore the object on a linear scale and without much degradation of its shape. The restored object will be an estimation of x'.

The restoration is obtained by minimizing a potential energy of the type

$$U(x' \mid y, y') \;=\; \theta \, U_1(x') + U_2(y, y' \mid x'),$$

where this energy is associated with the following Gibbs distribution

$$P(x' \mid y, y') \;\propto\; \exp -U(x' \mid y, y'),$$

which makes x' the occurrence of a Markov random field that is conditional on the pair (y, y'). The function U_1 quantifies the regularity of x'. Its expression can be chosen from among those presented in Section 10.1. For the images in Figure 10.13, which have indistinct edges, an energy with implicit edges is chosen. We use U_2 to denote the fidelity term. Its expression is a direct consequence of the fact that ϵ and ϵ' are white noises assumed to be Gaussian. More specifically, we have

$$P(y, y' \mid x') \;\propto\; \exp - U_2(y, y' \mid x'),$$

where the energy U_2 is written as

$$U_2(y, y' \mid x') \;=\; \frac{1}{2\sigma^2} \sum_s \left[y_s - H^{-1}(x'_s) \right]^2 + \frac{1}{2\sigma'^2} \sum_s \left[y'_s - T_s - x'_s \right]^2 .$$

This expression shows that, to perform the restoration, we must estimate H and T, as well as σ^2 and σ'^2.

10.4.2 Transfer Function Estimation

[4] We now wish to establish a relationship between Y and Y'. From equation (10.40), we obtain

$$Y'_s = H(Y_s - \epsilon_s) + T_s + \epsilon'_s.$$

After a Taylor expansion performed on H about Y_s, we obtain

$$Y'_s = H(Y_s) + T_s + W_s, \quad \forall s \in S, \tag{10.42}$$

and then, if we write $\dot{H}$ for the derivative of H, the noise W_s is written as

$$W_s = -\dot{H}(Y_s)\epsilon_s + \epsilon'_s.$$

This white noise is the linear combination of two independent white noises. The assumption $\mathrm{Var}(\epsilon'_s) \gg \mathrm{Var}(\epsilon_s)$ leads to the approximation $\mathrm{Var}(W_s) \approx \mathrm{Var}(\epsilon'_s)$. We therefore have

$$\mathrm{Var}(W_s) \approx \sigma'^2. \tag{10.43}$$

To introduce the estimation method gradually, we will consider the simple case without stains. The estimation for the general case will be an extension of this case.

• ESTIMATION OF H WHEN $T = 0$

In the absence of stains, for all $\ell \in \mathrm{L}$ and $u \in S$ such that $y_u = \ell$, it follows from equation (10.41) that y'_u is equal to $H(\ell)$ degraded by a noise w_u that is an occurrence of W_u. To reduce the noise with respect to $H(\ell)$, we can take the mean of the values y'_u on the sites u for which $y_u = \ell$:

$$m(\ell) = \frac{1}{|\mathcal{N}(\ell)|} \sum_{u \in \mathcal{N}(\ell)} y'_u = H(\ell) + \frac{1}{|\mathcal{N}(\ell)|} \sum_{u \in \mathcal{N}(\ell)} w_u, \tag{10.44}$$

where $\mathcal{N}(\ell) = \{u \in S : y_u = \ell\}$.

Therefore, for all ℓ, we calculate a value $m(\ell)$ that can be considered as a noisy version of $H(\ell)$, with the noise $\frac{1}{|\mathcal{N}(\ell)|} \sum_{u \in \mathcal{N}(\ell)} w_u$ that has a smaller variance $\sigma'^2 / |\mathcal{N}(\ell)|$. The mean $(m(\ell), \ell \in \mathrm{L})$ constitutes the pertinent data extracted from (y, y') for the estimation of H. This estimation consists in modeling H with a nondecreasing cubic spline. This model is adjusted to the data $\{m(\ell)\}$ by the method of least squares.

• DATA EXTRACTION FOR THE ESTIMATION OF $H + T$

In the presence of stains, the mean (10.44) no longer allows us to approximate H. But because the stains are locally constant according to assumption 10.1, the idea is to consider the mean (10.44) locally at each

---
[4]This subsection can be omitted at the first reading.

site s. Let us therefore write $m_s = (m_{s,\ell}, \ell \in L)$ for the values $(m(\ell), \ell \in L)$ calculated according to equation (10.44), but remaining within a small neighborhood of s. The mean $m_{s,\ell}$ is thus a noisy version of $H(\ell) + T_s$.

Let us look at $m_{s,\ell}$ in detail. We take a site $s \in S$ and choose a small, square window F_s centered on s. For all $\ell \in L$, we set $\mathcal{N}_s(\ell) = \{u \in F_s : y_u = \ell\}$, which is the restriction of $\mathcal{N}(\ell)$ to F_s. The local definition of (10.44) is then

$$m_{s,\ell} = \frac{1}{|\mathcal{N}_s(\ell)|} \sum_{u \in \mathcal{N}_s(\ell)} y'_u. \tag{10.45}$$

Using the expression (10.42) for y', we obtain the following breakdown of $m_{s,\ell}$:

$$m_{s,\ell} = H(\ell) + \overline{T}_s(\ell) + \overline{w}_{s,\ell}, \tag{10.46}$$

where we have set

$$\overline{T}_s(\ell) = \frac{1}{|\mathcal{N}_s(\ell)|} \sum_{u \in \mathcal{N}_s(\ell)} T_u, \quad \overline{w}_{s,\ell} = \frac{1}{|\mathcal{N}_s(\ell)|} \sum_{u \in \mathcal{N}_s(\ell)} w_u. \tag{10.47}$$

We write (s_1, s_2) for the coordinates of the site s in S. The main part $H(\ell) + \overline{T}_s(\ell)$ of equation (10.46) is a function of three variables s_1, s_2 and ℓ, known as the stained transfer function and written as G:

$$G(s_1, s_2, \ell) = H(\ell) + \overline{T}_s(\ell). \tag{10.48}$$

We can then rewrite (10.46) as follows:

$$m_{s,\ell} = G(s_1, s_2, \ell) + \overline{w}_{s,\ell}. \tag{10.49}$$

Figures 10.12(a) and 10.12(b) show how much the shape of this stained transfer function varies according to the position s. This demonstrates that the estimation based on (10.44) without taking the stains into account is very wrong.

The variance of the noise $\overline{W}_{s,\ell}$ is $\sigma'^2/|\mathcal{N}_s(\ell)|$. The noise $\overline{W}_{s,\ell}$ is not a white noise, because the windows F_s overlap. Inside a window F_s, certain gray levels ℓ can be rare or even absent. To work around this problem, we impose the condition that $m_{s,\ell}$ must have the same accuracy for all s, meaning that $\mathrm{Var}(\overline{W}_{s,\ell})$ must be independent of s, which gives

$$|\mathcal{N}_s(\ell)| = c(\ell).$$

The $c(\ell)$ are chosen to be proportional to the histogram values $h(\ell)$ calculated on y. To do this, at any site s and for any gray level ℓ, the size of the window F_s is either decreased or increased until $|\mathcal{N}_s(\ell)|$ is sufficiently close to $c(\ell)$. We therefore have

$$\mathrm{Var}\left(\overline{W}_{s,\ell}\right) \approx \frac{\sigma'^2}{c(\ell)}. \tag{10.50}$$

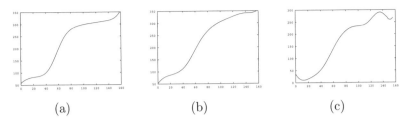

Figure 10.12. Stained transfer function. (a) and (b) $\widehat{G}(s_1, s_2, .)$ at two particular sites s. (c) Estimation of H disregarding stains.

Finally, let us look at how the data extracted from $m_{s,\ell}$ are pertinent for $H(\ell) + T_s$. In equation (10.46), $m_{s,\ell}$ is a noisy version of $H(\ell) + T_s$ if $\overline{T}_s(\ell) \approx T_s$. There are two situations. Firstly, when a gray level ℓ occurs very frequently around s in y, then the dimension of F_s is smaller than κ (i.e., $F_s(\ell)$ is contained in the window of dimension κ centered on s). In this case, according to Hypothesis 10.1, if $u \in \mathcal{N}_s(\ell)$ then $T_u \approx T_s$, and the mean $\overline{T}_s(\ell)$ is close to T_s. The second situation is the reverse. When a gray level ℓ is rare around s, $\mathcal{N}_s(\ell)$ can contain sites very far from s, and therefore $\overline{T}_s(\ell)$ and T_s can be very different. To distinguish between these two situations, we introduce the following definition.

Definition 10.2. *For any $s \in S$, we call $\mathbf{L}_s$ the set of gray levels $\ell \in \mathbf{L}$ such that the corresponding windows F_s have dimension smaller than κ.*

A consequence of this definition is

$$\forall\, \ell \in \mathbf{L}_s, \quad \overline{T}_s(\ell) \approx T_s. \tag{10.51}$$

• ESTIMATION OF H

The estimations of H and T are based on (10.48) and (10.51). Let us assume G to be known. In fact, G is replaced by its estimation $\widehat{G}$, as described below. In equation (10.48), G is defined, and the mean $\overline{T}_s(\ell)$ is given by equation (10.47). If $\ell \in \mathbf{L}_s$, the mean $\overline{T}_s(\ell)$ is close to T_s (cf. (10.51)). We thus obtain a linear system that relates the variables $\{H(\ell),\ \ell \in \mathbf{L}\}$ and $\{T_s,\ s \in S\}$ to G, for all $s \in S$:

$$\begin{aligned} H(\ell) \ + \ \tfrac{1}{|\mathcal{N}_s(\ell)|} \sum_{u \in \mathcal{N}_s(\ell)} T_u &= G(s_1, s_2, \ell), \quad \forall \ell \in \mathbf{L} \backslash \mathbf{L}_s, \\ H(\ell) \ + \ T_s &= G(s_1, s_2, \ell), \quad\quad\ \forall \ell \in \mathbf{L}_s, \end{aligned} \tag{10.52}$$

where $\mathbf{L} \backslash \mathbf{L}_s$ is the set of gray levels that are not in $\mathbf{L}_s$. We must solve this system to determine the values $H(\ell)$ and T_s. Because this system contains $|\mathbf{L}| \times |S|$ equations for only $|\mathbf{L}| + |S|$ variables, it will be solved by the method of least squares. This solution is defined to within an additive constant: If $(\hat{H}, \hat{T})$ is a solution of equation (10.52), then for any constant

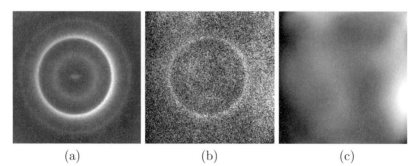

(a)   (b)   (c)

Figure 10.13. x-ray imaging for nondestructive testing: (a) and (b) Image pair (Y, Y'). (c) Estimated stains $\widehat{T}$.

b, $(\widehat{H}+b, \widehat{T}-b)$ is also a solution. To remove this uncertainty, we impose the constraint $H(1) = a$, where a is a fixed value. Solving this system requires a large amount of effort. To simplify it, its dimension can be reduced [178].

● EXPERIMENTAL RESULTS

Figure 10.11 shows how the pair of images 10.10(a) and 10.10(f) is processed. This processing of artificial images is useful because the results of processing can be compared to the real entities: transfer function, stains, and variance of the noise. We can therefore see that the estimation of H (Figure 10.10(b)) is very accurate because H and $\widehat{H}$ are almost superimposed. We can see a big increase in accuracy compared to the estimation of H without considering the stains (Figure 10.12). A visual comparison of the image of the stains T (Figure 10.10(d)) to its estimation $\widehat{T}$ (Figure 10.11(b)) shows how accurate the estimation is. The pair (Y, Y') in Figures 10.13(a) and 10.13(b) comes from nondestructive testing. These images raise exactly the same problem as above: They are x-ray images obtained with very short exposure times. The purpose of restoration is not to obtain a visually legible image, but to obtain an image with pertinent gray levels. Here, we no longer have a reference to check the estimation of H and T. Nonetheless, we can see that the estimation of the stains (Figure 10.13(c)) matches the stains that are suggested in Figure 10.13(b).

10.4.3 Estimation of Stained Transfer Function

[5] Let us consider the model (10.49): $m_{s,\ell} = G(s_1, s_2, \ell) + \overline{w}_{s,\ell}$. Because G is the result of a smoothing procedure, it is appropriate to represent it

---

[5]This subsection can be omitted at the first reading.

with a smooth function, in this case a tricubic spline function on $S \times L$ (see Chapter 3). To do this, the grid is partitioned by placing equidistant knots on each of the coordinates t, u, and ℓ. Let $\mathcal{T}_1$, $\mathcal{T}_2$, and $\mathcal{T}_3$ be these sets of knots, and B^1, B^2, and B^3 the bases of respective B-splines. The stained transfer function G is modeled by a tricubic spline defined on $S \times L$ with the set of knots $\mathcal{T}_1 \times \mathcal{T}_2 \times \mathcal{T}_3$. For all $(s, \ell) \in S \times L$, this representation is

$$G(s_1, s_2, \ell) = \sum_{i,j,k} B_i^1(s_1) B_j^2(s_2) B_k^3(\ell) \beta_{i,j,k}, \qquad (10.53)$$

where $\beta_{i,j,k}$ are the parameters of the model. Although the $\overline{W}_{s,\ell}$ are not independent, we will estimate G using the weighted least-squares criterion, disregarding the autocorrelation structure of the noise, because taking this structure into account would not significantly improve our estimation of a polynomial surface [37]. We therefore calculate the following:

$$\hat{\beta} = \arg\min_{\beta} \sum_{s,\ell} [m_{s,\ell} - G(s_1, s_2, \ell)]^2 \frac{1}{\operatorname{Var}(\overline{W}_{s,\ell})}.$$

This minimization was discussed in Section 3.2. Note that $\hat{\beta}$ cannot be calculated in vector form because it is too large. (Typically, we place seven knots on each of the three coordinates, which produces a model with 1331 parameters). For the case of separable variance $\operatorname{Var}(\overline{W}_{s,\ell}) = p_{s_1} q_{s_2} r_\ell$, we have seen that $\hat{\beta}$ could be calculated in matrix form. This holds true here, with $p_{s_1} = 1$, $q_{s_2} = 1$, and $r_\ell = \sigma'^2/c(\ell)$. The separability requirement is one of the reasons why $|\mathcal{N}_s(\ell)|$ was chosen independently from s in (10.50).

- SIGNAL-DEPENDENT VARIANCE

In the introduction we mentioned that the variance of ϵ'_s depended on the signal X'_s. The above estimation of G can be improved if we take this assumption of heteroscedasticity into account. This dependence can be expressed in terms of X_s because $X'_s = H(X_s)$. We therefore write

$$\operatorname{Var}(W_s) = \sigma'^2(X_s).$$

Applying this assumption, (10.50) remains valid. In more specific terms, the variance of $\overline{W}_{s,\ell}$ is approximately a function of ℓ:

$$\operatorname{Var}(\overline{W}_{s,\ell}) \approx \frac{\sigma'^2(\ell)}{c(\ell)}. \qquad (10.54)$$

From equation (10.47), we obtain

$$\operatorname{Var}(\overline{W}_{s,\ell}) = \frac{1}{|\mathcal{N}_s(\ell)|^2} \sum_{u \in \mathcal{N}_s(\ell)} \operatorname{Var}(W_u),$$

where

$$\text{Var}(W_u) = \sigma'^2(X_u) = \sigma'^2(\ell - \epsilon_u) \approx \sigma'^2(\ell)\frac{\partial}{\partial\ell}\sigma'^2(\ell)\epsilon_u$$

because on $\mathcal{N}_s(\ell)$, we have $Y_u = \ell$. This last expression comes from the Taylor expansion of σ'^2 around ℓ. Because ϵ is a centered white noise, we have $\sum_{u \in \mathcal{N}_s} \epsilon_u/\mathcal{N}_s(\ell) \approx 0$, and (10.54) is therefore established.

It is still possible to estimate G, by a least-squares estimation of a tricubic spline whose adjustment variance has unknown anisotropy. This problem is examined in Section 3.2.

11
Detection of Filamentary Entities

One of the main tasks of nondestructive testing is to detect defects in materials. Here, we discuss the automatic detection of the projection of defects on radiographic images. We will focus in particular on gammaradiographic images in highly degraded situations that lead to barely visible defects. In this type of image, defects can easily be confused with artifacts. In fact when digitally processing a single image, it is virtually impossible to distinguish between artifacts and defects. Both entities appear as dark, elongated regions on a more or less uniform, noisy background. We will use the generic term "valleys" to describe them. This chapter is devoted to the detection of such valleys. We should emphasize that this detection problem is found in fields other than nondestructive testing [62, 82, 144]. We will conclude this chapter with an extension to the situation where we have a matched pair of images. The fusion of the two images allows us to distinguish between defects and artifacts.

These tasks were undertaken by Coldefy et al. [55, 12]. Because of severe image degradation, we will estimate the valley bottom lines instead of the valleys themselves. The reason for this choice is that the signal-to-noise ratio is at its maximum on the valley bottom lines. This is another problem of searching for discretized curves, but the situation is more difficult than that in Chapter 9. The resulting, more complicated, energy will show how the Markovian–Bayesian framework allows sophisticated models to be built. The model is made up of a variety of components: parametric spline

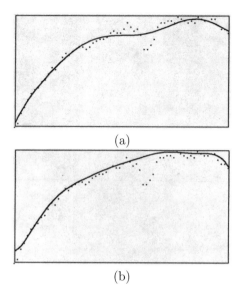

(a)

(b)

Figure 11.1. Profile of a γ image with estimation of the gradient μ. (a) Rigid method. (b) Adaptive method.

for the luminosity gradient, and statistical hypothesis tests and curvature indicator lines to detect the valley bottom lines.

11.1   Valley Detection Principle

11.1.1   Definitions

Let $y = \{y_s, s \in S\}$ be the image to be processed (see the profile given in Figure 11.1). We assume that this image is the occurrence of a random field Y, resulting from the degradation of the surface of the valleys x^p by an independent noise $\epsilon = \{\epsilon_s\}$ of variance σ^2 and varying mean $\mu = \{\mu_s\}$. We can also see x^p as the result of a random field $X^p = \{X_s^p, s \in S\}$. The degradation model is

$$Y = \mathcal{O}(X^p, \epsilon) = X^p + \epsilon = X^p + \mu + W, \tag{11.1}$$

where W is the white noise that comes from the decentering of ϵ: $\mathrm{E}(W_s) = 0$. We write μ to represent a smooth surface that is characteristic of a luminosity gradient. Its physical meaning is the variable thickness of material through which the γ radiation passes. The image of the valleys is denoted by x^p. Figure 11.1 gives the profile of such an image, i.e., the graph of a line of y. An estimation of μ is shown superimposed. We should point out that because of the gradient, the defects are almost invisible to anyone other than a specialist. They become obvious on $y - \widehat{\mu}$ when the gradient is

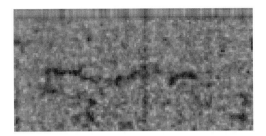

Figure 11.2. A γ image after suppression of the gradient μ.

suppressed (Figure 11.2), although this operation greatly accentuates the noise.

For any site $s \in S$, we set $x_s^p \leq 0$, with

$$x_s^p \quad < \quad 0, \quad \text{if } s \text{ is the site of a valley,}$$
$$x_s^p \quad = \quad 0, \quad \text{otherwise.}$$

Given a configuration x^p, we encode the binary field x^ℓ of valley bottom lines with

$$x_s^\ell = \begin{cases} 1 & \text{if } s \text{ is the site of a valley bottom line,} \\ 0 & \text{otherwise.} \end{cases}$$

Valley bottom lines are defined with respect to the following three types of feature:

- ($f1$) curvature feature in x^p,

- ($f2$) depth feature,

- ($f3$) geometrical regularity feature.

The field x^ℓ is also seen as the occurrence of a random field X^ℓ. This field is said to be virtual because it is defined according to the hidden field X^p. At this stage, we note an analogy between this model and the model from the previous chapter involving degradation by a noise, $Y = X^p + W$. In that model, the virtual hidden field was the field of edge elements X^b.

One special feature of the model using valley bottom lines is the presence of the unknown surface μ. Because this surface comes from a luminosity gradient, it is smooth enough to be modeled by a bicubic spline surface (Section 3.2.1). This representation is built on a subgrid of S whose resolution is deduced from the experimental conditions. As a result, μ is written as follows:

$$\mu = B' \, \alpha \, B, \tag{11.2}$$

where B is the matrix of B-splines calculated at the chosen resolution. Note that a coarser resolution gives an estimation of the surface that is less sensitive to the presence of valleys. Finally, the degradation model includes the pair of hidden fields

$$X = (X^p, X^\ell)$$

and the unknown parameters (α, σ^2).

11.1.2 Bayes–Markov Formulation

This approach consists in defining an energy function $U(x^p, x^\ell \mid y, \theta)$ that, given y, expresses what we know about the set of configurations (x^p, x^ℓ) and then calculates the configuration $(\widehat{x^p}, \widehat{x^\ell})$ that minimizes U. In the Markov context, U is the energy associated with the posterior Gibbs distribution on (x^p, x^ℓ):

$$P(x^p, x^\ell \mid y) \quad \propto \quad \exp{-U(x^p, x^\ell \mid y, \theta)},$$
$$U(x^p, x^\ell \mid y, \theta) = U_1(x^p, x^\ell, \theta_1) + U_2(y \mid x^p, x^\ell, \theta_2). \qquad (11.3)$$

Using the Bayes formula, we write

$$P(x^p, x^\ell \mid y) \propto P(y \mid x^p, x^\ell) P(x^p, x^\ell) \propto P(y \mid x^p) P(x^p, x^\ell). \quad (11.4)$$

It is very important to consider the pair (x^p, x^ℓ) because of the dependent relationship between x^p and x^ℓ modeled by the joint energy $U_1(x^p, x^\ell, \theta_1)$. Because of this, the estimation of x^ℓ will be constrained by the estimation of x^p. This is similar to the denoising situation with which we made a comparison above, where smoothing and edge detection were modeled together. We should also note that in the denoising problem associated with $Y = X^p + W$, the field of edge elements x^b was introduced only to improve the smoothing. The estimation of x^b was merely a secondary task. In our present case, the situation is reversed, because the primary task is now to detect x^ℓ.

Let us take the expression for $P(y \mid x^p)$. If W is a white noise, it follows that, subject to x^p, the random field Y is an independent Gaussian field. The prior distribution $P(y_s \mid x_s^p)$ is defined by the Gaussian distribution $LG(\mu_s + x_s^p, \sigma^2)$, and therefore

$$P(y \mid x^p) \quad \propto \quad \exp{-U_2(y \mid x^p, \theta_2)},$$
$$U_2(y \mid x^p, \theta_2) = \sum_{s \in S} \frac{(y_s - \mu_s - x_s^p)^2}{\sigma^2},$$

where $\theta_2 = (\alpha, \sigma^2)$. This expression shows that $P(y \mid x^p, x^\ell)$ can be reduced to $P(y \mid x^p)$, as was suggested in (11.4). The energy $U_2(y \mid x^p, \theta_2)$ expresses the distance between the configuration x^p and the data y.

Looking again at the general expressions (Chapter 6), we can write $P(x^p, x^\ell)$ as follows:

$$P(x^p, x^\ell) \quad \propto \quad \exp -U_1(x^p, x^\ell, \theta_1),$$

$$U_1(x^p, x^\ell, \theta_1) = U_{1,1}(x^\ell, \theta_{1,1}) + U_{1,2}(x^p, x^\ell, \theta_{1,2}) + U_{1,3}(x^p, \theta_{1,3}).$$

$$\tag{11.5}$$

Of all the parameters, the only ones that must be estimated at the same time as the hidden fields are $(\alpha, \sigma^2) = \theta_2$. We will see that θ_2 will also appear in $U_{1,2}$, such that we can rewrite (11.3) and (11.5) as follows:

$$U^\gamma(x^p, x^\ell \mid y) = U_1^\gamma(x^p, x^\ell) + U_2^\gamma(y \mid x^p), \tag{11.6}$$

$$U_1^\gamma(x^p, x^\ell) = U_{1,1}(x^\ell) + U_{1,2}^\gamma(x^p, x^\ell) + U_{1,3}(x^p), \tag{11.7}$$

where, to simplify the notation, we have set

$$(\alpha, \sigma^2) \doteq \gamma$$

instead of θ_2. The energy $U_{1,1}(x^\ell) + U_{1,2}^\gamma(x^p, x^\ell)$ expresses the features $f1$, $f2$, and $f3$, and the energy $U_{1,3}(x^p)$ takes into account the low occurrence rate of valleys in the images x^p. To implement this Markov approach, we must first build $U_1^\gamma(x^p, x^\ell)$, and then minimize the posterior energy (11.6). Note that U is nonlinear with respect to γ. This approach raises some difficulties because of the unknown parameters γ and the level of optimization.

11.2   Building the Prior Energy

Ideally, the valley bottom lines in the low-energy configurations x^ℓ should be represented by discrete curves that are one pixel wide and satisfy the features $f1$, $f2$, and $f3$ that express our prior information. This would mean that the value of $U_{1,1}(x^\ell) + U_{1,2}^\gamma(x^p, x^\ell)$ is low when the valley bottom lines satisfy the features, and high when they do not. The regularization energy $U_{1,1}(x^\ell)$ expresses $f3$: It favors valley bottom lines that are regular, long, and one pixel wide. The detection energy $U_{1,2}^\gamma(x^p, x^\ell)$ expresses $f1$ and $f2$: It favors valley bottom lines of which every point is close to a curvature indicator line in x^p (see above) and that are located at a sufficient depth under the surface μ.

11.2.1   Detection Term

- CURVATURE INDICATOR LINES

For any site s, we wish to extract a curvature feature from x^p in order to define the feature $f1$. To do this, let us consider x^p to be a discrete

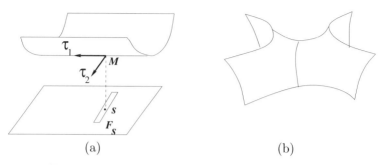

Figure 11.3. Two examples of curved surfaces. (a) $\rho_1 = 0$, $\rho_2 > 0$, (b) $\rho_1 < 0$, $\rho_2 > 0$.

surface obtained from the sampling of a continuous surface of the class $\mathcal{C}^2$, written as $\tilde{d}$. Let M be a point on $\tilde{d}$. We also write $\rho_1 \leq \rho_2$ for the main curvatures of $\tilde{d}$ in M, and τ_1, τ_2 for the principal axes associated with each of these curvatures [31].[1] We define the indicator lines as the set of points M on $\tilde{d}$ for which $\rho_2 > 0$ and τ_2 is a horizontal vector. For example, in Figure 11.3(a), it is the parabolic line that minimizes the surface.

For any site $s \in S$, we consider a $(2l + 1) \times 1$ window F_s whose direction is given by the orthogonal projection on S of the principal vector τ_2 calculated at the point M of x^p with coordinate s. (Here, ρ_2 and τ_2 are calculated discretely on a smoothed version of x^p). Such a point on x^p is considered to be close to an indicator line of $\tilde{d}$ if it is the local minimum of x^p in F_s. We thus define the discrete field Γ of curvature indicator lines:

$$\Gamma_s = \begin{cases} 1 & \text{if } x_s^p = \min_{u \in F_s} x_u^p, \\ 0 & \text{otherwise.} \end{cases}$$

When $\Gamma_s = 1$, the site s is called a site of feature $f1$. Note that $\Gamma = \Gamma(x^p)$ is a deterministic function of x^p. This is an efficient way to extract information for smooth surfaces. The sites of feature $f1$ constitute a set of irregular chains, one or two pixels wide, covering almost all the valley bottom lines (Figure 11.4). The field Γ can be interpreted as a field of valley bottom lines that needs to be regularized and cleansed of its many erroneous sites.

● CONSTRUCTION OF THE DETECTION TERM $U_{1,2}$

Let us assume that, subject to x^p, the field x^ℓ is an independent field: $P(x^\ell \mid x^p) = \prod_s P(x_s^\ell \mid x_s^p)$. If we then set $U_{1,2}^\gamma(x^p, x^\ell) = -\log P(x^\ell \mid x^p)$,

---

[1]See also the appendix at the end of this chapter.

the expression for this energy is

$$U_{1,2}^{\gamma}(x^p, x^\ell) = - \sum_s \mathbf{1}\left[x_s^\ell = 1\right] \log P(x_s^\ell = 1 \mid x_s^p) \qquad (11.8)$$

$$- \sum_s \mathbf{1}\left[x_s^\ell = 0\right] \log P(x_s^\ell = 0 \mid x_s^p).$$

After we have built this energy, its role will be clear. We can already say that $U_{1,2}^{\gamma}$ corresponds to a local analysis of x^p at any site s via $P(x_s^\ell \mid x_s^p)$, which gives an estimation of the presence of a valley at s. This energy is built to favor the detection of a valley at the site s when $\Gamma_s = 1$ (feature $f1$) and when the signal-to-noise ratio exceeds an empirical threshold k (feature $f2$). Thanks to this threshold, shallow valleys are treated as the low frequencies of noise and therefore ignored.

We will define $P(x_s^\ell = 1 \mid x_s^p)$ as the product of two probabilities P_1 and P_2. The relationship $x_s^\ell = 1$ is true only if s belongs to a valley $(s \in \mathcal{V})$. Let us examine $P(x_s^\ell = 1 \mid x_s^p)$ under this assumption. Because of the inclusion of the events $[x_s^\ell = 1] \subset [s \in \mathcal{V}]$, we obtain

$$P(x_s^\ell = 1 \mid x_s^p) = P(x_s^\ell = 1, s \in \mathcal{V} \mid x_s^p)$$

$$= P_1(x_s^\ell = 1 \mid s \in \mathcal{V}, x_s^p) \, P_2(s \in \mathcal{V} \mid x_s^p).$$

The probabilities P_1 and P_2 correspond to the features $f1$ and $f2$ respectively. Let us begin with the expression for P_1. Because this probability must express $f1$, it is appropriate to make the event $x_s^\ell = 1$ depend only on $\Gamma_s(x^p)$ when we observe $s \in \mathcal{V}$:

$$P_1(x_s^\ell = 1 \mid s \in \mathcal{V}, x_s^p) = P_1(x_s^\ell = 1 \mid \Gamma_s(x^p)) \doteq \Psi(\Gamma_s(x^p)). \qquad (11.9)$$

This probability is chosen empirically. Because a site s such that $\Gamma_s(x^p) = 1$ is considered to be a potential point of a valley bottom line, $\Psi(1)$ is chosen close to 1, and $\Psi(0)$ is chosen close to 0. The probability $\Psi(1)$ is in fact chosen to be strictly less than 1 to allow the regularization of the valley bottom lines; the value of this probability characterizes the distance between the field Γ and the field x^ℓ.

Let us now take the probability $P_2(s \in \mathcal{V} \mid x_s^p)$, which must express $f2$. We define it as follows:

$$P_2(s \in \mathcal{V} \mid x_s^p) = \Phi(-k - x_s^p/\sigma). \qquad (11.10)$$

The value k is the η-percentile of the Gaussian distribution $LG(0,1)$:

$$\eta = \Phi(-k) \doteq P[LG(0,1) < -k].$$

For example, for $k = 1.96$, we have $\Phi(-k) \approx 0.025$. For the same k, we obtain $P_2(s \in \mathcal{V} \mid x_s^p = 0) = \Phi(-k) \approx 0.025$ and $P_2(s \in \mathcal{V} \mid x_s^p = -4\sigma) = \Phi(-k + 4) \approx 0.975$.

This probability P_2 can be interpreted as the power of the hypothesis test

$$H_0 : \quad x_s^p = 0, \qquad \text{versus} \qquad H_1 : \quad x_s^p < 0.$$

Because the random variable $(Y_s \mid X_s^p = x_s^p)$ is governed by the Gaussian distribution $LG(\mu_s + x_s^p, \sigma^2)$, the test can be expressed as

$$H_0 : \quad \mathrm{E}(Y_s \mid x_s^p) = \mu_s, \qquad \text{versus} \qquad H_1 : \quad \mathrm{E}(Y_s \mid x_s^p) < \mu_s.$$

The classical decision rule accepts H_0 if $[(y_s - \mu_s)/\sigma < -k]$ and H_1 otherwise [23]. By definition, the power of the test is the probability of deciding H_1 when H_1 is true. In our case, this is $\Phi(-k - x_s^p/\sigma)$. Because k is defined by the level of the test (i.e., the probability of deciding H_1 when H_0 is true), it is easy to calibrate because it is related to the signal-to-noise ratio required to detect valleys.

Finally, with (11.9) and (11.10), we obtain

$$P(x_s^\ell = 1 \mid x_s^p) \;=\; \Psi(\Gamma_s(x^p)) \, \Phi(-k - x_s^p/\sigma). \qquad (11.11)$$

This means that $U_{1,2}^\gamma$ is dependent on σ^2 alone. We can now see why the low-energy configurations of (11.8) are those containing valley bottom lines close to indicator lines and located at a sufficient depth under μ. Taken alone, however, this energy provides poor detection because, it is performed independently at each site. The spatial interactions taken into account in $U_{1,1}$ will create interaction between the decisions and will provide greater accuracy.

11.2.2 Regularization Term

The purpose of $U_{1,1}(x^\ell)$ is to organize the valley bottom lines in x^ℓ according to closely related chains that are one pixel wide, in a sufficiently regular manner to satisfy $f3$. Because very small valleys are similar to artifacts, this energy must also tend to avoid chains that are too short. Its expression was given in Section 8.2.3:

$$U_{1,1}(x^\ell) \;=\; \sum_s x_s^\ell H_{\vartheta_s}(x^\ell),$$

$$H_{\vartheta_s}(x^\ell) \;=\; \beta_1 \mathbf{1}[n_s = 0] \;+\; \beta_2 \mathbf{1}[n_s = 1]$$
$$+\, \mathbf{1}[n_s = 2] \left(\beta_{3,1} \mathbf{1}[a_s \leq \pi/2] - \beta_{3,2} \mathbf{1}[a_s > \pi/2]\right)$$
$$+\, \mathbf{1}[n_s \geq 3] \, h(n_s),$$

with the following notation:

- ϑ_s is a 3×3 window centered on the site s (excluding s).
- $n_s = \mathrm{card}\{ t \in \vartheta_s \mid x_t^\ell = 1 \}$.

- a_s, which is defined only when $n_s = 2$, is the absolute value of the angle formed by $s - s'$ and $s'' - s$, where s' and s'' belong to ϑ_s and are such that $x^\ell_{s'} = 1$ and $x^\ell_{s''} = 1$.

- All the parameters β are positive and unknown.

- $h(n_s)$ is a positive function that strongly penalizes local configurations such as $n_s \geq 3$.

The parameters β and h play the role of $\theta_{1,1}$ in (11.5). The parameter β_1 penalizes isolated detections, and β_2 controls the length of the valley bottom lines in x^ℓ. Finally, $\beta_{3,2}$ favors valley bottom lines whose local variations of a_s are larger than $\pi/2$, unlike the $\beta_{3,1}$ term. The low-energy configurations x^ℓ are those with regular valley bottom lines that are not too short.

The parameters β and the values of $h(n_s)$ are calibrated according to the qualitative box method (Section 8.2.3) for the energy

$$U_{1,*}\big(x^p, x^\ell\big) \doteq U_{1,1}\big(x^\ell\big) + U^\gamma_{1,2}\big(x^p, x^\ell\big),$$

or, equivalently, by rewriting the $U^\gamma_{1,2}$ given in (11.8) to within an additive constant

$$U_{1,*}(d, x^\ell) = U_{1,1}(x^\ell) - \sum_s x^\ell_s d_s, \qquad (11.12)$$

$$\text{with } d_s = \log \frac{P\big(x^\ell_s = 1 \mid x^p_s\big)}{1 - P\big(x^\ell_s = 1 \mid x^p_s\big)}.$$

This is an expression for the energy (8.37) discussed in Section 8.2.3 (page 172). To perform the calibration, we take a set of typical valley bottom line local configurations associated with local values of d. For each of these configurations, we then set a minimum and maximum probability of occurrence, leading to a system of inequalities that surrounds the parameters β.

- ENERGY $U_{1,3}$

At this stage, it is not obvious that the local conditional probabilities $P(x^\ell_s \mid x^p_s)$ defined in (11.11) are consistent with those that would be calculated based on the distribution $P(x^p, x^\ell)$ given in (11.5) for the energy $U^\gamma_1(x^p, x^\ell)$. To be sure that they are, we would have to prove that there exists an energy $U_{1,3}$ such that the probability $P(x^p, x^\ell)$ satisfies

$$P\big(x^p, x^\ell\big) \propto \exp - \big[U_{1,1}\big(x^\ell\big) + U^\gamma_{1,2}\big(x^p, x^\ell\big) + U_{1,3}\big(x^p\big)\big],$$

$$\text{with } P\big(x^\ell_s = 1 \mid x^p_s\big) = \Psi\big(\Gamma_s(x^p)\big) \Phi\big(-k - x^p_s/\sigma\big), \ \forall\ s. \qquad (11.13)$$

It was shown in [55] that the following expression satisfies (11.13):

$$U_{1,3}(x^p) = -\sum_s \mathbf{1}\left[x_s^p = 0\right] \log(p) - \sum_s \mathbf{1}\left[x_s^p < 0\right] \log(1-p). \quad (11.14)$$

This energy is the same as $-\log P(x^p)$ for independent values of X_s^p governed by the following distribution

$$P\left(X_s^p = 0\right) = p, \quad P\left(X_s^p < 0\right) = 1 - p.$$

This problem of the consistency of the conditional probabilities with the overall probability is similar to the consistency problem seen in Chapter 5. Here we have a new example of this type of problem, which is not often encountered in the building of Markov models. Finally, with $\kappa^2 = \log[p/(1-p)]$, the expression (11.14) is equivalent to

$$U_{1,3}(x^p) = \sum_s \mathbf{1}\left[x_s^p < 0\right]\kappa^2, \quad (11.15)$$

where κ^2 replaces $\theta_{1,3}$ in equation (11.5). Note that (11.15) is not the only expression that satisfies (11.13), but it gives a very simple analytical expression for the estimation task that will now be performed.

11.3 Optimization

[2] We need to estimate (x^p, x^ℓ) and $\gamma = (\alpha, \sigma)$. The estimation is based on the energy given by equation (11.6), which is

$$
\begin{aligned}
U^\gamma\left(x^p, x^\ell \mid y\right) &= U_1^\gamma\left(x^p, x^\ell\right) + U_2^\gamma\left(y \mid x^p\right) \\
&= \left[U_{1,1}\left(x^\ell\right) + U_{1,2}^\gamma\left(x^p, x^\ell\right) + U_{1,3}\left(x^p\right)\right] + U_2^\gamma\left(y \mid x^p\right).
\end{aligned}
$$

If γ were known, our aim would be to calculate a local minimum of U^γ at x^p and x^ℓ simultaneously. We limit ourselves to an initial estimation that is performed in two stages, as follows:

1. Calculate $(\widehat{\gamma}, \widehat{x^p})$ while minimizing the energy $U_{1,3}(x^p) + U_2^\gamma(y \mid x^p)$ with the constraint $x_s^p \le 0$ for all s.

2. Calculate the estimation of the MPM of x^ℓ based on $U_{1,1}(x^\ell) + U_{1,2}^{\widehat{\gamma}}(\widehat{x^p}, x^\ell)$.

This estimation could be improved by the relaxation of (γ, x^p) and x^ℓ by working on the overall energy $U^\gamma(x^p, x^\ell \mid y)$, but the improvement obtained in this case would not be sufficiently worthwhile to justify the added complication.

---

[2]This section can be omitted at the first reading.

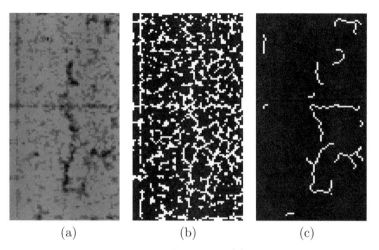

$$(a) \qquad\qquad (b) \qquad\qquad (c)$$

Figure 11.4. Estimation on a single image. (a) Estimation of valleys. (b) Curvature indicator lines. (c) Estimation of valley bottom lines.

● INITIAL ESTIMATION

We begin with Step 1. A necessary condition for a local minimum is obtained by making the gradient of $U_{1,3}(x^p) + U_2^\gamma(x^p \mid y)$ vanish at x^p and γ. As a result, the estimations $\widehat{\mu}$, $\widehat{x^p}$, and $\widehat{\sigma}$ are solutions of the following system:

$$\widehat{\mu} = \mathrm{P}\left(y - \widehat{x^p}\right),$$

$$\widehat{\sigma} = \sqrt{\frac{1}{|S|} \sum_s \left(y_s - \widehat{\mu}_s - \widehat{x_s^p}\right)^2}, \qquad (11.16)$$

$$\forall\, s, \quad \widehat{x_s^p} = y_s - \widehat{\mu}_s \;\; \text{if } y_s < \widehat{\mu}_s - 2\kappa\widehat{\sigma},$$
$$= 0 \qquad\qquad \text{otherwise.}$$

We write P to denote the projection on the spline space (11.2). This calculation is performed using the algorithm below, in which two relaxation algorithms are nested. The second loop is very fast, and converges in only three iterations. It is similar to the robust smoothing procedure (Section 2.2.3) and the method given by Carroll (Section 3.2.2). In general, the whole procedure requires about fifteen iterations.

Step 2 calculates the estimation of the MPM (Chapter 6) of x^ℓ:

$$\forall\, s \in S, \quad \widehat{x_s^\ell} = \arg\max_{\lambda=0,1} P_{\widehat{\gamma}}\left[X_s^\ell = \lambda \mid \widehat{x^p}\right],$$

where the probabilities $P_{\widehat{\gamma}}[X_s^\ell = \lambda \mid \widehat{x^p}]$ are estimated using the Gibbs sampler algorithm (Chapter 7).

Algorithm 11.1. (Detection of Filamentary Entities).

> **(R1)** First relaxation algorithm, acting on μ and (x^p, σ).
>
> - *Step 0 of* (R1): $\widehat{\mu}^0 = \mathrm{P}(y)$, $\widehat{x^p}^0 = 0$,
> $\widehat{\sigma}^0 = \sqrt{\frac{1}{|S|} \sum_s (y_s - \widehat{\mu}_s^0)^2}$,
>
> - *Step n of* (R1) : $\widehat{x^p}^{n-1}$ estimation of x^p at step $n - 1$.
>
>   – *Estimation of μ:* $\widehat{\mu}^n = \mathrm{P}\left(y - \widehat{x^p}^{n-1}\right)$.
>   – *Estimation of x^p and σ:*
>   **(R2)** Second relaxation algorithm on x^p and σ.
>   Let $\widehat{x^p}^{n,i-1}$ and $\widehat{\sigma}^{n,i-1}$ be the estimations of x^p and σ at Step $i - 1$ of (R2).
>   * *Step i of* (R2):
>
> $$\widehat{x^p}_s^{n,i} = y_s - \widehat{\mu}_s^n \quad \text{if } y_s < \widehat{\mu}_s^n - 2\kappa\widehat{\sigma}^{n,i-1}, \quad \forall\, s,$$
> $$= 0 \qquad \text{otherwise,}$$
>
> $$\widehat{\sigma}^{n,i} = \sqrt{\frac{1}{|S|} \sum_s \left(y_s - \widehat{\mu}_s^n - \widehat{x^p}_s^{n,i}\right)^2}.$$
>
>   If $\widehat{\sigma}^{n,i} \approx \widehat{\sigma}^{n,i-1}$, then $\widehat{\sigma}^n \doteq \widehat{\sigma}^{n,i}$ and $\widehat{x^p}_s^n \doteq \widehat{x^p}_s^{n,i}$
>   and end of (R2).
>   If $\widehat{\sigma}^n \simeq \widehat{\sigma}^{n-1}$, end of (R1).

- COMMENTS

The expression for $\widehat{x^p}$ in (11.16) provides fresh justification for the expression of $U_{1,3}$. Here, x^p is estimated with the weighted variance $\kappa^2\sigma^2$ taken into account. To ignore $U_{1,3}$ would be equivalent to choosing $\kappa = 0$ and therefore estimating meaningless valleys.

We now look at a simplified method that does not consider $U_{1,3}$ but that tries nonetheless to estimate x^p. As before, we start by minimizing $U_2^\gamma(x^p \mid y)$ at x^p and γ, and then $U_{1,1}(x^\ell) + U_{1,2}^{\widehat{\gamma}}(\widehat{x^p}, x^\ell)$ at x^ℓ. The first minimization is performed using a multiscale representation. On the coarse scale g, μ is represented by a bicubic spline (11.2): $\mu = B'_g \alpha_g B_g$. On the fine scale f, x^p is represented by $x^p = B'_f \alpha_f B_f$ ($g = 5f$ in the examples shown in Figure 11.1). The estimations are therefore

$$\widehat{\mu} = \mathrm{P}_g(y), \quad \widehat{x^p} = \mathrm{P}_f(y - \widehat{\mu}), \quad \widehat{\sigma}^2 = \sum_s \left(y_s - \widehat{\mu}_s - \widehat{x^p}_s\right)^2 / |S|.$$

This approach, which we call "rigid", gives good results, but it has the following drawback. Because of the valleys, the estimation of μ is biased, and these valleys are therefore underestimated. This is a classic robustness problem where the valleys are interpreted as outliers that are the cause of the poor estimation of μ [112, 134]. This type of problem was discussed in Chapter 2. The method that takes $U_{1,3}$ into account avoids this problem, as illustrated in Figure 11.1, which can be compared to Figure 2.2. This method is said to be "adaptive".

11.4 Extension to the Case of an Image Pair

Let us take the case of two images acquired by stacking two films, which have the same sensitivity in this example (in contrast to Section 10.4). The two resulting images, y^1 and y^2, are assumed to match. This means that at any site s we have two values (y_s^1, y_s^2) that come from the same radiation. With this pair of images it is possible to distinguish certain artifacts from real projections of defects. The model (11.1) is replaced by

$$Y^1 = \mu^1 + X^{p1} + W^1,$$
$$Y^2 = \mu^2 + X^{p2} + W^2.$$

The configurations x^{p1} and x^{p2} are seen as deformations of the same field x^p whose valley bottom lines are represented by the field x^ℓ. Our main unknown is the hidden random field X^ℓ of occurrence x^ℓ. Let $\left(\widehat{x^{p1}}, \widehat{\sigma}_1\right)$ be the estimations of $\left(x^{p1}, \sigma_1\right)$ and let $\left(\widehat{x^{p2}}, \widehat{\sigma}_2\right)$ be those of $\left(x^{p2}, \sigma_2\right)$ obtained by single-image processing using the method described in the preceding sections.

In fact, $\widehat{x^{p1}}$ and $\widehat{x^{p2}}$ are our new data. The field X^ℓ is a Markov random field with the posterior energy

$$U^{\widehat{\gamma}}\left(x^\ell \mid \widehat{x^{p1}}, \widehat{x^{p2}}\right) = U_1\left(x^\ell\right) + U_2\left(\widehat{x^{p1}}, \widehat{x^{p2}}, x^\ell\right),$$
$$U_1\left(x^\ell\right) = U_{1,1}\left(x^\ell\right),$$

where $\widehat{\gamma} = (\widehat{\sigma}_1, \widehat{\sigma}_2)$. The value U_1 is the prior energy defined in the case of only one image. The value U_2 takes on an expression similar to that of $U_{1,2}$. It characterizes the relationship between the data $(x^{p1}, \widehat{x^{p2}})$ and any configuration x^ℓ. This energy favors the detection of sites in the neighborhood of the curvature lines of the mean field $(\widehat{x^{p1}} + \widehat{x^{p2}})/2$ where the signal-to-noise ratio is sufficiently large in both images [55]. Minimizing this energy will give the configuration of valley bottom lines that best suits each of the two images. In the context of image analysis, we would call this a detection with information fusion.

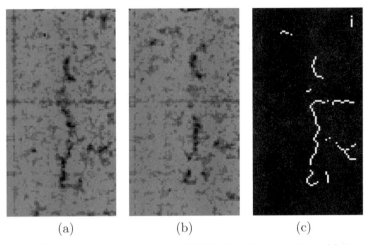

(a)   (b)   (c)

Figure 11.5. Estimation on a pair. (a) and (b) Estimation of valleys. (c) Detection of valley bottom lines with fusion of an image pair.

Processing example.
The size of the images γ is reduced by Gaussian filtering from 512×512 to 128×128. Figures 11.5 (a, b) show the valleys that were estimated separately. Note that the valleys look different in the two images. The erratic behavior of the indicator lines $\Gamma(\widehat{x^p})$ is explained by noise (Figure 11.4(b)). The final detection of valley bottom lines by fusion is shown in Figure 11.5(c).

- APPENDIX: CURVATURES

Let $r(s)$, $s \in \mathbb{R}$, be a regular curvature in $\mathbb{R}^3$ with derivative $\frac{d}{ds}r = \dot{r}$. Let us consider the unit tangent vector $T = \dot{r}/\|\dot{r}\|$. Because $\langle T, T \rangle = 1$, we obtain the following by differentiation: $\langle T, \dot{T} \rangle = 0$. We write $N = \dot{T}/\|\dot{T}\|$ for the unit normal vector. The component $\kappa = \|\dot{T}\|$ of $\dot{T}$ on N defines the curvature of r: $\dot{T} = \kappa N$. The function $\kappa(s)$ quantifies the variation of the tangent along the curve $r(s)$.

Let us now take a regular surface $\mathcal{S}$ in Cartesian coordinates:

$$r(u, v) = \begin{pmatrix} x = u \\ y = v \\ z = z(u, v) \end{pmatrix}.$$

For any point on $\mathcal{S}$ with coordinates (u, v), the tangent plane is defined by the vectors $\frac{\partial r}{\partial u}$ and $\frac{\partial r}{\partial v}$. We write $N = \dot{r}/\|\dot{r}\|$ for the unit normal vector where $\dot{r} = \frac{\partial r}{\partial u} \wedge \frac{\partial r}{\partial v}$. The variation of the normal along the surface is quantified by $\langle dN, dr \rangle$, which is known as the "second fundamental form" of the surface. Expansion gives

$$\langle dN, dr \rangle = \left\langle \frac{\partial N}{\partial u} du + \frac{\partial N}{\partial v} dv \, , \, \frac{\partial r}{\partial u} du + \frac{\partial r}{\partial v} dv \right\rangle$$
$$= (du, dv) \, H \, (du, dv)',$$

where, because $\langle N, \frac{\partial r}{\partial u} \rangle = 0$,

$$H_{1,1} = \left\langle \frac{\partial N}{\partial u}, \frac{\partial r}{\partial u} \right\rangle = \left\langle -N, \frac{\partial^2 r}{\partial u^2} \right\rangle = \frac{\partial^2 z}{\partial u^2}.$$

If we apply the same procedure to the other elements, H is the matrix of the Hessian

$$H = \begin{pmatrix} \dfrac{\partial^2 z}{\partial u^2} & \dfrac{\partial^2 z}{\partial u \partial v} \\ \dfrac{\partial^2 z}{\partial u \partial v} & \dfrac{\partial^2 z}{\partial v^2} \end{pmatrix}.$$

Let P be any point on $\mathcal{S}$, and let π be a plane that passes through P and contains the normal N at P. Let $\Gamma = \mathcal{S} \cap \pi$ be the resulting curve. When we rotate π around N by an angle θ, the curvature κ of the curve passes through a minimum and a maximum, called the principal curvatures. To calculate these curvatures, we consider a specific plane π with orientation θ. With respect to the basis $\{ \frac{\partial r}{\partial u}, \frac{\partial r}{\partial v} \}$ of the tangent plane, the unit tangent vector T at the curve Γ is written as

$$T = t_1 \frac{\partial r}{\partial u} + t_2 \frac{\partial r}{\partial v}, \quad \langle T, T \rangle = 1.$$

It is equivalent to take the angle θ or the vector $t = (t_1, t_2)$. The tangents T corresponding to the main curvatures are called the principal axes. Let us set

$$G = \begin{pmatrix} \left(\dfrac{\partial z}{\partial u}\right)^2 & \dfrac{\partial z}{\partial u} \dfrac{\partial z}{\partial v} \\ \dfrac{\partial z}{\partial u} \dfrac{\partial z}{\partial v} & \left(\dfrac{\partial z}{\partial v}\right)^2 \end{pmatrix}.$$

Proposition 11.1. *The principal curvatures and principal axes are solutions in λ and t of the equation $Ht = \lambda Gt$.*

Proof:
By definition, in the plane π with orientation t, we have $\kappa = \langle N, \dot{T} \rangle = \langle -\dot{N}, T \rangle$, and we can easily prove that $\kappa = t'Ht$. Furthermore, $\langle T, T \rangle = t'Gt$. Using the Lagrange multiplier λ, maximizing the curvature κ at t with the constraint $\langle T, T \rangle = 1$ allows us to calculate $\max_t [t'Ht - \lambda(t'Gt - 1)]$. We obtain $Ht = \lambda Gt$. If λ^* is the largest eigenvalue of $G^{-1}H$, and t^* is the associated eigenvector, then we deduce from $Ht^* = \lambda^* Gt^*$ that $t^{*'} Ht^* = \lambda^* t^{*'} Gt^*$, i.e., $\kappa^* = \lambda^*$. The procedure is exactly the same for the smallest κ. $\qquad\square$

12

Reconstruction from Projections

The purpose of reconstruction is to determine the internal structure of an object, given a set of views that are radiographic projections of the object. This technique was first used for medical applications, which are still the best-known examples of its use. It is also used for nondestructive testing in an industrial context, for example, to inspect welds or composite materials, or to monitor sensitive structures such as nuclear reactors. Here, we focus on tomographic techniques. These require the introduction of a regularization energy because the physics of the acquisition makes this an ill-conditioned problem.

12.1 Projection Model

This model depends on the type of radiation and the type of equipment used [110, 142].

12.1.1 Transmission Tomography

The most common case involves the reconstruction of a cross section from one-dimensional views. Let us consider the reconstruction of the cross section $\{x(s), s = (s_1, s_2) \in \Pi\}$ of an object in a plane Π, with respect to the absorption function of the radiation used (typically x-rays). To do this, the object is illuminated by a bar of point sources whose rays located in the plane Π are parallel. If we make the bar rotate around the object in the plane, we obtain a set of views. We now set $s = (s_1, s_2)$. Let us consider, for example, the bar positioned along the s_1 axis. If I_0 is the initial intensity of the ray at the output of the source at position s_1, then, after traveling in the object over an interval $T(s_1)$, it emerges with the following intensity I:

$$I(s_1) = I_0 \exp\left(-\int_{T(s_1)} x(s_1, s_2)\, ds_2\right).$$

Instead, we consider

$$\mathcal{R}_x(s_1) = -\log\left(\frac{I(s_1)}{I_0}\right) = \int_{T(s_1)} x(s_1, s_2)\, ds_2, \qquad (12.1)$$

known as the Radon transform of x in s_1. The other bar positions can be deduced from this initial position according to rotations whose angles a are measured with respect to the s_1-axis. Let $s_1' = s_1 \cos a + s_2 \sin a$ and $s_2' = -s_1 \sin a + s_2 \cos a$ be the resulting coordinates. The Radon transform of x with angle a is therefore the function

$$\mathcal{R}_x(s_1', a) = \int_{T(s_1', a)} x(s_1, s_2)\, ds_2', \quad \forall s_1'.$$

The set of projections $\{\mathcal{R}_x(s_1', a),\ s_1' \in \mathbb{R}\}$ defines the view of angle a. It is theoretically possible to invert the Radon transform in x when it is continuously defined on (s_1, a). The following theorem states this in more specific terms.

Theorem 12.1. *The one-dimensional Fourier transform of $\mathcal{R}_x(., a)$ is equal to the two-dimensional Fourier transform of x restricted to the line of angle a.*

This theorem forms the basis of all classic inversion methods, but for our purposes this inversion is unrealistic if the number of views is too limited or if a noise is added to the projections.

Another approach is to work with a discrete version of the section to be reconstructed on a rectangular grid S in the plane II. We thus write $x = \{x_s,\ s \in S\}$ for any discrete configuration of a section on S. The configuration x is a piecewise plane surface on the grid elements. An element of the grid is a square surface called a pixel, and is written as $\bar{s}$, where s is the node grid situated at the center of $\bar{s}$. Let us further assume that the bar is partitioned into n intervals of width δ, each corresponding to a source emitting a beam of width δ. For any angle a_ℓ ($\ell = 1, \ldots, L$), the view $\tilde{y}^\ell$ is therefore discrete, and each of its components $\tilde{y}_d^\ell$ ($d = 1, \ldots, n$), is the result of the Radon transform via the beam written as R_d^ℓ of width δ. The instrumentation required for this includes a bar of n detectors placed opposite the bar and on the other side of the object to receive the n beams. The source bars and detector bars form a single assembly and rotate in unison. We write $\mathcal{H}_{d,s}^{\ell} = \text{surface}\big(\bar{s} \cap R_d^{\ell}\big)$ for the contribution of $\bar{s}$ to the projection $\tilde{y}_d^{\ell}$ (see Figure 12.1).

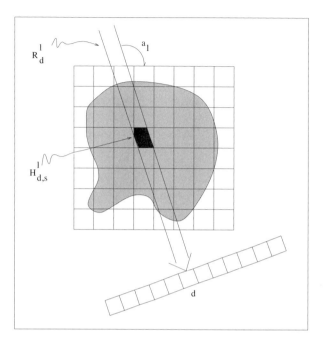

Figure 12.1. Transmission tomography apparatus.

Finally, in the absence of noise, the discretization of the Radon transform (12.1) gives

$$\tilde{y}_d^{\ell} = \sum_{s:\ \bar{s} \cap R_d^{\ell} \neq \emptyset} \mathcal{H}_{d,s}^{\ell}\, x_s, \quad \ell = 1,\dots,L, \quad d = 1,\dots,n, \qquad (12.2)$$

or, in vector form,

$$\tilde{y}^{\ell} = \mathcal{H}^{\ell} x, \quad \ell = 1,\dots,L.$$

If we write $\mathcal{H}$ for the block column matrix whose ℓth block is $\mathcal{H}^{\ell}$, and $\tilde{y}$ for the corresponding block vector, these L equations can be represented as

$$\tilde{y} = \mathcal{H} x.$$

With this notation, we can omit the index ℓ and simply write

$$\tilde{y}_d = \sum_s \mathcal{H}_{d,s}\, x_s,$$

where $d = 1, \ldots, nL$ is now the set of indices of the detector positions. In the presence of a noise, the projection model is

$$Y = \mathcal{H}x + W, \tag{12.3}$$

where the W_d are independent random variables with the same zero expectation distribution, giving

$$E(Y \mid x) = \mathcal{H}x, \tag{12.4}$$

which means that the Radon transform is now only approximate. Each of the projections y_d is now seen as the occurrence of a random variable Y_d. We write V for the diagonal covariance matrix of Y. In the Gaussian noise situation, we have

$$P(y \mid x) \; \propto \; \exp -\frac{1}{2}\|y - \mathcal{H}x\|_{V^{-1}}^2 \,, \tag{12.5}$$

where $\|z\|_V^2 = z'Vz$. From now on, we will use this discrete formulation. Note that $\mathcal{H}$ is known from physical laws and the configuration of the apparatus. The reconstruction problem involves estimating x from the observed projections $\{y_d, \; d = 1, \ldots, nL\}$.

12.1.2 Emission Tomography

In transmission tomography, the sources are outside the object, and the detectors measure the attenuation of rays caused by the object. In emission tomography, the object constitutes the set of sources. Typically, $x(s)$ represents the concentration of an isotope, and the detectors, which are γ-cameras, count the photons emitted by the isotopes (see Figure 12.2). For example, in nondestructive testing, radioactive waste containers are inspected in this way [130]. As before, we wish to reconstruct a cross section of the object. To do this, we use a bar of n detectors, evenly spaced and placed in the cross-sectional plane in L positions obtained by rotation about the object. Also as before, we wish to reconstruct a discrete section $\{x_s, \; s \in S\}$. We write y_d for the number of photons counted at the dth detector position. Using both mathematical and physical arguments, we can show that y_d is the occurrence of a random Poisson variable Y_d whose mean is linearly dependent on the concentration x of isotopes and that the Y_d are independent conditionally on X. As in (12.4), we therefore have

$$E(Y_d \mid x) = \sum_s \mathcal{H}_{d,s}x_s,$$

$$E(Y \mid x) = \mathcal{H}x,$$

where the element $\mathcal{H}_{d,s}$ of the matrix $\mathcal{H}$ is proportional to the probability that a photon emitted at $\bar{s}$ will be received by the detector d (see Section 10.3). The matrix $\mathcal{H}$ is assumed to be known. The conditional distribution of the observations is given by the Poisson distribution

$$P(y \mid x) \ = \ \prod_{d=1}^{n\mathrm{L}} \frac{\left(\sum_s \mathcal{H}_{d,s} x_s\right)^{y_d}}{(y_d)!} \ \exp\left(-\sum_s \mathcal{H}_{d,s} x_s\right).$$

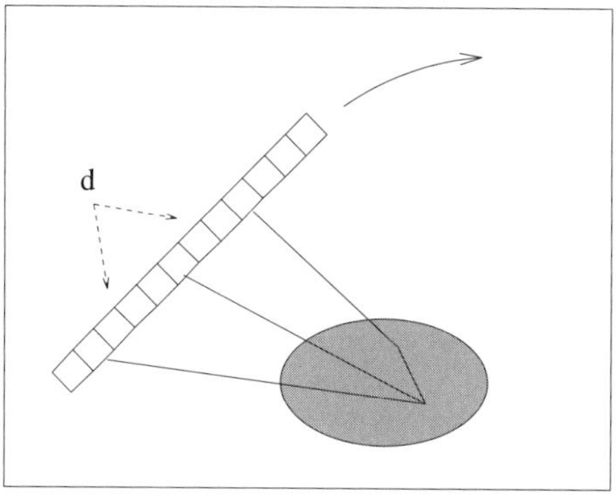

Figure 12.2. Emission tomography apparatus.

12.2 Reconstruction with Regularity Constraints

In both types of tomography described above, the Radon transform is only approximate in the sense that $\mathrm{E}(Y \mid x) = \mathcal{H}x$. The specific nature of each type lies in the definition of $\mathcal{H}$ and the distribution $P(y \mid x)$. Note that the model $Y = \mathcal{H}x + W$ is formally the same as the model for degradation by blurring given in (10.17). We have seen the ill-conditioned nature of the inverse problem that attempts to reconstruct x from y. Here we have an ill-conditioned problem of the same type. A naive approach would be to solve the inverse problem using the maximum likelihood principle $\max_x P(y \mid x)$. This is the method suggested in [162, 176] for emission tomography; the standard EM algorithm (Section 8.2.1) is used to estimate the hidden field x, but this provides a poor solution. In particular, the variance of the estimator $\widehat{X}$ increases with the discretization of X. This means that, here again, we must introduce a regularity constraint on x via a prior distribution $P(x)$. We then perform the maximum likelihood calculation under this constraint, and we find ourselves in the Bayesian context. Here, the specific nature of the inverse problem lies in the distribution $P(y \mid x)$, not in the distribution $P(x)$, which contains prior information about regularity similar to that used in situations other than tomography. The problem here is to reconstruct the cross section of an object composed of several distinct homogeneous regions, where a region is homogeneous if it consists of a single material. As a result, the two types of tomography can be handled with similar prior distributions. As in the case of image restoration (Section 10.1), we can work with either implicit discontinuities or explicit discontinuities in x.

Regularization with implicit discontinuities was first suggested for emission tomography by S. Geman and S. McLure [88]. If we consider that the isotope concentration is more or less constant within a region and that the transition between two regions is clear, i.e., that the isotope concentration has a sharp gradient at the borders between regions, then the proposed energy is

$$U_1(x) = \theta \sum_{[s,t]} \phi(x_s - s_t) + \frac{\theta}{\sqrt{2}} \sum_{<s,t>} \phi(x_s - s_t),$$

where $[s,t]$ are horizontal or vertical neighboring sites, and $< s,t >$ are diagonal neighboring sites. The chosen potential function ϕ, already used in Section 10.1.2, is

$$\phi(u) = \frac{-1}{1 + (u/\beta)^2}.$$

For this function, the low-energy configurations x are those containing smooth regions with a nearly constant gradient, separated by clear borders. This type of function has given rise to variants for tomography. In particular, convex potential functions ϕ have been considered in order to obtain a convex energy $U(x \mid y)$ whose minimization is better justified. This is particularly true of the Green potential, written as $\phi(u) = \log \cosh(u/\beta)$. With these choices, however, Theorem 10.1 on page 199 does not necessarily remain valid [47].

12.2.1 Regularization with Explicit Discontinuities

We now write X^p for the hidden field that was previously written as X. We have seen (Section 10.1) that the introduction of explicit discontinuities requires the introduction of a dual field, written as X^b, that encodes the discontinuities. Reconstruction is then performed simultaneously on the pair of fields

$$(X^p, X^b) \doteq X,$$

because by taking the field X^b into consideration, we improve the reconstruction of X^p. Furthermore, X^p was, conditionally to X^b, a Gaussian Markov random field, while X^b was a Bernoulli Markov random field. Instead of dealing with the discontinuities in terms of edges, we now deal with them in terms of regions. This means that x^b is the encoding of a partition of S. For this, X^b has values in $E^b = \{1, \ldots, m\}$, where m denotes the number of materials: $x_s^b = i$ when s belongs to a region of type i material. The field X^b is an m-ary field. The field X^p has values in E^p, for example, $E^p = \{0, \ldots, 255\}$. Note that the fields X^p and X^b are defined on the same grid S. The field X is modeled by a Markov random field with neighborhood system $\{N_s, \ s \in S\}$ and prior energy $U_1(x)$, making $(X \mid Y = y)$ a Markov random field with the energy

$$U\left(x^p, x^b \mid y\right) \;=\; U_1\left(x^p, x^b\right) + U_2\left(y \mid x^p\right).$$

For transmission tomography, U_2 is deduced from the following conditional distribution (12.5):

$$U_2\left(y \mid x^p\right) \;=\; \left\| y - \mathcal{H}\, x^p \right\|_{V^{-1}}^2.$$

- DEFINITION OF PRIOR ENERGY

The approach using explicit discontinuities was adopted by Dinten [70] in the context of nondestructive testing to solve the problem of transmission tomography for a small number of views L < 10. This approach requires the energy U_1 to express two pieces of prior information. The first indicates that the materials are arranged in wide, well-structured regions. The second indicates that inside each region, the image x^p of the reconstructed section is smooth, as follows:

$$U_1\left(x^p, x^b\right) \;=\; U_{1,1}\left(x^b\right) + U_{1,2}\left(x^p, x^b\right),$$

where this energy follows from the Bayesian breakdown

$$P\left(x^p, x^b\right) \;=\; P\left(x^b\right) P\left(x^p \mid x^b\right).$$

The first prior information is expressed by the local specifications associated with $P(x^b)$:

$$P\left[X_s^b = i \mid N_s(x)\right] \;\propto\; \exp\left(-\sum_{t \in N_s} \beta_{i,x_t^b}\right),$$

which is a more general version of the definition given in (7.15), page 139. The parameters $\beta_{i,i'}$ are defined for all $(i, i') \in E^b \times E^b$. In nondestructive testing, the arrangement of the materials with respect to each other is often known. The values of the parameters are modulated according to our prior knowledge. In particular, we choose $\beta_{i,i} = 0$ as a reference, and $\beta_{i,i'}$ a very large positive value to make the juxtaposition of the materials i and i' very improbable. The above specifications describe the Gibbs distribution $P(x^b)$ with the energy

$$U_{1,1}\left(x^b\right) = \sum_{<s,s'>} \sum_{i,i'} \beta_{i,i'} \mathbf{1}\left[x_s^b = i,\ x_{s'}^b = i'\right] = \sum_{i,i'} \nu_{i,i'}\left(x^b\right) \beta_{i,i'},$$

where $\nu_{i,i'}(x^b) = \sum_{<s,s'>} 1[x^b_s = i, x^b_{s'} = i']$ is the number of cliques whose elements are in the materials i and i' respectively. The energy $U_{1,1}$ is comparable to a "segmentation" energy.

The second prior information is expressed by the local specifications associated with $P(x^p \mid x^b)$. These are chosen to be Gaussian for each material. Their moments are as follows:

$$\mathrm{E}\big[X^p_s \mid N_s(x^p, x^b), x^b_s = i\big] = \mu_i + \gamma_i \sum_{t \in N_s} (x^p_t - \mu_i)\, 1\big[x^b_s = x^b_t\big],$$

$$\mathrm{Var}\big[X^p_s \mid N_s(x^p, x^b), x^b_s\big] = \sigma^2, \tag{12.6}$$

where μ_i is the mean for the material i, and γ_i is its isotropic autocorrelation. This expression for the expectation, which takes into account only neighboring sites made of the same material as s, is similar to that of Proposition 10.1 (Section 10.1.1, page 195). In that proposition, the expectation did not include the neighboring sites of s located on the other side of a neighboring edge. These are the specifications of the Gibbs distribution $P(x^p \mid x^b)$ with the following energy:

$$U_{1,2}(x^p, x^b) = \sum_s \frac{1}{2\sigma^2}\big(x^p_s - \mu_{x^b_s}\big)^2$$
$$- \sum_{<s,t>} \frac{\gamma_{x^b_s}}{\sigma^2}\big(x^p_s - \mu_{x^b_s}\big)\big(x^p_t - \mu_{x^b_t}\big)\, 1\big[x^b_s = x^b_t\big].$$

This expression is an immediate extension of the Gaussian energy given in Proposition 5.1 (page 109). Note that σ^2 and γ_i quantify the regularity in the ith homogeneous region. More concisely, we can write

$$U_{1,2}(x^p, x^b) = (x^p - \mu_{x^b})' Q_{x^b}(x^p - \mu_{x^b}) \doteq \big\|x^p - \mu_{x^b}\big\|^2_{Q_{x^b}}.$$

• RECONSTRUCTION

The posterior distribution

$$P(x^p, x^b \mid y) \propto P(x^b) P(x^p \mid x^b) P(y \mid x^p, x^b) \tag{12.7}$$

has the energy

$$U(x^p, x^b \mid y) = U_{1,1}(x^b) + U_{1,2}(x^p, x^b) + U_2(y \mid x^p)$$
$$= \sum_{i,i'} \nu_{i,i'}(x^b)\, \beta_{i,i'} + \big\|x^p - \mu_{x^b}\big\|^2_{Q_{x^b}} + \big\|y - \mathcal{H}x^p\big\|^2_{V^{-1}}.$$

Note that the local specification of the field (X^p, X^b) at any site s involves all the sites t encountered by the rays passing through s. This long-range dependence explains why it is so difficult to minimize U. In [70], a local minimum of $U(x^p, x^b \mid y)$ is obtained using a block-by-block relaxation algorithm applied alternately to x^p and x^b (Section 7.4.2). For fixed x^b,

$U(x^p, x^b \mid y)$ is quadratic, and for fixed x^p, the minimization can be performed by an ICM algorithm.

Another approach consists in adapting the Gibbs EM algorithm (Section 8.2.2). In this case, $X^p \mid X^b$ is Gaussian, and X^b is m-ary. This algorithm makes it possible to estimate the hidden field and the parameters at the same time, as follows. The kth iteration of the algorithm solves equations (8.24) and (8.25), which are

$$\mathbf{E}_{\theta(k)}\left[\frac{d}{d\theta_1}\log\mathcal{P}_{\theta_1}(X)\,\Big|\,Y=y\right] = 0, \qquad (12.8)$$

$$\mathbf{E}_{\theta(k)}\left[\frac{d}{d\theta_2}\log\mathcal{P}_{\theta_2}(y\mid X^p)\,\Big|\,Y=y\right] = 0. \qquad (12.9)$$

The distribution $\mathcal{P}_{\theta_1}(x)$ is the pseudolikelihood of the field x:

$$\begin{aligned}\mathcal{P}_{\theta_1}(x) &= \mathcal{P}_{\theta_{1,1}}(x^b)\,\mathcal{P}_{\theta_{1,2}}(x^p\mid x^b)\\ &= \prod_s P(x_s^b\mid N_s(x^b))\,\prod_s P(x_s^p\mid N_s(x^p,x^b),x_s^b)\,,\end{aligned}$$

and $\mathcal{P}_{\theta_2}(y\mid x^p)\propto\exp-\|y-\mathcal{H}x^p\|_{V^{-1}}^2$ is the distribution (12.5). The vector θ_2 is the vector of unknown parameters describing V, here assumed to be equal to $\sigma_W^2\mathrm{Id}$. The set of unknown parameters of the prior distribution is written as $\theta_1 = (\theta_{1,1},\theta_{1,2})$. The parameter $\theta_{1,1}$ contains the $\beta_{i,j}$, and $\theta_{1,2}$ contains the parameters of the homogeneous regions $(\mu_i,\gamma_i,\sigma^2)$. The expectation $\mathbf{E}_{\theta(k)}$ refers to the posterior distribution $\mathcal{P}_{\theta(k)}(x^p,x^b\mid y)$ given in (12.7). Equation (12.8) therefore breaks down into the following two equations:

$$\mathbf{E}_{\theta(k)}\left[\frac{d}{d\theta_{1,1}}\log\mathcal{P}_{\theta_{1,1}}(X^b)\,\Big|\,Y=y\right] = 0, \qquad (12.10)$$

$$\mathbf{E}_{\theta(k)}\left[\frac{d}{d\theta_{1,2}}\log\mathcal{P}_{\theta_{1,2}}(X^p\mid X^b)\,\Big|\,Y=y\right] = 0. \qquad (12.11)$$

For (12.10), let us look again at the second formulation of the m-ary case discussed in Section 8.2.2. Let $E^{(j)}$ be the local neighborhood configurations, $p_{ij} = P[X_s^b = i \mid N_s(x^b) \in E^{(j)}]$ the associated specifications, and $n_{ij}(x^b) = \sum_s \mathbf{1}[x_s^b = i,\ N_s(x^b)\in E^{(j)}]$. If we overparametrize by replacing $\theta_{1,1}$ with the p_{ij}, the solution of (12.10) is similar to (8.29):

$$p_{ij}(k+1) = \frac{\mathbf{E}_{\Theta(k)}\left[n_{ij}(X^b)\mid Y=y\right]}{\sum_i \mathbf{E}_{\Theta(k)}\left[n_{ij}(X^b)\mid Y=y\right]}.$$

The symbol Θ is used instead of θ to denote the overparametrization. Equation (12.11) is handled in a similar way to the Gaussian case (Section 8.2.2). Using the expression for the local specifications (12.6), the log-pseudolikelihood is

$$\log \mathcal{P}_{\theta_{1,2}}(x^p \mid x^b)$$

$$= -\frac{|S|}{2}\log(2\pi\sigma^2) - \frac{1}{2\sigma^2}\sum_{i=1}^{m}\sum_{s}\mathbf{1}_{x_s^b=i}\left(x_s^p - \mu_i - \gamma_i\sum_{t\in N_s}\mathbf{1}_{x_t^b=i}(x_t^p - \mu_i)\right)^2.$$

After the gradient in (12.11) vanishes, we obtain the estimated mean

$$\gamma_i(k+1)$$

$$= \frac{\mathbf{E}_{\Theta(k)}\left[\sum_s \mathbf{1}_{X_s^b=i}(X_s^p - \mu_i(k+1))\left(\sum_{t\in N_s}\mathbf{1}_{X_t^b=i}(X_t^p - \mu_i(k+1))\right)\Big| y\right]}{\mathbf{E}_{\Theta(k)}\left[\sum_s \mathbf{1}_{X_s^b=i}\left(\sum_{t\in N_s}\mathbf{1}_{X_t^b=i}(X_t^p - \mu_i(k+1))\right)^2\Big| y\right]},$$

and the estimated variance

$$\sigma^2(k+1) = \frac{1}{|S|}\sum_{i=1}^{m}\mathbf{E}_{\Theta(k)}\left[S_i^2 \mid Y=y\right],$$

$$S_i^2 = \sum_s \mathbf{1}_{X_s^b=i}\left(X_s^p - \mu_i(k+1) - \gamma_i(k+1)\sum_{t\in N_s}\mathbf{1}_{X_t^b=i}(X_t^p - \mu_i(k+1))\right)^2,$$

and, as an approximation,

$$\mu_i(k+1) = \frac{\mathbf{E}_{\Theta(k)}\left[\sum_s \mathbf{1}\left[X_s^b=i\right]X_s^p \mid Y=y\right]}{\mathbf{E}_{\Theta(k)}\left[\sum_s \mathbf{1}\left[X_s^b=i\right] \mid Y=y\right]}.$$

Finally, $\mathbf{E}_{\Theta(k)}$ is approximated by an empirical mean calculated on a sample $(x(1),\ldots,x(N))$ of $X = (X^p, X^b)$ obtained according to the Gibbs sampler for the distribution $P_{\Theta(k)}(x^p, x^b \mid Y=y)$. At each iteration, the hidden field is estimated according to the TPM criterion for X^p and the MPM for X^b (Section 6.2).

12.2.2   *Three-Dimensional Reconstruction*

We now give an example of three-dimensional reconstruction from a small number of projections that are two-dimensional radiographic images. In nondestructive testing, we might wish to reconstruct a three-dimensional image of the internal structure of the material being inspected in order to view any defects or cracks, which we will call "inclusions". A classic reconstruction is not very effective when there is only a small number of projections available, and so several authors have suggested reconstructing only certain geometric features of the inclusions, such as their polygonal outline [136] or their skeleton [43]. This robust approach is not valid in the

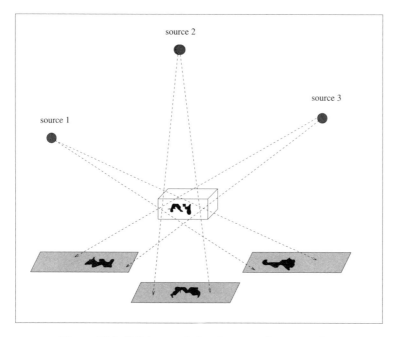

Figure 12.3. 3-D transmission tomography apparatus.

case of occlusion, i.e., when the material contains inclusions whose projections overlap. In nondestructive testing, however, we can often assume that the material is approximately binary, with more or less contrasting inclusions on a more or less homogeneous background. A reasonable goal, then, would be to determine the number of inclusions and their shape, without trying to achieve a finely shaded reconstruction with the exact gray levels.

The apparatus used here is a generalization of the two-dimensional case described in Section 12.1.1, except that the sources are now conical. It consists of L sources (γ or X), typically $L \leq 9$, where the maximum view angle between sources is small. Each source emits a conical beam of rays [165]. As a result, L two-dimensional projections are obtained and sampled on a grid D_ℓ. Let S be the discretized three-dimensional volume in which the reconstruction is performed, and let $(x(s), s \in S)$ be the image to be reconstructed. The volume S is a subset of a three-dimensional grid and is the discretized intersection of the L cones whose bases are the D_ℓ (see Figure 12.3). The cubic elements of the grid are written as $\bar{s}$. We write y^ℓ for the projections, R_d^ℓ for the elementary cone connecting the ℓth source and the two-dimensional element $\bar{d}$, and finally, $\mathcal{H}_{d,s}^\ell = \text{volume}\ (\bar{s} \cap R_d^\ell)$ is the contribution of $\bar{s}$ to the projection y_d^ℓ.

As in (12.3), y_d^ℓ is the occurrence of the following random variable:

$$Y_d^\ell = \sum_{s:\ \bar{s} \cap R_d^\ell \neq \emptyset} \mathcal{H}_{d,s}^\ell x_s + W_d^\ell, \quad \ell = 1, \dots, \mathrm{L}.$$

Here again, it is natural to think of an energy containing a fidelity term $\sum_\ell \|y^\ell - \mathcal{H}^\ell x\|^2$ and a regularization term such as $\|x - \bar{x}\|^2$, where $\bar{x}_s$ denotes the mean of the values of the $3^3 - 1$ neighboring sites of s. Unfortunately, these terms are not of the same order of magnitude, which makes them difficult to compare. To alleviate this problem, we suggest using

$$U_1(x) = -\frac{\langle x, \bar{x} \rangle^2}{\|x\|^2 \|\bar{x}\|^2}, \tag{12.12}$$

$$U_{2,\ell}(y^\ell \mid x) = -\frac{\langle y^\ell, \mathcal{H}^\ell x \rangle^2}{\|y^\ell\|^2 \|\mathcal{H}^\ell x\|^2}, \quad \ell = 1, \dots, \mathrm{L}, \tag{12.13}$$

where $\langle a, b \rangle$ is the usual scalar product. These functional expressions are of the form $\left[\frac{\langle y, z \rangle}{\|y\| \|z\|} \right]^2$, or $\cos^2(y, z)$. Let us set

$$U(x \mid y) = \alpha\, U_1(x) + \frac{1}{\mathrm{L}} \sum_\ell U_{2,\ell}(y^\ell, x),$$

where α is the regularization parameter. Unlike all the energies encountered until now, this is not a Markov energy. The energies (12.12) and (12.13) cannot be expressed as a *sum* of neighborhood potentials [44].

Figures 12.4 and 12.5 illustrate the use of this type of energy on simulated data. The original inclusions (Figure 12.5 (a)) were simulated under nondestructive testing conditions. The $\mathrm{L} = 4$ sources are placed in a plane above the two inclusions at a height of $z = 482$ mm. The first source is above the inclusions and the three others are equally spaced on a circle of radius 125 mm centered on the first source. The projection plane is at $z = 0$ mm. The mean position of the inclusions is $z = 443$ mm. Each voxel $\bar{s}$ is a cube with side 0.75 mm. The total volume of the inclusions is 4024 voxels, and the volume of S is 19,000 voxels. A Gaussian noise of variance 4 was added to the projections of these two inclusions. We note that on the projections the inclusions overlap.

The results come from a computer program written by F. Coldefy. After the initial stage of detection, inherent in all nondestructive testing, we have a segmentation of the images in which we keep only the segments that are the presumed projections of the inclusions (Chapter 11), and the rest are set to zero. This is equivalent to assuming that the density μ of the material outside of inclusions is zero. More specifically, if L is the total path length of a ray and ℓ_o is the length restricted to inclusions when present, then the value of the projection is $(L - \ell_o)\mu + \ell_o\mu'$, where μ' is the density of the inclusions. We initialize the energy minimization with

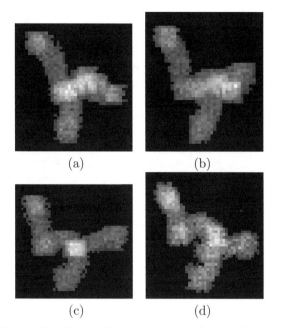

Figure 12.4. Four radiographic views of two inclusions.

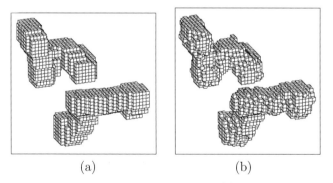

Figure 12.5. 3-D tomographic reconstruction. (a) Original inclusions. (b) Reconstructed inclusions.

the configuration ($x_s = 1$, $\forall s$). The volume of the reconstruction, 4086 voxels, is accurate. There are 776 incorrect voxels, but most of them are located on the surface of the inclusions, a fact that halves the error rate.

12.3   Reconstruction with a Single View

With only one view, the multiview tomography projection model can no longer be used. The problem is too ill-posed to make reconstruction possible without introducing a large amount of additional information. Here, this information comes from a simple generic shape of the objects to be reconstructed. There are two possible situations. The first, and simpler, situation involves a single projected object. The second situation, which we plan to discuss, is more difficult, because it involves the presence of two unknown objects in the radiographic field. The projections of these two objects can be superimposed in the image. This type of occlusion was described in the above section for the case of objects with nearly constant density in a material with nearly constant density. This assumption about the densities is not sufficient, however. We must also apply some simplifying assumptions concerning the shape of the object. In the presence of occlusion, we consider one of the objects to be deformable such that we can estimate its deformations.

12.3.1   Generalized Cylinder

• DEFINITIONS

A generalized cylinder (GC) is a volume characterized by an axial curve C from which closed curves located in the planes that are transverse to this curve generate a surface G [16, 39, 170]. This model can be expressed in many ways. We will now show two of them.

An example of an explicit analytical expression is [148]:

$$G(u, v) = v\mathbf{e}_1 + f(u, v)\cos(u)\mathbf{e}_2 + f(u, v)\sin(u)\mathbf{e}_3.$$

The function $f(u, v)$ is regular and u-periodic, and $(u, v) \in [0, 2\pi[\times [0, 1]$. The vectors $\{\mathbf{e}_i\}$ constitute the orthonormal reference of (x, y, z) space.

An implicit expression is formulated in the form of a spline-type energy. The axis is a curve with parameter $t \in [0, 1]$, and $C(t) = (x(t), y(t), z(t))'$ is a spline function with the following prior energy (Chapter 2):

$$U_{1,1}(C) = \int_0^1 \beta_1 \|D^2 C(t)\|^2 dt.$$

The parameters of the surface G of the GC are $(u, v) \in [0, 1]^2$, and $G(u, v) = (x(u, v), y(u, v), z(u, v))'$ is a spline with the following prior energy:

$$U_{1,2}(G) = \int_0^1 \int_0^1 \left(\beta_2 \|D^{2,0}G\|^2 + \beta_3 \|D^{1,1}G\|^2 + \beta_4 \|D^{0,2}G\|^2 \right) du\, dv.$$

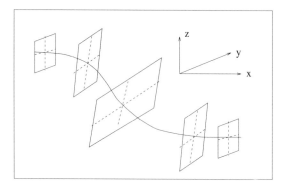

Figure 12.6. Generalized cylinder with horizontal rectangular section.

The following constraints guarantee that the surface is closed:

$$G(u,0) = G(u,1) \quad D^{0,1}G(u,0) = D^{0,1}G(u,1).$$

A special choice of parameter would be $t \equiv u$, where t is the parameter of the axis. This means that at fixed u_0, $\{G(u_0, v), v \in [0,1]\}$ describes a curve, written as $G(u_0, .)$, that is transverse to $C(u)$. We must, however, build an interaction energy between C and G, for example so that C will be an axis inside G. The center of mass of the curve $G(u_0, .)$ is

$$\bar{G}(u_0) = \frac{1}{l} \int_0^1 G(u_0, v) \left\| D^{0,1}G(u_0, v)\right\| dv,$$

where l is the length of the curve:

$$l(u_0) = \int_0^1 \left\| D^{0,1}G(u_0, v)\right\| dv.$$

The interaction energy expressing the fact that C is approximately the axis of the GC is

$$U_{1,3}(C, G) = \int_0^1 \beta_5 \left\| \bar{G}(u) - C(u)\right\|^2 du.$$

The use of this type of formulation, whether explicit or implicit, requires the building of a fidelity energy, which depends on the application. We now examine a special case.

● AN EXAMPLE AND ITS RADIOGRAPHIC PROJECTION

Let us consider a GC that is principally oriented in the x direction. Let C be the axial curve with the coordinates

$$y = A(x),$$
$$z = B(x),$$

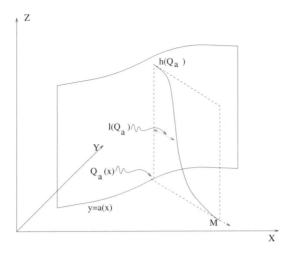

Figure 12.7. Radiographic projection of a generalized cylinder with rectangular section.

and let Π be the surface that is orthogonal to the (x, y)-plane and contains C. The surface describing the GC is generated by moving a closed curve centered on C in a plane orthogonal to Π. The simplest examples of such curves are ellipses [42] and rectangles [95]. A cylinder with a horizontal rectangular section with equation

$$\left((y - A(x))^2 - \frac{L(x)^2}{4}\right)\left((z - B(x))^2 - \frac{H(x)^2}{4}\right) = 0 \qquad (12.14)$$

is therefore defined by the *characteristic curves* A, B, L, H, where L is the width of the rectangles and H is their height, as illustrated in Figure 12.6.

We now examine the radiographic projection on the $z = 0$ plane along rays that are orthogonal to this plane. The result is a surface of gray levels $z = \mathcal{G}(x, y)$, which can itself by described by sections that are orthogonal to a vertical surface $\widetilde{\Pi}$ described by the equation $y = a(x)$. For the cylinder with a rectangular section (12.14), the sections of the projection are modeled by bell curves defined by $t \mapsto h \exp\left[-K_q \left|\frac{t}{\ell}\right|^q\right]$ where q quantifies the blurring introduced by the radiographic acquisition system. With $K_q = 2^q \log 2$, ℓ represents the width of the section at its mid-height, and h represents its height. For any point $M = (x, y)$ we use $Q_a(x)$ to denote the point nearest to M on the curve $y = a(x)$ (Figure 12.7). The projection at M is

$$\mathcal{G}(M) = h(Q_a) \exp\left[-K_q \left(\frac{\|Q_a M\|}{\ell(Q_a)}\right)^q\right]. \qquad (12.15)$$

The projections constitute a new GC, whose characteristic curves are $a(x), h(Q_a(x))$, and $\ell(Q_a(x))$. This is not a classic GC because its sections are not defined by closed curves.

12.3.2   Training the Deformations of a Generalized Cylinder

We examine a GC defined by a set of characteristic curves. Furthermore, we assume that this GC represents a deformable object. In formal terms, this GC is either G or its projection $\mathcal{G}$. For the application being considered, $\mathcal{G}$ is more appropriate. We wish to model the deformations of this type of GC from a set of observations obtained during the deformation of this GC. This is equivalent to modeling the deformations of each of the characteristic curves of the GC. We therefore consider a specific characteristic curve, called h, and assume that we have an N-sample $\{h^j\}_{1 \leq j \leq N}$ of its deformations, each discretized at n points: $h^j = (h_1^j, \ldots, h_n^j)'$. These discretizations h^j are performed on the same interval and are considered to be occurrences of a random vector $\vec{h}$. This assumption is difficult to put into practice, because it requires the observations to be *matched*. This means that, for example, when the curves have not been realigned, and their misalignment is not synonymous with natural deformation, the training will highlight these misalignments rather than the deformations of interest (see Section 4.1.1).

When $\vec{h}$ is a random Gaussian vector, it is appropriate to apply a model that uses linear decomposition on the principal modes of deformation as described by principal component analysis (Section 4.1.1) [152]. In the non-Gaussian case, AAC-type modeling as described in Section 4.2 should be used. For the remainder of the present chapter, we consider the Gaussian case.

• GAUSSIAN DEFORMATIONS

PCA modeling (Section 4.1.1) seeks the principal directions of the scattering of the cloud $\mathcal{X} = \{h^j\}$ with respect to the mean vector

$$\bar{h} = \frac{1}{N} \sum_{j=1}^{N} h^j.$$

The scattering of the cloud is given by the empirical covariance matrix V^h, which quantifies the linear variations of the $\{h^j\}$ about the mean

$$V^h = \frac{1}{N} \sum_{j=1}^{N} (h^j - \bar{h})(h^j - \bar{h})'.$$

The matrix V^h is symmetric and can be diagonalized on an orthonormal basis of eigenvectors

$$V^h = \mathsf{H}\Lambda^h\mathsf{H}',$$

where $\Lambda^h = \mathrm{diag}(\lambda_1^h, \ldots, \lambda_n^h)$ is the diagonal matrix of eigenvalues in decreasing order, and $\mathsf{H} = [\bar{h}^1, \ldots, \bar{h}^n]$ is the matrix of associated eigenvectors. The inertia of the $\{h^j\}$ in the $\bar{h}^i$ direction is $\lambda_i^h = {}^t\bar{h}^i V^h \bar{h}^i$. Any occurrence h of $\vec{h}$ therefore breaks down on eigenvectors according to the following relationship:

$$h = \bar{h} + \sum_{i=1}^{n} \beta_i \bar{h}^i, \qquad (12.16)$$

where $\{\beta_i,\ i = 1, \ldots, n\}$ constitutes the coordinates on the $\bar{h}^i$. We must emphasize that in the context of curves or images, the eigenvectors $\bar{h}^i$ often correspond to specific deformation modes that are known as "natural modes" because they have a particular physical interpretation such as elongation or torsion.

Let us consider the n_h first eigenvectors such that

$$\frac{\sum_{i=1}^{n_h} \lambda_i^h}{\sum_{i=1}^{n} \lambda_i^h} \approx 1,$$

meaning that the subspace they generate represents almost all the inertia. The model used to approximate the deformations is as follows:

$$h \approx \bar{h} + \sum_{i=1}^{n_h} \beta_i \bar{h}^i \doteq \bar{h} + \check{\mathsf{H}}\,\check{\beta}, \qquad (12.17)$$

where $\check{\mathsf{H}}$ comprises the first n_h columns of H. For any vector h, equation (12.16) gives

$$\beta_i = (h - \bar{h})'\bar{h}^i \ \text{ which gives } \ \check{\beta} = \check{\mathsf{H}}'(h - \bar{h}),$$

and, in particular, for each h^j of the sample,

$$\{h^j,\ j = 1, \ldots, N\} \longmapsto \{\check{\beta}^j = \check{\mathsf{H}}'(h^j - \bar{h}),\ j = 1, \ldots, N\}. \quad (12.18)$$

To deal with the inverse problem of reconstruction in the presence of occlusion, we need a probability distribution for $\vec{h}$. Assuming a Gaussian distribution, for fixed $\bar{h}^i$, β_i is governed by the Gaussian distribution $LG(0, \lambda_i^h)$. Equation (12.17) therefore comes from the stochastic model whose random components are the β_i:

$$\vec{h} = \bar{h} + \check{\mathsf{H}}\,\vec{\beta}.$$

The parameters of this model are the λ_i^h, the $\bar{h}^i$, and $\bar{h}$, which are estimated on the observed deformations $\{h^j,\ j = 1, \ldots, N\}$. Simulating a deformation h therefore amounts to simulating the n_h Gaussian coordinates $\{\beta_i,\ i = 1, \ldots, n_h\}$.

The deformations of a GC are induced by the deformations of its characteristic curves. Thus, for example, deformations of the GC of the radiographic projection (12.15) are obtained by simulating the random vectors $\vec{a}$, $\vec{h}$, and $\vec{\ell}$ according to

$$a = \bar{a} + \sum_{i=1}^{n_a} \alpha_i \bar{a}^i, \quad \alpha_i \sim LG(0, \lambda_i^a),$$

$$h = \bar{h} + \sum_{i=1}^{n_h} \beta_i \bar{h}^i, \quad \beta_i \sim LG(0, \lambda_i^h), \tag{12.19}$$

$$\ell = \bar{\ell} + \sum_{i=1}^{n_\ell} \gamma_i \bar{\ell}^i, \quad \gamma_i \sim LG(0, \lambda_i^\ell).$$

where the natural modes $\bar{a}^i$, $\bar{h}^i$, and $\bar{\ell}^i$ are calculated on the N samples $\{a^j\}$, $\{h^j\}$ and $\{\ell^j\}$ respectively. Using the notation from (12.17), for all set values of $(\check{\alpha}, \check{\beta}, \check{\gamma})$, we obtain the discrete characteristic curves a, h, and ℓ such that

$$a - \bar{a} = \check{\mathsf{A}}\,\check{\alpha},$$
$$h - \bar{h} = \check{\mathsf{H}}\,\check{\beta},$$
$$\ell - \bar{\ell} = \check{\mathsf{L}}\,\check{\gamma},$$

or, more concisely, by setting $\theta' = (\check{\alpha}', \check{\beta}', \check{\gamma}')$, $\Im = \mathrm{diag}(\check{\mathsf{A}}, \check{\mathsf{H}}, \check{\mathsf{L}})$ and $I' = (a' - \bar{a}', h' - \bar{h}', \ell' - \bar{\ell}')$, we have

$$I = \Im\theta. \tag{12.20}$$

Here, $\Im$ is an $n \times n_\theta$ matrix with $n_\theta = n_a + n_h + n_\ell$. The expression (12.20) is general and is not related to the special case of the GC in our example. We should emphasize that here, the deformations of the characteristic curves are independent. This restrictive assumption will later be relaxed.

12.3.3   Reconstruction in the Presence of Occlusion

The radiographic projection is the sum of two projections we wish to estimate, one of which is a GC. The model of the projection at the point $M = (x, y)$ is

$$P(M) = \mathcal{G}_\theta(M) + \mathcal{O}(M) + W(M), \tag{12.21}$$

$$\text{with the constraint } \mathcal{O}(M) > 0.$$

Here, $\mathcal{G}_\theta$ denotes a GC whose characteristic curves have θ as their parameters. This cylinder is obscured by $\mathcal{O}$ up to a noise W. The GC $\mathcal{G}_\theta$ is represented by (12.20). Under certain conditions, the estimation of the two components $\mathcal{G}_\theta$ and $\mathcal{O}$ can recreate the GC of the three-dimensional object whose projection is $\mathcal{G}_\theta$, but we will not pursue this here. Once again, we have a sample of the deformations of each characteristic curve.

● BAYESIAN ESTIMATION

First we wish to estimate $\mathcal{G}_\theta$, and the estimation of $\mathcal{O}$ is obtained from this by subtraction from P:

$$\hat{\mathcal{O}} = P - \mathcal{G}_{\hat{\theta}}. \tag{12.22}$$

The estimation of the GC is equivalent to estimating the parameters θ from the characteristic curves. We adopt a Bayesian approach by building an energy on θ with the following expression:

$$U(\theta \mid P) = \sum_{i=1}^{2} U_{1,i}(\theta) + \sum_{i=1}^{2} \rho_i U_{2,i}(P \mid \theta).$$

The first term $U_{1,1}(\theta)$ of the prior energy comes from the Gaussian distribution introduced to simulate the deformations. Similarly to (2.22), this term is written as

$$U_{1,1}(\theta) = \theta'\Omega_1\theta,$$

where Ω_1 is deduced from the expressions in (12.19):

$$\Omega_1 = \mathrm{diag}(\Lambda^a, \Lambda^h, \Lambda^\ell) \quad \text{with} \quad \Lambda^* = \mathrm{diag}\Big(\ldots, \frac{1}{\lambda_i^*}, \ldots\Big).$$

Note that the characteristic curves were modeled independently from each other. It is, however, more realistic to consider that the deformations between characteristic curves are dependent. When these dependencies are linear, they can be estimated by "canonical analysis", giving the linear system $\Omega_2\theta \approx 0$. This is explained in the appendix at the end of this section. The second prior energy is then

$$U_{1,2}(\theta) = \|\Omega_2\theta\|^2. \tag{12.23}$$

The second term $U_{2,2}$ of the posterior energy, which comes from the constraint (12.21), represents the "negative" reconstructed volume

$$U_{2,2}(P \mid \theta) = \sum_{M} \|P(M) - \mathcal{G}_\theta(M)\|^2 \, \mathbf{1}[(P(M) - \mathcal{G}_\theta(M)) \le 0].$$

Although the prior energies $U_{1,i}(\theta)$ are not directly related to the application, the fidelity energy $U_{2,1}(P \mid \theta)$ is strongly dependent on the application. There is one situation in which this energy has a general expression: when the occlusion is not total. In this case, a few points of the characteristic curves of $\mathcal{G}_\theta$ can sometimes be extracted from P. Let $\check{n}$ be the number of these points $(n > \check{n} > n_\theta)$. We write $\check{I}$ and $\check{\mathfrak{I}}$ for the corresponding components of (12.20). The occlusion is negligible at these points, and therefore (12.21) is reduced to $P(M) = \mathcal{G}_\theta(M) + W(M)$. Assuming W to be Gaussian, we can choose the following expression as the fidelity energy:

$$U_{2,1}(\mathrm{P} \mid \theta) = \left\| \check{I} - \check{\mathfrak{J}}\theta \right\|^2. \tag{12.24}$$

The value of $\check{I}$ depends on P, while the $\check{\mathfrak{J}}$ comes from the training stage. When $\check{n}$ is small, i.e., close to n_θ, this energy is insufficient. We must then seek another form of energy, as is illustrated in the example below.

• EXAMPLE

[1] This is an example of nondestructive testing by x-ray radiography described in Girard et al. [95]. In this application, P represents the projection of a soldered tab on a printed circuit, and the aim is to estimate the solder $\mathcal{O}$ under the tab $\mathcal{G}_\theta$. Here, the estimation of $\mathcal{O}$ contains all the relevant information, and it is not necessary to recreate its full three-dimensional representation, as Figure 12.9 shows. The tabs being examined are gull-wing tabs, see Figure 12.8. They are modeled by the GC (12.14), and their radiographic projections are modeled by equation (12.15).

Although there is total occlusion in this application, we are able to determine some linear equations with respect to θ, which are written as $\check{I} = \check{\mathfrak{J}}\theta$ (a notation that is not strictly correct). We will give an overview of how this is done [94]. Considering the physical phenomena involved in forming the image, we can obtain the abscissa x_1 at which the function h reaches a maximum: $Dh(x_1) = 0$. We can also obtain the abscissa x_2 of the point of inflection of h at the free end of the tab: $D^2h(x_2) = 0$, where the raised level of the weld is clearly identifiable. We therefore obtain two linear equations with respect to β,

$$D\bar{h}(x_1) + \sum_{i=1}^{n_h} \beta_i D\bar{h}^i(x_1) = 0, \quad D^2\bar{h}(x_2) + \sum_{i=1}^{n_h} \beta_i D^2\bar{h}^i(x_2) = 0,$$

and this notation assumes that we have a continuous version $\bar{h}^i(x)$ of $\bar{h}^i$. Note that in this case $\bar{h}$ and $\bar{h}^i$ are known, because they come from the training on the deformation samples. By performing the same procedure on the other characteristic curves, we obtain four new linear equations.

In the absence of the energy term $U_{2,2}$, we saw that the failure to satisfy the constraint (12.21) at the output of the calculation (12.22) was due to errors in the estimation of the dimensions of the GC $\mathcal{G}_\theta$. The best approach now is to try to control certain cylinder dimensions. To do this, we set $h_1 = h(x_1)$ as the maximum value of h. For fixed h_1, we have the

---

[1]This paragraph can be omitted at the first reading.

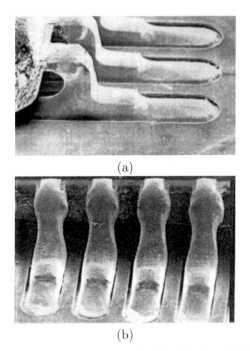

(a)

(b)

Figure 12.8. Electron microscope views of soldered tabs. (a) Profile view. (b) Frontal view.

following equation:

$$h_1 - \bar{h}(x_1) - \sum_{i=1}^{n_h} \beta_i \bar{h}^i(x_1) = 0.$$

We also have an equation for the maximum width ℓ_1 of ℓ. We write $d = (h_1, \ell_1)$ to denote these unknowns. All the $6 + 2$ linear equations obtained can be summarized by $\check{I} - \check{\Im}\theta = 0$. Because $\check{I}$ has two unknown components, $d = (h_1, \ell_1)$, we write $\check{I}(d)$ instead of $\check{I}$. Finally in this example, the fidelity energy (12.24) is rewritten as

$$U_{2,1}(\mathrm{P} \mid \theta, d) = \left\| \check{I}(d) - \check{\Im}\theta \right\|^2,$$

and the posterior energy is therefore

$$U(\theta, d \mid \mathrm{P}) = \sum_{i=1}^{2} U_{1,i}(\theta) + \rho_1 U_{2,1}(\mathrm{P} \mid \theta, d) + \rho_2 U_{2,2}(\mathrm{P} \mid \theta).$$

For a joint optimization in (θ, d), a very large value of ρ_2 would have to be chosen to favor the constraint (12.21). We will decouple the θ and d

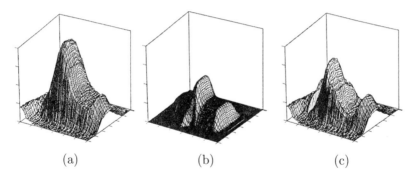

Figure 12.9. Soldered tab. (a) Radiographic projection P. (b) Estimation of the projection of the tab $\mathcal{G}_\theta$. (c) Estimation of the projection of the weld $\mathcal{O}$.

optimizations, however. Noting that the solution $\widehat{\theta}(d)$ of the expression

$$\min_\theta \left[\sum_{i=1}^{2} U_{1,i}(\theta) + \rho_1 U_{2,1}(\mathrm{P} \mid \theta, d) \right]$$

has an explicit analytical expression that comes from setting the gradient to zero:

$$(\Omega_1 + \Omega_2'\Omega_2 + \rho_1 \check{\mathfrak{F}}'\check{\mathfrak{F}})\widehat{\theta}(d) - \rho_1 \check{\mathfrak{F}}'\check{I}(d) = 0,$$
$$\text{i.e.,} \quad \widehat{\theta}(d) = (\Omega_1 + \Omega_2'\Omega_2 + \rho_1 \check{\mathfrak{F}}'\check{\mathfrak{F}})^{-1}\rho_1 \check{\mathfrak{F}}'\check{I}(d).$$

We obtain the estimation of d from

$$\min_d U_{2,2}\big(\mathrm{P} \mid \widehat{\theta}(d)\big),$$

whose solution $\widehat{d}$ gives the estimation $\widehat{\theta}(\widehat{d})$ of θ. Clearly, the introduction of the unknowns d has allowed us to decouple $U_{2,2}$ from the rest of the energy.

● APPENDIX: CANONICAL ANALYSIS

Let $\vec{A}$ and $\vec{B}$ be two random vectors with values in R^p and R^q, respectively. Let $\mathbf{A} = [a^1, \ldots, a^N]'$ and $\mathbf{B} = [b^1, \ldots, b^N]'$ be the $N{\times}p$ and $N{\times}q$ matrices composed of N samples of these random vectors arranged in rows. We also write $\mathbf{A} = [\underline{a}_1, \ldots, \underline{a}_p]$, where $\underline{a}_i$ is the ith column of $\mathbf{A}$ and represents an N-sample of the ith coordinate A_i of the random vector $\vec{A}$ (the same type of notation is used for $\mathbf{B}$).

The canonical analysis seeks the linear relationships between $\vec{A}$ and $\vec{B}$. The first step of this analysis assumes that there exist $\alpha_1 \in \mathrm{R}^p$ and $\beta_2 \in \mathrm{R}^q$ such that the following random variables have maximum correlation:

$$\tilde{A}_1 = \vec{A}'\alpha_1 \quad \text{and} \quad \tilde{B}_1 = \vec{B}'\beta_1. \tag{12.25}$$

This is interpreted as a measure of the mutual information between $\vec{A}$ and $\vec{B}$. We must then estimate α_1 and β_1. To do this, we note that for all α and β, the vectors $\underline{a} = \mathbf{A}\alpha$ and $\underline{b} = \mathbf{B}\beta$ are N-samples of the random variables $\vec{A}'\alpha$ and $\vec{B}'\beta$. The estimation of α_1 and β_1 then consists in maximizing the following empirical correlation:

$$(\hat{\alpha}_1, \hat{\beta}_1) = \arg\max_{\alpha,\beta} \widehat{\mathrm{corr}}(\mathbf{A}\alpha, \mathbf{B}\beta). \tag{12.26}$$

The second step assumes the existence of two random variables

$$\tilde{A}_2 = \vec{A}'\alpha_2 \quad \text{and} \quad \tilde{B}_2 = \vec{B}'\beta_2$$

with maximum correlation under the following constraint:

$$\mathrm{corr}(\tilde{A}_1, \tilde{A}_2) = 0 \quad \text{and} \quad \mathrm{corr}(\tilde{B}_1, \tilde{B}_2) = 0.$$

The analysis continues in this manner until the step $n_c = \min(p, q)$.

There are two approaches for implementing this analysis. The first looks for the solutions in $\mathsf{R}^p \times \mathsf{R}^q$ as above. The second looks for the solutions in $\mathsf{R}^N \times \mathsf{R}^N$ by replacing (12.26) with

$$\left(\hat{\underline{a}}_1, \hat{\underline{b}}_1\right) = \arg\max_{E_A \times E_B} \widehat{\mathrm{corr}}(\underline{a}, \underline{b}),$$

where E_A and E_B are the subspaces of R^N generated by the samples of $\vec{A}$ and $\vec{B}$, respectively:

$$E_A = \{\underline{a} \mid \underline{a} = \mathbf{A}\alpha, \ \alpha \in \mathsf{R}^p\},$$
$$E_B = \{\underline{b} \mid \underline{b} = \mathbf{B}\beta, \ \beta \in \mathsf{R}^q\}.$$

Solutions in $\mathsf{R}^N \times \mathsf{R}^N$. Assuming the two samples $\{a^j\}$ and $\{b^j\}$ to be centered, the $\underline{a}$ and $\underline{b}$ are also centered, and in this case the empirical correlation is identified with the cosine: $\widehat{\mathrm{corr}}(\underline{a}, \underline{b}) = \cos(\underline{a}, \underline{b})$. The solution is then geometric. The first stage of the analysis seeks two vectors $\hat{\underline{a}}_1 \in E_A$ and $\hat{\underline{b}}_1 \in E_B$ with the smallest possible angle. The second stage seeks the $\hat{\underline{a}}_2 \in E_A$ and $\hat{\underline{b}}_2 \in E_B$ with the smallest possible angle with the constraint $\hat{\underline{a}}_1 \perp \hat{\underline{a}}_2$ and $\hat{\underline{b}}_1 \perp \hat{\underline{b}}_2$, and so on until the step $n_c = \min(p, q)$. The variables $(\hat{\underline{a}}_i, \hat{\underline{b}}_i)$ are called canonical variables. Let us now give details of these solutions. We write P_A and P_B for the orthogonal projection operators onto E_A and E_B:

$$\mathrm{P}_A = \mathbf{A}(\mathbf{A}'\mathbf{A})^{-1}\mathbf{A}', \quad \mathrm{P}_B = \mathbf{B}(\mathbf{B}'\mathbf{B})^{-1}\mathbf{B}'.$$

The solutions in $\mathsf{R}^N \times \mathsf{R}^N$ are characterized in this way.

Proposition 12.1. *The matrices* $\mathrm{P}_A\mathrm{P}_B$ *and* $\mathrm{P}_B\mathrm{P}_A$ *have the same eigenvalues. The canonical variables* $\hat{\underline{a}}_i$ *and* $\hat{\underline{b}}_i$ *are the eigenvectors of* $\mathrm{P}_A\mathrm{P}_B$ *and* $\mathrm{P}_B\mathrm{P}_A$*, respectively, associated with the same eigenvalue* λ_i:

$$\mathrm{P}_A\mathrm{P}_B \ \hat{\underline{a}}_i = \lambda_i \ \hat{\underline{a}}_i, \quad \mathrm{P}_B\mathrm{P}_A \ \hat{\underline{b}}_i = \lambda_i \ \hat{\underline{b}}_i.$$

They satisfy $\hat{\underline{a}}'_i \, \hat{\underline{a}}_j = 0$ and $\hat{\underline{b}}'_i \, \hat{\underline{b}}_j = 0$ for $i \neq j$, and $\hat{\underline{a}}'_i \, \hat{\underline{b}}_j = 0$ for $i \neq j$. We also have $\lambda_i = \cos^2\left(\hat{\underline{a}}_i, \hat{\underline{b}}_i\right)$.

Proof:
The vector of E_A nearest to $\hat{\underline{b}}_1$ is the orthogonal projection of $\hat{\underline{b}}_1$ on E_A, and $P_A\hat{\underline{b}}_1$ is therefore collinear with $\hat{\underline{a}}_1$, i.e., $P_A\hat{\underline{b}}_1 \propto \hat{\underline{a}}_1$. Similarly, $P_B\hat{\underline{a}}_1 \propto \hat{\underline{b}}_1$ and therefore $P_B P_A \hat{\underline{b}}_1 = \lambda_1 \hat{\underline{b}}_1$. Similarly, we immediately have $P_A P_B \hat{\underline{a}}_1 = \lambda_1 \hat{\underline{a}}_1$.

The matrix $P_B P_A$ can be diagonalized. Because the eigenvectors of $P_B P_A$ belong to E_B, we simply need to show that the restriction of $P_B P_A$ to E_B is symmetric. Because of the symmetry of P_A and P_B, we have, for all b and b^* belonging to E_B, $\langle b, P_B P_A b^* \rangle = \langle P_B b, P_A b^* \rangle = \langle b, P_A b^* \rangle = \langle P_A b, b^* \rangle = \langle P_A b, P_B b^* \rangle = \langle P_B P_A b, b^* \rangle$. This proves that $P_B P_A$ is symmetric. The eigenvectors of $P_B P_A$ are orthogonal. The λ_i are positive because P_A and P_B are positive semidefinite.

The value λ_1 is the square of the cosine of the angle between $\hat{\underline{a}}_1$ and $\hat{\underline{b}}_1$:

$$\lambda_1 = \frac{\|P_B P_A \hat{\underline{b}}_1\|}{\|\hat{\underline{b}}_1\|} = \frac{\|P_B P_A \hat{\underline{b}}_1\|}{\|P_A \hat{\underline{b}}_1\|} \frac{\|P_A \hat{\underline{b}}_1\|}{\|\hat{\underline{b}}_1\|} = \cos^2\left(\hat{\underline{a}}_1, \hat{\underline{b}}_1\right).$$

$\square$

Justification of (12.23). Equation (12.23) quantifies the degree of the linear relationship between the deformations of the characteristic curves of the GC. For the case of two centered and reduced random variables X and Y, this relationship is given by the following correlation: $\operatorname{corr}(X, Y) = \arg\min_\rho \mathbf{E}\left[(Y - \rho X)^2\right]$. As a result, when $\lambda_1 \approx 1$, assuming $\widehat{\operatorname{corr}}(\hat{\underline{a}}_1, \hat{\underline{b}}_1) \approx \operatorname{corr}(\vec{A}'\hat{\alpha}_1, \vec{B}'\hat{\beta}_1)$, we have the approximation in the sense of the quadratic error $\vec{A}'\hat{\alpha}_1 \approx \sqrt{\lambda_1}\vec{B}'\hat{\beta}_1$ or

$$\hat{\alpha}'_1 \vec{A} - \sqrt{\lambda_1}\hat{\beta}'_1 \vec{B} \approx 0. \tag{12.27}$$

As an example, let us take the curves h and ℓ. From a PCA of their samples, $\{h^j\}$ and $\{\ell^j\}$, we obtain the vectors that break them down into the natural modes (12.18):

$$\{h^j\} \longmapsto \{a^j\} \quad \text{and} \quad \{\ell^j\} \longmapsto \{b^j\}$$

(where we have changed the notation of the decomposition vectors for the expression (12.18), which were written as $\{\beta^j\}$ and $\{\gamma^j\}$). These decomposition vectors $\{a^j\}$ and $\{b^j\}$ are seen as N-samples of the random vectors $\vec{A}$ and $\vec{B}$. A relationship between the deformations of h and ℓ is then given by the relationship between the random decomposition vectors $\vec{A}$ and $\vec{B}$, shown in equation (12.27). If $\vec{A}$ and $\vec{B}$ are Gaussian, a prior energy on the pair $(\vec{A}, \vec{B})$ is

$$\left\|\hat{\alpha}'_1 \vec{A} - \sqrt{\lambda_1}\hat{\beta}'_1 \vec{A}\right\|^2.$$

By applying the same procedure to the deformable curves a and h and then a and ℓ, we finally obtain (12.23). As an illustration, in our example we calculated $\sqrt{\lambda_1} = 0.64$, 0.88, and 0.93 for the pairs (a, ℓ), (a, h), and (h, ℓ), respectively.

13
Matching

In this chapter we discuss the problem of recognizing the two-dimensional outline ℵ of an object in an image (or in several images), based on an approximate knowledge of its shape. This problem of pattern recognition is complicated, and we could justify discussing it at great length. An excellent introduction to pattern recognition is given in Younes [189].

The shape of the outline (or silhouette) is ideally described by a planar, closed curved with parameter C called the "template":

$$t \in [0,1] \longmapsto C(t) = (C_1(t), C_2(t)), \text{ with } C(0) = C(1). \qquad (13.1)$$

The object present in the image $I = (I(s), s \in S)$ that we wish to find is not necessarily described in the form of a contour. We only have features extracted from the image that are more or less suggestive of the presence of a contour (see Section 9.1.1). The object is thus *hidden* in the image. We write $\tilde{S}$ to denote the continuous domain of which S is the discretization, for example $\tilde{S} = [0,1]^2$. The most immediate feature is the gradient

$$\tilde{I}(s) = \|\nabla I(s)\|^2, \quad \forall s \in \tilde{S}, \qquad (13.2)$$

where ∇ is the continuous gradient operator. In practice, this means that the discrete field $\{\nabla I(s),\ s \in S\}$ is interpolated on $\tilde{S}$. Because the information obtained from the gradient contains a high level of error concerning the presence of an edge, it is better to perform a more sophisticated preprocessing on I to extract the significant edges from it. Let B be the resulting binary image of edges. For reasons related to the matching method described below, we must transform B into an image in gray levels by convolution, for example using the non-normalized Gaussian blur function (10.21) as follows:

$$\tilde{I}(s) \;=\; \int_{R^2} \exp - \left(\frac{\|u\|^2}{2\alpha^2} \right) B(s-u) \; du, \quad \forall \; s \in \tilde{S}. \qquad (13.3)$$

Here, this type of field is called a potential field.

In general, the position, orientation, and scale of the hidden object are unknown. Furthermore, its shape can be deformed with respect to the shape of the template. Matching the template and the hidden object therefore involves estimating the rigid transformations Θ (similarities in the plane) and the elastic transformations ξ (deformations) which, when applied to the template, make it match the features of the image $\tilde{I}$ associated with the hidden object. The template transformed in this way is an estimation of the outline of the object.

Rigid transformations are related to the concept of invariance, which is fundamental to pattern recognition. Note that the term "recognition" is not strictly accurate in this context; it is preferable to talk about "matching". After the matching procedure, there is a "distance" between the initial template and the transformed template. This distance can be used to decide whether or not the estimated outline is the outline of the object to be recognized [188]. A related question, which goes beyond the scope of this discussion, concerns the relationship between the object and its outline. In the case of retinal vision, the outline is defined by the rays leaving the center of vision that are tangential to the surface of the object. The outline can have a very different appearance depending on the angle of sight [147]. This calls the theory of aspect graphs into play.

Section 13.1 introduces rigid transformations and elastic deformations separately. The elastic deformations described have a small amplitude. Radiographic examples are given as illustrations. These examples do not show the simplest type of matching; a radiographic projection operator is involved in the imaging process. Section 13.2 describes an elastic deformation model that applies to any $\tilde{S}$ in a continuous manner and preserves the overall structure of the template.

13.1   Template and Hidden Outline

13.1.1   Rigid Transformations

To introduce this topic, we discuss a real example from the field of radiographic nondestructive testing. We have several images at our disposal.

In one image, the two-dimensional outline $\aleph$ of the object is theoretically the external outline of the projection of the object, i.e., the outline delin-

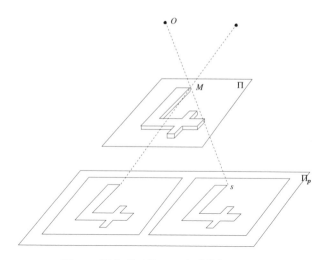

Figure 13.1. Outline and visible contour.

eating the projection of its background. It is associated with the visible contour of the three-dimensional object. Let there be a source O and any point $s \in \aleph$ on the outline (see Figure 13.1). Let us assume that the ray Os is tangential to the surface of the object at a single point M. When s runs along $\aleph$, the point M then defines the visible contour of the object. The visible contour varies from one source to another. Let us take the situation in which the sources are very close to each other and we therefore assume there to be a unique template giving an approximate representation of the visible contour independently of the source. This is the case shown, for example, in Figure 13.1, which shows L = 2 points of view. This is also the case in Figure 13.2(a), which shows L = 3 x-ray images of an object made of lead representing the number 4, and the associated template projected into the plane of the films.

Our example deals with the problem of realigning radiographic images. Let us consider Figures 13.1 and 13.2. In this situation, the position of the films in their common plane Π_p and the shape C of the template are not known with any precision. Realignment consists in determining the position of the films and the shape of the template so that the projection of C matches the projection of the object. In other words, after placing C in its plane Π, we must estimate the scaling parameter ς and the realignment parameters (ρ, τ) that achieve this matching. The film realignment parameters are $\{(\rho_\ell, \tau_\ell), \ell = 1, \ldots, L\}$, where ρ_ℓ denotes a rotation and $\{\tau_\ell = (\tau_{\ell,1}, \tau_{\ell,2})\}$ denotes a translation. These transformations are defined in the plane of the films. Note that for L = 1, the situation is exactly the same as classic matching, which means that all the parameters $\Theta = (\rho, \tau, \varsigma)$ could relate to the template C only.

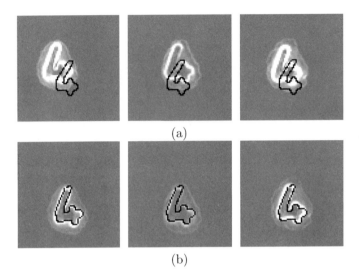

(a)

(b)

Figure 13.2. Rigid realignment of three images using a template: (a) Images of blurred edges $\tilde{I}_\ell$ and projection of the template. (b) Final realignment.

The parameter estimation is Bayesian. We write the following: P_ℓ is the projection corresponding to the ℓth source, $\tilde{I}_\ell^{(\rho,\tau)}$ is the image $\tilde{I}_\ell$ realigned according to (ρ,τ), and C^ς is the template C adjusted to the scale ς. The prior distribution on Θ is chosen to be uniform, and the posterior energy is therefore equal to the fidelity energy. The energy is written as follows:

$$U\left(\rho,\tau,\varsigma \mid \{\tilde{I}_\ell\}\right) \;=\; -\sum_{\ell=1}^{L}\int_0^1 \tilde{I}_\ell^{(\rho,\tau)}\left(P_\ell C^\varsigma(t)\right) dt. \qquad (13.4)$$

It has its minimum value when the projections $\{P_\ell C^\varsigma\}$ pass through high-intensity sites in the chosen potential fields $\tilde{I}_\ell$. The realignment shown in Figure 13.2 is the result of minimizing U using a relaxation algorithm operating alternately on (ρ,τ) and ς (Section 7.4.2).

13.1.2 Spline Model of a Template

Let us now imagine that the parameter Θ is known and that the matching of the template with the image amounts to an elastic deformation of the image. In the above example, once Θ has been estimated, this would require us to allow deformations of the template "4" to perfect the matching.

The deformation of a planar curve C in a potential field $\tilde{I}$ has been widely studied. Classically, the matching of C and $\tilde{I}$ is quantified by an energy using the potential field (13.2) and a convoluted image (13.3) [120]:

$$U(C \mid \tilde{I}) = U_1(C) + U_2(\tilde{I} \mid C), \tag{13.5}$$

$$U_1(C) = \beta_1 \int_0^1 \|D^1 C(t)\|^2 dt + \beta_2 \int_0^1 \|D^2 C(t)\|^2 dt,$$

$$U_2(\tilde{I} \mid C) = - \int_0^1 \tilde{I}(C(t)) dt,$$

applying the closure constraint (13.1). The parameters β_1 and β_2 are the elasticity and stiffness parameters. The energy U_1 belongs to the class of spline energies discussed in Chapter 2 (equation (2.8) page 26), where we looked at minimizing U using the Euler–Lagrange equation (Section 2.2.2).

Let us now see how to minimize this energy using the finite elements method. This method was introduced in Section 2.5 for the quadratic energy (2.29). In the present situation, (13.5) has the following form for all curves C and K belonging to Sobolev space $\mathrm{H}^2([0, 1])$:

$$U(C \mid \tilde{I}) = \frac{1}{2} A(C, C) - F(C), \tag{13.6}$$

$$A(C, K) = \beta_1 \int_0^1 \langle D^1 C(t), D^1 K(t) \rangle dt + \beta_2 \int_0^1 \langle D^2 C(t), D^2 K(t) \rangle dt,$$

$$F(C) = \int_0^1 \tilde{I}(C(t)) dt.$$

However, if A is still a positive symmetric continuous bilinear form, then F is not a linear form, in contrast to Section 2.4. Because of this, the minimum of U is not guaranteed to be unique. Here, the finite elements constitute a partition of the interval $[0, 1]$ into n intervals $\{e_j\}$ of length l: $[0, 1] = \bigcup_i e_j$ with $e_j = [jl, (j+1)l]$. The minimization is restricted to the space $\mathcal{E}$ consisting of piecewise polynomial curves of class $\mathcal{C}^1$. This space has a basis of functions $\{\phi_i\}$ with local support on the $\{e_i\}$ such that for all $C \in \mathcal{E}$, we have the breakdown $C = \sum_{i=1}^n c^i \phi_i$, where $c^i = (c_1^i, c_2^i)$. As a result, the functional expression U can be expressed as a function of $c = (c^1, \ldots, c^n)$. A curve $C^* \in \mathcal{E}$ is therefore a minimum of $U(C \mid \tilde{I})$ if c^* is a minimum of $U(c \mid \tilde{I})$. A necessary condition for a minimum is for the gradient of U to vanish: $\nabla_c U(c^* \mid \tilde{I}) = 0$, i.e.,

$$\frac{\partial}{\partial c_k^i} U(c^* \mid \tilde{I}) = 0, \quad \forall\, i = 1, \ldots, n, \quad k = 1, 2.$$

Taking the equation (13.6) into account, the gradient at c^i leads to the variational equation for all $i = 1, \ldots, n$:

$$\nabla_{c^i} U(c \mid \tilde{I}) = \beta_1 \sum_j \int_0^1 c^j \langle D^1 \phi_i(t), D^1 \phi_j(t) \rangle dt$$

$$+ \beta_2 \sum_j \int_0^1 c^j \langle D^2 \phi_i(t), D^2 \phi_j(t) \rangle dt$$

$$- \int_0^1 \nabla_{c^i} \tilde{I} \left(\sum_{j=1}^n c^j \phi_j(t) \right) dt$$

$$= \sum_j c^j A(\phi_i, \phi_j) - F^i(c),$$

$$\nabla_c U(c \mid \tilde{I}) \doteq A\, c - F(c).$$

The last equation is the formal equivalent of the matrix expression (2.34). Because $F(c)$ is not linear, we do not have a linear system as we had in (2.34). The solution is therefore the result of a gradient descent algorithm (Section 7.4.2). The $(k+1)$th iteration of the algorithm gives the following equation:

$$c(k + 1) = c(k) - \epsilon_k \left(A\, c(k) - F(c(k)) \right). \tag{13.7}$$

This algorithm is initialized with a specific position of the template $C^{(0)}$. The sequence of deformed templates obtained in this way has given the name *active contour* to this technique [120, 54]. We must emphasize that this method is very sensitive to the initial position of the template and to the "false edges" located between the initial position and the hidden outline. This method should therefore be recommended when we are sure that the initialization is in the neighborhood of the outline. Now we can understand why the blurring is introduced (13.3) on the image of the edges; without it, outside of the edges the dynamic (13.7) could not evolve toward the outline, because the potential field would remain constant [54].

Another difficulty is the weighting of the terms U_1 and U_2 composing the energy. We note, however, that it can be advantageous to replace a nonparametric spline with a parametric spline in which the number of knots represents a smoothing factor [127], or a regularization parameter (see the note in Section 3.1.1). Thus, the finite-elements representation with a small number of elements is partially redundant with the energy U_1. The idea then is not to take U_1 into account and to represent C with the following parametric spline:

$$C(t) = \sum_{i=1}^N c^i B_i(t),$$

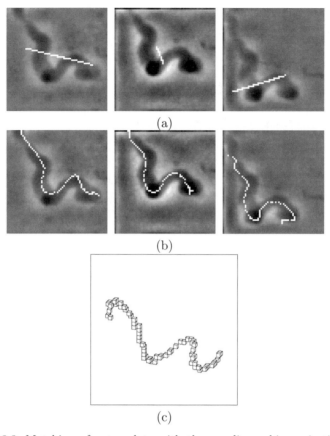

(a)

(b)

(c)

Figure 13.3. Matching of a template with three radiographic projections: (a) Images of blurred edges $\{\tilde{I}_\ell\}$ and projection of the rigid template. (b) Projections of the spline template after matching. (c) 3-D representation of the deformed template after matching.

where $B_i(t)$ denotes a B-spline (Section 3.1). The energy becomes

$$U\left(c \mid \tilde{I}\right) = -\int_0^1 \tilde{I}\left(\sum_{i=1}^N c^i B_i(t)\right) dt,$$

and we therefore have $U\left(c \mid \tilde{I}\right) = U_2\left(\tilde{I} \mid c\right)$.

Figure 13.3 illustrates this approach. The radiographic protocol is the same as in the example of Figure 13.2, except that the images are realigned in this case. A curve must be positioned in space (rather than in the plane) so that its radiographic projection matches three blurred views of a three-dimensional filamentary object [43]. Matching takes place in two stages. First, the initial template $C^{(0)}$ is a segment of a straight line that is matched according to a rigid transformation. Next, starting from this matching, the segment is represented by a spline, which after its energy has been minimized, refines the matching.

13.2   Elastic Deformations

13.2.1   Continuous Random Fields

The following approach provides an alternative to the one described above. We note that the template spline model contains no overall structure. Although the user defines the initial template $C^{(0)}$ based on an overall structure (the figure "4" in the above example), as the deformed templates $\{C^{(k)}, \ k = 1, 2, \dots\}$ evolve, nothing guarantees that the initial overall structure will be preserved, except when $C^{(0)}$ is very close to the hidden outline. We should point out that the term "deformable template model" would not quite be appropriate in this case, because the deformations that are solely the result of a minimization dynamic are not modeled. We then face the problem of defining a real deformation model whose simulation would give deformed templates that preserve the overall structure of the initial template. This problem has already been encountered (Chapter 4 and Chapter 12), but based on a sample of observable deformations. Here, we have no such sample.

Let us use the approach suggested by Amit, Grenander, and Piccioni [3]. The template C is included in the *continuous* domain $\tilde{S} = [0, 1]^2$. We can imagine that C is a drawing on the sheet of paper $\tilde{S}$. Instead of deforming the template by defining deformations on C only, we define a field $\vec{\mathcal{G}}$ of deformations that is continuous on the entire domain $\tilde{S}$, and that does not *globally* change the edges of $\tilde{S}$. The field $\vec{\mathcal{G}} = (G_1, G_2)$ is a bivariate vector field, indexed by $\tilde{S}$ for which the components G_1 and G_2 are the fields of the horizontal and vertical components of the deformations respectively. Any site $s = (s_1, s_2) \in \tilde{S}$ therefore undergoes a displacement $\vec{\mathcal{G}}(s)$:

$$(s_1, s_2) \longmapsto (s_1, s_2) + (G_1(s_1, s_2), G_2(s_1, s_2)).$$

The conditions for the nondisplacement of the edge of $\tilde{S}$ are as follows:

$$G_1(0, s_2) = G_1(1, s_2) = 0, \quad \frac{\partial G_1}{\partial s_2}(s_1, 0) = \frac{\partial G_1}{\partial s_2}(s_1, 1) = 0,$$

$$G_2(s_1, 0) = G_2(s_1, 1) = 0, \quad \frac{\partial G_2}{\partial s_1}(0, s_2) = \frac{\partial G_2}{\partial s_1}(1, s_2) = 0. \quad (13.8)$$

We can imagine that $\tilde{S}$ is a planar elastic sheet that deforms under the action of $\vec{\mathcal{G}}$; $\vec{\mathcal{G}}(\tilde{S})$ remains a planar sheet with the same edges as $\tilde{S}$ (Figure 13.4). This produces a deformation of the drawing $C = (C(t), t \in [0, 1])$ that follows the deformation of the sheet. For all $t \in [0, 1]$, we have $C(t) \in \tilde{S}$

and therefore

$$C(t) = (C_1(t), C_2(t)) \longmapsto C(t) + (G_1(C(t)), G_2(C(t))).$$

In more formal terms, $\vec{\mathcal{G}}$ is seen as a continuous field of random deformations.

● GAUSSIAN MODELS

Random fields were used a great deal in the previous chapters, but in a discrete Markov context. Now we need to define the stochastic processes when the random fields are indexed on a continuous domain such as $\tilde{S}$. Although we will choose a Gaussian field model, in Section 13.2.2 we will present second-order continuous processes. A scalar field $\{G(s), s \in \tilde{S}\}$ with *zero* mean is of second order if it has a covariance function $V(s, u)$ on $\tilde{S} \times \tilde{S}$:

$$V(s, u) = \mathrm{Cov}(G(s), G(u)) = \mathrm{E}[G(s)G(u)].$$

If, furthermore, for any set $(s^1, \dots, s^n)$ of n points in $\tilde{S}$ the random vector $(G(s^1), \dots, G(s^n))$ is Gaussian for all n, then the field is said to be Gaussian. The probability distribution of this vector is proportional to

$$\exp - \left[\frac{1}{2} \sum_{i,j=1}^{n} g(s^i) \, \Gamma_{ij} \, g(s^j) \right],$$

where Γ is the inverse of the $n \times n$ covariance matrix $(V(s^i, s^j))$. The matrix $V(s, u)$ is positive semidefinite in the sense that

$$\sum_{i,j=1}^{n} z_i V(s^i, s^j) z_j \geq 0, \tag{13.9}$$

for all $\{z_i\}$ and all $\{s^i\}$. By setting $X = \sum_{i=1}^{n} z_i G(s^i)$, we obtain $0 \leq \mathrm{Var}(X) = \sum_{i,j=1}^{n} z_i V(s^i, s^j) z_j$. The process is stationary when $V(s, t)$ is dependent on $s - t$ only.

For our application, we have a bivariate vectorial process $\{\vec{\mathcal{G}}(s), s \in \tilde{S}\}$ that has zero mean and is Gaussian. Its distribution is defined by the following covariance function

$$\mathrm{E}\left[\vec{\mathcal{G}}(s) \, {}^t\vec{\mathcal{G}}(u)\right] = \begin{pmatrix} \mathrm{Cov}(G_1(s), G_1(u)) & \mathrm{Cov}(G_1(s), G_2(u)) \\ \mathrm{Cov}(G_2(s), G_1(u)) & \mathrm{Cov}(G_2(s), G_2(u)) \end{pmatrix}.$$

To model deformations, we assume $\mathrm{Cov}(G_1(s), G_2(u)) = 0$, which means that the scalar fields G_1 and G_2 are independent. For any set $(s^1, \dots, s^n)$ of n points in $\tilde{S}$, $(\vec{\mathcal{G}}(s^1), \dots, \vec{\mathcal{G}}(s^n))$ is Gaussian, and its density is proportional to

$$\exp - \left[\frac{1}{2} \sum_{i,j=1}^{n} g_1\left(s^i\right) \, \Gamma_{ij}^{(1)} \, g_1\left(s^j\right) \right] \times \exp - \left[\frac{1}{2} \sum_{i,j=1}^{n} g_2\left(s^i\right) \, \Gamma_{ij}^{(2)} \, g_2\left(s^j\right) \right],$$

where $\left(\Gamma_{ij}^{(k)}\right) = \left(\mathrm{Cov}(G_k(s_i), G_k(s_j))\right)^{-1}$ for $k = 1, 2$. Modeling $\vec{G}$ consists in breaking down the horizontal and vertical deformations G_1 and G_2 on a restricted basis of orthogonal trigonometric functions $\{\phi^{n,m}\}$.

Model 13.1. *The vector field of deformations $\vec{G} = (G_1, G_2)$ is represented by the Gaussian model*

$$G_i(s) \;=\; \sum_{n,m=1}^{N} \frac{1}{\lambda_{n,m}} \phi_i^{n,m}(s) \, \xi_i^{n,m}, \quad i = 1, 2 \qquad (13.10)$$

$$\phi_1^{n,m}(s_1, s_2) \;=\; 2 \, \sin(\pi n s_1) \, \cos(\pi m s_2),$$
$$\phi_2^{n,m}(s_1, s_2) \;=\; 2 \, \sin(\pi n s_2) \, \cos(\pi m s_1),$$
$$\lambda_{n,m} \;=\; -\pi^2 \left(n^2 + m^2\right),$$

where the $\{\xi_i^{n,m}\}$ are independent Gaussian random variables $LG(0, \sigma^2)$.

Equation (13.10) satisfies the boundary conditions (13.8). For example, for $s_2 = 0$, the points $(s_1, 0)$ located on the abscissa axis s_1 are forced to remain there and move along that axis: $G_2(s_1, 0) = 0$ and possibly $G_1(s_1, 0) \neq 0$, $\frac{\partial G_1}{\partial s_1}(s_1, 0) \neq 0$. On the other hand, we have $\frac{\partial G_1}{\partial s_2}(s_1, 0) = 0$, which explains why the deformation lines are perpendicular to the abscissa axis when they meet it (Figure 13.4). If we set

$$\vec{e}_1^{\,n,m}(s) \;=\; (\phi_1^{n,m}(s), \; 0),$$
$$\vec{e}_2^{\,n,m}(s) \;=\; (0, \; \phi_2^{n,m}(s)),$$

we obtain a more concise expression for $\vec{G}$:

$$\vec{G}(s) \;=\; \sum_{n,m=1}^{N} \frac{1}{\lambda_{n,m}} \left(\xi_1^{n,m} \, \vec{e}_1^{\,n,m} + \xi_2^{n,m} \, \vec{e}_2^{\,n,m}\right). \qquad (13.11)$$

Note that we had a similar model in Chapter 12. The deformations were expressed on a vector basis, and the coefficients were governed by a Gaussian distribution. The main difference between that and the present approach lies in the fact that the basis vectors were estimated on a set of deformation examples but without a template. Here, we have no such set of examples, and we must choose a prior covariance function for the deformation field.

The template (Figure 13.4(a)) is the template from Section 13.1.1. The deformations shown in Figures 13.4(b) through 13.4(e) were simulated with $N = 3$ and $\sigma = 1$. We can see what a wide variety of deformations has been obtained; they are diverse in scale, orientation, and curvature. When the parameters N and σ are "well" chosen, as they are here with $N = 3$ and $\sigma = 1$, the deformations respect the overall structure of the template. If N and/or σ is too large, however, the deformations become increasingly complicated, and we can no longer be sure of preserving the overall structure. This is illustrated in Figure 13.4(f) for a simulation with $N = 3$ and $\sigma = 1.5$. We can see in this figure that bends and intersections have appeared. For more on this topic, consult [172], in which the linear model (13.11) is generalized for better modeling of large deformations.

- BAYESIAN MATCHING

As in Section 13.1, matching is equivalent to estimating the parameters of the model. These parameters are $\xi = \{\xi_i^{n,m}; \ i = 1, 2; \ n, m = 1, \ldots, N\}$ for elastic deformations and $\Theta = (\rho, \tau, \varsigma)$ for rigid transformations (rotation, translation, homothetic transformation). We again use $C = (C(t), t \in [0, 1])$ to denote the template in its initial position, and $C^{\Theta, \xi}$ for the deformed template. For all $t \in [0, 1]$, we have $C(t) \in \tilde{S}$ and therefore

$$C^{\Theta, \xi}(t) \ = \ \mathrm{T}_\varsigma \circ \mathrm{T}_\rho \left(C(t) + \vec{G}(C(t)) \right) + \mathrm{T}_\tau,$$

where $(\mathrm{T}_\rho, \mathrm{T}_\tau, \mathrm{T}_\varsigma)$ denotes the rigid transformation of the parameter $\Theta = (\rho, \tau, \varsigma)$.

The prior distribution of ξ is Gaussian, and the prior distribution of Θ is uniform on an *appropriate* interval, which means that $C^{\Theta, \xi}$ remains in the domain $\tilde{S}$. The prior energy is

$$U_1(\Theta, \xi) \ = \ \sum_{n,m=1}^{N} \frac{1}{\sigma^2} \left[\left(\xi_1^{n,m} \right)^2 + \left(\xi_2^{n,m} \right)^2 \right].$$

The fidelity energy is similar to (13.4) in the sense that it is minimized when the curve $C^{\Theta, \xi}$ passes through edge points in the image I. As in Section 13.1, we have a binary image B of the edges of I and its blurred version $\tilde{I}$ given by (13.3). Furthermore, for all $s \in \tilde{S}$, we write $s^B \in S$ to denote the site nearest to s where there is an edge, and $\vec{\nabla} I(s^B)$ to denote the unit gradient at s^B. We write $\overset{\perp}{C}(t)$ to denote the unit normal vector to the curve. We then take the following energy [127, 117]:

$$U_2(I \mid \Theta, \xi) \ = \ \frac{1}{\ell} \int_0^1 \left[1 - \tilde{I}\left(C^{\Theta, \xi}(t) \right) \left\langle \overset{\perp}{C}{}^{\Theta, \xi}(t), \vec{\nabla} I\left(C^{\Theta, \xi}(t)^B \right) \right\rangle^2 \right] dt,$$

$$(13.12)$$

where ℓ is the length of the curve C. This energy favors deformed templates that pass close to the edges and have the same local orientation as the edges. In addition, it takes its values from $[0, 1]$, which facilitates the

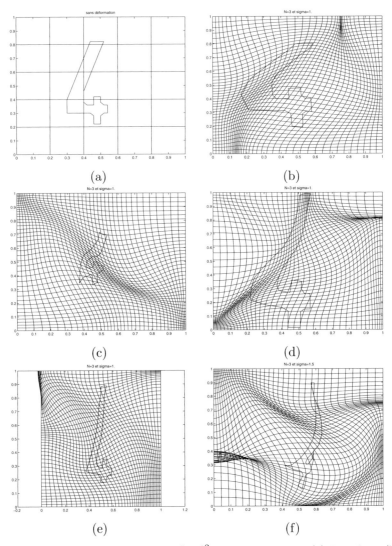

Figure 13.4. Elastic deformations on $[0,1]^2$ of the template 4. (a) Template. (b - e) $N = 3$, $\sigma = 1$. (f) $N = 3$, $\sigma = 1.5$.

empirical choice of the weighting parameter β between the energies U_1 and U_2. Matching is performed for the parameters $\left(\widehat{\Theta}, \widehat{\xi}\right)$ that minimize the posterior energy

$$U(\Theta, \xi \mid I) = \beta\, U_1(\Theta, \xi) + U_2(I \mid \Theta, \xi). \tag{13.13}$$

In [3], the minimization of a neighborhood energy that is limited to the parameters ξ is examined, based on a "stochastic gradient"-type algorithm.

This algorithm is implemented *hierarchically*, which significantly enhances its performance. The hierarchy is based on the degree N of the Gaussian model (13.10). The algorithm starts with $N = 1$ and continues by gradually increasing N. With $N = 1$ the deformations are small, and convergence is therefore faster than it would be if the algorithm started directly with a larger N. The N-level result is then used to initialize the level $N + 1$.

In [117], the Amit–Grenander–Piccioni approach is again tried, but with the introduction of the Θ parameters and working with an energy similar to that in equation (13.13). A dually hierarchical gradient descent minimization procedure is used. It operates on N as well as on the blurring parameter α from equation (13.3). The variation of this parameter is synonymous with multiresolution. For a large value of α, for example $\alpha = 8$, the field $\tilde{I}$ is very blurred and the details of the edges are lost. For this coarse resolution, the number of local minima is reduced and the convergence is improved. Useful information on this subject is found in the references [100, 118, 132].

We can illustrate this approach by matching the template "4" (Figure 13.4(a)) with the radiographic image in Figure 13.5(a). In this experiment we have concentrated on estimating ξ with a fixed position parameter Θ to analyze the performance of the deformation model. Figure 13.5(b) shows such an initial position. The fidelity energy is of the type given in equation (13.12) but without the scalar product, because in our application we have only the blurred image of the edges (Figure 13.5(a)): $U_2(I \mid \xi) = \frac{1}{M} \sum_i I(C^\xi(t_i))$ where $\{C^\xi(t_i)\}$ is the deformed version of the discretized initial template $\{C(t_i), i = 1, \ldots, M\}$ (Figure 13.5(b)). Matching was performed by a gradient descent algorithm (7.18): $\xi(n + 1) = \xi(n) - \epsilon_n \nabla U(\xi(n) \mid I)$, implemented coordinate by coordinate and in a simulated annealing version with fast cooling: $T_n = 0.9^n$ (Section 7.2). The gradient written as $2\beta\xi + \nabla(U_2(I \mid \xi))$ is calculated discretely: $\nabla(U_2(I \mid \xi)) \approx \frac{1}{h} [U_2(I \mid \xi + h) - U_2(I \mid \xi)]$. Figure 13.5(e) is the result of the matching, and Figure 13.5(f) is obtained from the resulting deformations. In this example, $N = 2$ and $[0, 1]^2$ is discretized on a grid with elements of side length 0.02. The algorithm converged in about fifty iterations.

A few comments on this approach: Because of the variety and amplitudes of the simulated deformations in Figure 13.4, we might think that matching is possible with large deformations. In fact, however, this method is effective for small deformations only. The example in Figure 13.5 is a special case where matching was achieved in spite of an initial position that was far from the hidden entity. This example does, however, show the specific nature of the overall deformation model: Although the slanted branch of the template "4" initially encloses a high-intensity zone in I,

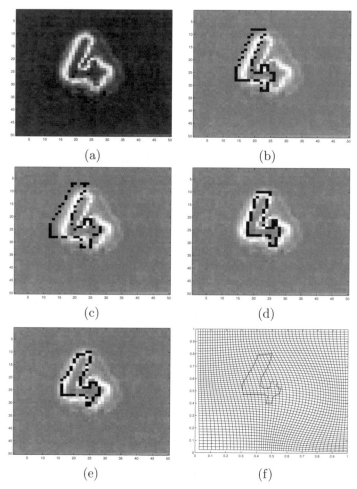

Figure 13.5. Elastic matching. (a) Image. (b) Initial position of template. (c, d) Intermediate positions. (e) Final position. (f) Deformations of the matching.

this branch was finally displaced and deformed. Finally, let us emphasize that this algorithm is difficult to adjust, particularly for the choice of ϵ_n.

13.2.2   Probabilistic Aspects

[1] Here we ask two questions: How can a continuous scalar process be expressed based on its covariance function? How should the covariance function be chosen? To answer these questions, we require an introduction

---

[1]This subsection can be omitted at the first reading.

to the second-order process [159], which will be presented in one dimension, i.e., for $s \in [0, 1]$.

● SPECTRAL REPRESENTATION

For a series of random variables $\{X_i, \; i = 1, 2, \ldots \}$, the quadratic mean (abbreviated as q.m.) convergence toward a random variable X is defined by

$$X = \lim_n (\text{q.m.}) X_n \quad \Longleftrightarrow \quad \lim_{n \to \infty} \mathbf{E}\left[(X_n - X)^2\right] = 0.$$

For a process X indexed by $[0, 1]$ and a deterministic function f, we define the following stochastic integral:

$$\int_0^1 f(t) \, dX(t) = \lim_n (\text{q.m.}) \sum_{i=1}^n f(t_i) \left[X(t_i) - X(t_{i-1})\right],$$

where $t_0, \ldots, t_n$ induces a partition of $[0, 1]$ such that $\max(t_i - t_{i-1}) \to 0$. Similarly, the derivative $D^1 X(t_0)$ of X at $t = t_0$ is defined, provided that it exists, by

$$D^1 X(t_0) = \lim_{h \to 0} (\text{q.m.}) \left[\frac{X(t_0 + h) - x(t_0)}{h}\right].$$

Having briefly set the above conditions, we now introduce the extended fundamental Karuhnen–Loève theorem, which gives the expression for a stochastic process.

Theorem 13.1. *Let there be a centered second-order process $\{G(s), s \in [0, 1]\}$ whose covariance function is as follows:*

$$V(s, t) = \int_0^1 \Phi(s, u) \Phi(t, u) \, d\mu(u), \tag{13.14}$$

where $\mu(u)$ is a positive measure on $[0, 1]$ with respect to which $\Phi(s, .)$ is square integrable. The process G then allows the following spectral representation:

$$G(s) = \int_0^1 \Phi(s, t) \, dZ(t), \tag{13.15}$$

where Z is a centered process with independent increments satisfying

$$\mathbf{E}[|dZ(t)|^2] = d\mu(t), \tag{13.16}$$
$$\text{Cov}(dZ(t), dZ(u)) = 0, \quad \forall t \neq u. \tag{13.17}$$

Reciprocally, if G allows the representation (13.15) with the properties (13.16) and (13.17), then its covariance function is written according to (13.14).

This theorem is expressed in exactly the same way for two dimensions [61], i.e., for stochastic processes indexed by $\tilde{S} = [0, 1]^2$.

Proof.
This topic is discussed in an excellent book by Priestley [151], among others. For the reciprocal case, we have

$$V(s, t) = \mathbf{E}[G(s)G(t)] = \iint \Phi(s, u)\Phi(t, u')\mathbf{E}[dZ(u)dZ(u')],$$

which, because of the noncorrelation of the $\{dZ(t)\}$, can be simplified to

$$V(s, t) = \int \Phi(s, u)\Phi(t, u)\mathbf{E}[|dZ(u)|^2] = \int \Phi(s, u)\Phi(t, u)\, d\mu(u).$$

$\square$

The property (13.17) means that the increments of Z are noncorrelated. A special choice for $Z(t)$ would be Brownian motion, i.e., a Gaussian process $LG(0, \sigma_Z^2\, t)$ such that

$$\mathbf{E}[|dZ(t)|^2] = \sigma_Z^2 dt. \tag{13.18}$$

Another special case is that of stationary G processes, for which $\Phi(s, t) = \exp(ist)$ are the Fourier components and $d\mu(t)$ is the associated spectrum.

• GREEN'S FUNCTION AND COVARIANCE FUNCTION

The relationship between these two functions [71, 122] allows us to choose the covariance function (13.14). The Karuhnen–Loève theorem left this choice unresolved. We will introduce this relationship in one dimension. It is based on the following mth-order linear differential operator

$$\mathsf{L} = D^m + a_{m-1}D^{m-1} + \cdots + a_1 D^1 + a_0, \tag{13.19}$$

where the a_i are deterministic functions. For all fixed $t \in [0, 1]$ we write $\phi(., t) = \{\phi(s, t), s \in [0, 1]\}$, the solution of the homogeneous differential equation

$$\mathsf{L}\phi(., t) = 0, \tag{13.20}$$

such that

$$\frac{\partial^j}{\partial s^j}\phi(s, t)\,|_{s=t} = 0, \quad j = 0, \ldots, m - 2,$$

$$\frac{\partial^{m-1}}{\partial s^{m-1}}\phi(s, t)\,|_{s=t} = 1.$$

We write $\mathcal{N} = \{g : D^j g(0) = 0, \ j = 0, \ldots, m - 1\}$.

Lemma 13.1. *For all $g \in \mathcal{N}$ and $f \in L^2$, we have $\mathsf{L}g = f \Leftrightarrow g(s) = \int_0^s \phi(s, t)f(t)dt$.*

We prefer to work with

$$\Phi(s,t) = \phi(s,t)\, \mathbf{1}_{s \geq t}.$$

This is called a Green's function. The equivalent of the above lemma for a Green's function is written as follows:

$$\mathsf{L}g = f \iff g(s) = \int_0^1 \Phi(s,t) f(t)\, dt.$$

One of the simplest physical examples of Green's function involves the equation for the speed g of a particle (of unit mass in a medium with drag a_0) subjected to a force f. It is given by the first-order differential equation $D^1 g + a_0 g = f$, where the initial speed is $g(0) = 0$. The solution, for all $s \in [0,1]$, is

$$g(s) = \int_0^s e^{-a_0(s-t)} f(t) dt,$$

which we rewrite as

$$g(s) = \int_0^1 \Phi(s,t) f(t) dt \quad \text{with} \quad \Phi(s,t) = e^{-a_0(s-t)} \mathbf{1}_{s>t}.$$

For all t, $\Phi(s,t)$ is interpreted as the response function (here, the speed) at the instant t. From the point of view of the theory of distributions, Φ satisfies the following equation:

$$\frac{d\Phi}{ds}(s,t) + a_0 \Phi(s,t) = \delta(s-t),$$

where δ denotes the Dirac function in t. The value g is interpreted as the superposition of the speeds resulting from the impulse forces.

Lemma 13.2. *The function $V_0(s,t)$ defined on $[0,1]^2$ by*

$$V_0(s,t) = \int_0^1 \Phi(s,u)\Phi(t,u) du, \tag{13.21}$$

is positive definite and symmetric.

The function V_0 therefore has the property (13.9) of a covariance function. Furthermore, V_0 is a specific expression of (13.14). In short, the choice of a differential equation lets us define a covariance function via a Green's function.

EXAMPLE.
This example involves the selection of a Green's function in two dimensions, chosen to model deformations. It comes from the partial differential equation known as the Poisson equation:

$$\Delta g(s_1, s_2) = f(s_1, s_2), \quad \text{with} \quad \Delta g = \frac{\partial^2 g}{\partial s_1^2} + \frac{\partial^2 g}{\partial s_2^2},$$

with the boundary conditions (13.8). The Laplacian Δ is the same Laplacian used in Section 10.2.1. The following characteristic functions and eigenvalues are the solutions of $\Delta\phi = \lambda\phi$:

$$\phi^{n,m}(s) = 2\sin(\pi n s_1)\cos(\pi m s_2), \tag{13.22}$$
$$\lambda_{n,m} = -\pi^2(n^2 + m^2), \quad n, m = 1, 2, \dots.$$

As in the previous example, for all $t = (t_1, t_2)$, Φ is a solution in the sense of distributions of

$$\Delta\Phi(s,t) = \delta(s_1 - t_1)\delta(s_2 - t_2) \doteq \delta(s - t).$$

By decomposing Φ with respect to the characteristic functions

$$\Phi(s,t) = \sum_{n,m} a_{n,m}(t)\phi^{n,m}(s),$$

we obtain $\Delta\Phi(s,t) = \sum_{n,m}\lambda_{n,m}a_{n,m}(t)\phi^{n,m}(s) = \delta(s - t)$. Similarly, if $\delta(s-t) = \sum_{n,m} b_{n,m}(t)\phi^{n,m}(s)$, then $b_{n,m}(t) = \int\int \delta(s-t)\phi^{n,m}(s)ds = \phi^{n,m}(t)$. Consequently, $\lambda_{n,m}a_{n,m}(t) = \phi^{n,m}(t)$, which gives the expression for the coefficients $\{a_{n,m}\}$ and from this the expression for Φ:

$$\Phi(s,t) = \sum_{n,m}\frac{1}{\lambda_{n,m}}\phi^{n,m}(s)\phi^{n,m}(t). \tag{13.23}$$

• STOCHASTIC DIFFERENTIAL EQUATIONS

Let us look again at the differential operator L in one dimension. Given a differentiable process Z, we consider the following stochastic differential equation:

$$\mathsf{L}\,G\,ds = dZ(s) \quad \text{with} \tag{13.24}$$
$$D^j G(0) = 0, \quad j = 0, \dots, m - 1.$$

A stochastic differential equation is equivalent to a stochastic integral. For the first-order case, we therefore have the following equivalence:

$$dG(s) = a(s)ds + dZ(s),$$
$$G(s) - G(0) = \int_0^s a(t)dt + \int_0^s dZ(t).$$

The relationship between the Green's function and the spectral representation, and therefore the covariance function, is given as follows [159].

Theorem 13.2. *The solution to the stochastic differential equation (13.24) is the following representation (13.15):*

$$G(s) = \int_0^1 \Phi(s,t)\,dZ(t), \tag{13.25}$$

where $\Phi(s,t) = \phi(s,t) \, \mathbf{1}_{s \geq t}$ is the Green's function associated with the solution to the deterministic differential equation (13.19):

$$\mathsf{L} \, \phi(.,t) = 0, \quad \text{such that}$$

$$\frac{\partial^j}{\partial s^j} \phi(s,t) \mid_{s=t} = 0, \quad j = 0, \dots, m-2,$$

$$\frac{\partial^{m-1}}{\partial s^{m-1}} \phi(s,t) \mid_{s=t} = 1,$$

Proof.
The proof uses the following lemma.

Lemma: If $F(s) = \int_0^s \psi(s,t) dZ(t)$, then $dF(s) = \left[\int_0^s \frac{\partial}{\partial s} \psi(s,t) dZ(t)\right] ds + \psi(s,s) \, dZ(s)$.

Let us check that (13.25) is a solution of (13.24), $\left(D^m + a_{m-1}D^{m-1} + \cdots + a_1 D^1 + a_0\right) G ds = dZ(s)$. The lemma, if applied to (13.25), gives

$$dG(s) = \left[\int_0^s \frac{\partial}{\partial s} \phi(s,t) \, dZ(t)\right] ds + \phi(s,s) dZ(s),$$

which is simplified by the assumption that $\phi(s,s) = 0$. We then obtain

$$D^1 G(s) = \int_0^s \frac{\partial}{\partial s} \phi(s,t) \, dZ(t).$$

By applying the lemma to $D^1 G$, and by continuing in this manner, we obtain

$$D^j G(s) = \int_0^s \frac{\partial^j}{\partial s^j} \phi(s,t) \, dZ(t), \quad \forall \, j \leq m-1. \tag{13.26}$$

Finally, by applying the lemma to the $(m-1)$th derivative, we obtain

$$d(D^{m-1}G)(s) = \left[\int_0^s \frac{\partial^m}{\partial s^m} \phi(s,t) dZ(t)\right] ds + \frac{\partial^{m-1}}{\partial s^{m-1}} \phi(s,s) dZ(s), \tag{13.27}$$

in which the $(m-1)$th partial derivative has the value 1 in accordance with the assumption. By considering $\mathsf{L} \, \phi(.,t) = 0$ in (13.27), we rewrite

$$\frac{\partial^m}{\partial s^m} \phi(s,t) = -\sum_{j=0}^{m-1} a_j \frac{\partial^j}{\partial s^j} \phi(s,t), \tag{13.28}$$

Finally, using (13.26), (13.27), and (13.28) again, we obtain

$$d(D^{m-1}G)(s) + \left[\sum_{j=0}^{m-1} D^j G(s)\right] ds = dZ(s).$$

This is equation (13.24). $\square$

• BACK TO MODELING

Seen in this light, the modeling of a stochastic process involves choosing
the linear differential operator L. For the application concerning deformations discussed in Section 13.2.1, and following [3], the operator L is the
Laplacian whose Green's function is (13.23). The field Z is a Brownian
field, an extension of Brownian motion (13.18). Formally, the covariance
matrix is then obtained according to (13.14) expressed in two dimensions:

$$V(s,t) = \int_0^1 \int_0^1 \Phi(s,u)\Phi(t,u)d\mu(u) = \sum_{n,m} \frac{1}{\lambda_{n,m}^2} \phi^{n,m}(s)\phi^{n,m}(t)\, \sigma_Z^2.$$

The spectral representation is

$$
\begin{aligned}
G(s) &= \int_0^1 \int_0^1 \Phi(s,t)\, dZ(t) \\
&= \int_0^1 \int_0^1 \left[\sum_{n,m} \frac{1}{\lambda_{n,m}} \phi^{n,m}(s)\phi^{n,m}(t)\right] dZ(t) \\
&= \sum_{n,m} \frac{1}{\lambda_{n,m}} \phi^{n,m}(s) \left[\int_0^1 \int_0^1 \phi^{n,m}(t)\, dZ(t)\right] \\
&\doteq \sum_{n,m} \frac{1}{\lambda_{n,m}} \phi^{n,m}(s)\, \xi^{n,m}, \quad\quad\quad\quad (13.29)
\end{aligned}
$$

where we have set $\xi^{n,m} = \int_0^1 \int_0^1 \phi^{n,m}(t)\, dZ(t)$. Because the $dZ(t)$ are
centered, we have $E[\xi^{n,m}] = 0$. We also show that the $\{\xi^{n,m}\}$ are uncorrelated by directly expanding the expectation $E[\xi^{n,m}\xi^{n',m'}]$ and taking
into account the orthogonality of the $\phi^{n,m}$ and the noncorrelation of the
$\{dZ(t)\}$, expressed in (13.17):

$$\mathrm{Cov}(\xi^{n,m}, \xi^{n',m'}) = \begin{cases} 0 & \text{if } (n,m) \neq (n',m'), \\ \sigma_Z^2 & \text{otherwise.} \end{cases}$$

Finally, the Gaussian nature of the Brownian field means that the $\{\xi^{n,m}\}$
are Gaussian random variables. This concludes the presentation of the
fundamentals of the model (13.10) on page 278.

Note that this model constitutes a prior distribution for matching. As in
the discrete case, other prior models are, of course, possible. For example,
in [118], the differential operator used comes from the theory of elasticity. It
is written as $L = a_2\Delta + a_1\nabla(\nabla.)$ and gives rise to a spectral representation
similar to that of the Laplacian. When applied in multiresolution, this
operator can be used to deal with large deformations [118, 50].

References

[1] Agresti A., *Categorical Data Analysis*, Wiley, 1990.

[2] Ahlberg J.H., Nilson E.N., Walsh J.L., *The Theory of Splines and Their Applications*, Academic Press, 1967.

[3] Amit Y, Grenander U., Piccioni M., "Structural image restoration through deformable template", *Journal of the American Statis. Ass.*, 86(414), 376-387, 1991.

[4] Ash J.M., *Studies in Harmonic Analysis*, American Mathematical Society, Providence RI., 1975.

[5] Aubert G., Kornprobst P., *Mathematical Problems in Image Processing: Partial Differential Equations and the Calculus of Variations*, Springer-Verlag, AMS series, 147, 2001.

[6] Azencott R., "Image analysis and Markov field", *Proc. of Int. Conf. on Ind. Appl. Math. SIAM*, (invited paper), 1987.

[7] Azencott R., "Simulated annealing", *Séminaire Bourbaki*, 40ème année, vol. 697, 1-15, 1988.

[8] Azencott R., Chalmond B., Julien Ph., "Bayesian 3-D path search and its application to focusing seismic data", *Lecture Notes in Statistics, "Stochastic Models, Statistical Methods and Algorithms in Image Analysis"*, 74, 46-74, Springer-Verlag, 1992.

[9] Azencott R., Graffigne C., Labourdette C., "Edge detection and segmentation of textured plane images", *Lecture Notes in Statistics, "Stochastic Models, Statistical Methods and Algorithms in Image Analysis"*, 74, 75-88, Springer-Verlag, 1992.

[10] Azencott R., (Ed.), *Simulated Annealing : Parallelization Techniques*, John Wiley, 1992.

[11] Azencott R., Chalmond B., Wang J.P., "Non-homogeneous transfer function estimation for film fusion using tricubic splines", Preprint 9336, *CMLA, Ecole Normale Supérieure, Cachan*, France, 1993.

[12] Azencott R., Chalmond B., Coldefy F., "Markov fusion of a pair of noisy images to detect intensity valleys", *International J. of Computer Vision*, 16, 135-145, 1995.

[13] Azencott R., Chalmond B., Wang J.P., "Transfer function estimation, film fusion and image restoration", *CVGIP : Graphical Models and Image Processing*, 58, 65-74, 1996.

[14] Azencott R., Wang J.P., Younes L., "Texture classification using windowed Fourier filters", *IEEE Trans. on Pattern Anal. Machine Intell.* , 19(2), 148-152, 1997.

[15] Baldi P. and Hornik K., "Neural network and principal component analysis : learning from examples without local minima", *Neural Networks*, 58, 53-58, 1989.

[16] Ballard D.H., Brown C.M., *Computer Vision*, Prentice-Hall, 1983.

[17] Barret H., Swindell W., *Radiological Imaging, The Theory of Image Formation, Detection, and Processing*, Academic Press, New York, 1981.

[18] Baum L., Petrie T., Soules G., Weis N., "A maximization thechnique occuring in the statistical analysis of probabilistic functions of Markov chains", *Ann. Math. Satist*, 41, 164-171, 1972.

[19] Bates R.H.I., McDonnel M.J., *Image Restoration and Reconstruction*, Clarendon, Oxford, 1986.

[20] Basseville M., "Distance measures for signal processing and pattern recognition", *Signal Processing*, 349-369, 1989.

[21] Besag J., "Spatial interaction and the statistical analysis of lattice systems", *Journal Royal Statis. Soc.*, B-36, 192-236, 1974.

[22] Besag J., "On the statistical analysis of dirty pictures", *Journal Royal Statis. Soc.*, B-148, 1988.

[23] Bickel P.J., Dosksum K.A., *Mathematical Statistics*, Holden Day, 1977.

[24] Bonin A., Chalmond B. Lavayssière B., "Monte Carlo simulation of industrial radiography images and experimental designs", *NDT & E*, 2002.

[25] Bookstein F.L., *The Measurement of Biological Shape and Shape Change*, Springer-Verlag, 1978.

[26] Bookstein F. L., "Principal warps : Thin-plate splines and the decomposition of deformations", *IEEE Trans. on Pattern Anal. Machine Intell.*, 11(6), 567-585, 1989.

[27] Boulard H., Kamp Y., "Autoassociative memory by multilayer perceptron and singular values decomposition", *Biological Cybernetics*, 59, 291-294, 1989.

[28] Burq C., *Degradation Models in Radiography and Image Restoration*, (in French), PhD thesis, University of Paris-Sud, Orsay, Mathematics, 1992.

[29] Butkov E., *Mathematical Physics*, Addison-Wesley, 1968.

[30] Carasso A.S., Sanderson J.G., Hyman J.M., "Digital removal of random media image degradations by solving the diffusion equation backwards in time", *Siam J. Numer. Anal*, 15, 344-367, 1978.

[31] do Carno M. P., *Differential Geometry of Curves and Surfaces*, Prentice Hall, 1976.

[32] Carroll R. J., Wu C. F. J., Ruppert D., "The effect of estimating weights in weighted least-sqares", *Journal of the Amer. Statist. Ass.*, 83, 1045-1054, 1988.

[33] Cea J., Glowinski R., "On optimization methods by relaxation", *Rev. Autom. Inform. Recher. Operat.* R-3, 5-32, 1973.

[34] Černy V., "Thermodynamical approach to the travelling salesman problem : an efficient simulation algorithm", *JOTA*, 45, 41-51, 1985.

[35] Ciarlet P. G., *The Finite Element Method for Elliptic Problems*, North-Holland, 1987.

[36] Chalmond B., "A Bayesian detection-estimation for a shift of the mean in an autoregressive process", *Proc. Comput. Statist. II*, Wien, Physica Verlag, 1982.

[37] Chalmond B., "Regression with autocorrelated residuals and spatial trend estimation", (in French), *Stat. et Anal. des Donn.*, 11(2), 1-25, 1986.

[38] Chalmond B., "Image restoration using an estimated Markov model", *Signal Processing*, 15, 115-129, 1988.

[39] Chalmond B., "Individual hip prosthesis design from CT images", *Pattern Recogn. Letters*, 8, 203-208, 1988.

[40] Chalmond B., "An iterative Gibbsian technique for reconstruction of m-ary images", *Pattern Recogn.*, 22, 747-761, 1989.

[41] Chalmond B., "PSF estimation for image deblurring", *CVGIP : Graph. Models and Image Processing*, 53(4), 364-372, 1991.

[42] Chalmond B., "Automatic fitting of custom femoral stem prostheses : Ten years of research and clinical experience", *Proc. of the 14th Conf. of the IEEE Engineering in Medicine and Biology Society*, 2066-56, Paris, 1992.

[43] Chalmond B., Coldefy F., Lavayssière B., "3D curve reconstruction from degraded projections", *Wavelets, Image and Curve Fitting*, In Laurent P.J., Le Méhauté A., Schumaker L.L. (Eds.), AKPeters, 113-119, 1994.

[44] Chalmond B., Coldefy F., Lavayssière B., "Tomographic reconstruction from non-calibrated noisy projections in non-destructive evaluation", *Inverse Problems*, 15, 399-411, 1999.

[45] Chalmond B., Girard S.C., "Non linear modeling of scattered multivariate data and its application to shape change", *IEEE Trans. on Pattern Anal. Machine Intell.*, 21(5), 422-432, 1999.

[46] Chalmond B., Graffigne G., Prenat M., Roux M., "Contextual performance prediction for low-level image analysis algorithm", *IEEE Trans. on Image Processing*, 10(7), 2001.

[47] Charbonnier P., *Image Reconstruction : Regularization with Discontinuities*, (in French), PhD thesis, University of Nice Sophia-Antipolis, 1994.

[48] Chellappa R., Jain A. (Eds.), *Markov Random Fields: Theory and Applications*, Academic Press, 1993.

[49] Chen H., "Estimation of a projection-pursuit type regression model", *The Annals of Statistics*, 19(1), 142-157, 1991.

[50] Christensen G., Rabbitt R., Miller M., "Deformable templates using large deformation kinematics", *IEEE Trans. on Image Processing*, 5, 1435-1447, 1996.

[51] Cinquin P., *Splines for Image Processing*, (in French), Thèse d'Etat, University of Grenoble, Mathematics, 1987.

[52] Cinquin P., Tricubic spline functions, an efficient tool for 3D image modeling, *Proc. of Int. Conf. on Ind. Appl. Math.*, Invited Paper, SIAM, 1987.

[53] Clearbout J.F., *Imaging the Earth's Interior*, Blackwell Scientific Publishing, Palo Alto, 1985.

[54] Cohen L.D., Cohen I., "Finite-element methods for active contour models and balloons for 2D and 3D images", *IEEE Trans. on Pattern Anal. Machine Intell.*, 15(11), 131-147, 1993.

[55] Cofdefy F., *Markovian Random Fields and Variational Methods for Low Level Image Processing*, (in French), PhD thesis, University of Paris-Sud, Orsay, Mathematics, 1993.

[56] Comon P., "Independent Component Analysis - a new concept?", *Signal Processing*, 36, 287-314, 1994.

[57] Cootes T., Taylor C., Cooper D.H., Graham J., "Active shape models, their training and application", *Computer Vision and Image Understanding*, 61(1), 38-59, 1995.

[58] Courant R., "Variational methods for the solution of problems of equilibrium and vibrations", *Bull. American Mathematical Society*, 49, 1-23, 1943.

[59] Courant R., Hilbert D., *Methods of Mathematical Physics*, vol. 1, John Wiley, 1989.

[60] Craven P., Wahba G., "Smoothing noisy data with spline functions", *Numer. Math.* 31, 377-403, 1971.

[61] Dacunha-Castelle D., Duflo M., *Probability and Statistics*, (in French), Vol. II, Masson, Paris, 1983.

[62] David C., Zucker S.W., "Potentials, valleys, and dynamic global coverings", *Int. Journal of Comp. Vision*, 5, 219-238, 1990.

[63] DeBoor C., "Bicubic spline interpolation", *J. Math. Phys.*, 41, 212-218, 1962.

[64] DeBoor C., "On calculation with B-splines", *Journal of Approximation Theory*, 6, 50-62, 1972.

[65] DeBoor C., *A practical Guide to Splines*, Springer-Verlag, 1978.

[66] Dempster A., Laird N., Rubin D., "Maximum likelihood from incomplete data via the EM algorithm", *Journal Royal Statis. Soc.*, B-39, 1-38, 1977.

[67] Derin H., Elliott H., "Modeling and segmentation of noisy and textured images using Gibbs random fields", *IEEE Trans. on Pattern Anal. Machine Intell.*, 9(1), 39-55, 1987.

[68] Diaconis P., Shashahani M., "On nonlinear functions of linear combinations", *SIAM Journal of Scientifical Statis. Comput.*, 5(1), 175-191, 1984.

[69] Dierckx P., *Curve and Surface Fitting with Splines*, Oxford Science Publications, 1993.

[70] Dinten J.M., *Tomography From a Limited Number of Projections*, (in French), PhD thesis, University of Paris-Sud, Orsay, Mathematics, 1990.

[71] Dolph C.L., Woodbury M.A., "On the relation between Green's function and covariances of certain stochastic processes", *Trans. Amer. Math. Soc.*, 72, 519-550, 1952.

[72] Doob J.L., *Stochastic Processes*, John Wiley, 1953.

[73] Draper N., Smith H., *Applied Regression Analysis*, John Wiley, New-York, 1966.

[74] Duflo M., *Stochastic Algorithms*, (in French), Springer Verlag, 1996.

[75] Eubank P., *Spline Smoothing and Nonparametric Regression*, Decker, 1990.

[76] Feller W., *An Introduction to Probability Theory and Its Applications*, vol. I, John Wiley, 1968.

[77] Figueiredo M., Leitao J.M. and Jain A.K., *Introduction to Bayesian Theory and Markov Random fFields for Image Analysis*, Springer Verlag, 2001.

[78] Fishman G.S., *Monte Carlo*, Springer-Verlag, 1995.

[79] Friedman J.H., "Exploratory projection pursuit", *Journal of the American Statis. Ass.*, 82(397), 249-266, 1987.

[80] Friedman J.H., Stuetzle W., "Projection pursuit regression", *Journal of the American Statis. Ass.*, 76(376), 817-823, 1981.

[81] Friedman J.H., "Multivariate adaptive regression splines", *The Annals of Statistics*, 19(1), 1-141, 1991.

[82] Gauch J.M., Pizer S.M., "Multiresolution analysis of ridges and valleys in grey-scale images", *IEEE Trans. on Pattern Anal. Machine Intell.*, 635-646, 1993.

[83] Geiger D., Girosi F., "Parallel and deterministic algorithms from MRF's : Surface reconstruction", *IEEE Trans. on Pattern Anal. Machine Intell.*, 13(5), 401-412, 1991.

[84] Geman S., Geman D., "Stochastic relaxation, Gibbs distribution, and the Bayesian restoration of images", *IEEE Trans. on Pattern Anal. Machine Intell.*, 6, 721-741, 1984.

[85] Geman S., McLure D.E., "Bayesian image analysis" : An application to single photon emission tomography", In *Proc. of the Statistical Computing Section, American Statistical Association*, 12-18, 1985.

[86] Geman S., Graffigne C. "Markov random fields image models and their applications to computer vision", *Proc. ICM86, American Society*, Ed. A.M. Gleason, Providence 1987.

[87] Geman S., Geman D. Graffigne C., Dong P., "Boundary detection by constrained optimization", *IEEE Trans. on Pattern Anal. Machine Intell.*, 12, 609-628, 1990.

[88] Geman S., Mc-Clure S., Geman D., "A nonlinear filter for film restoration and other problems in image processing", *CVGIP : Graph. models and Image processing*, 54, 281-289, 1992.

[89] Geman D.,"Random fields and inverse problems in imaging", *Lecture Notes in Mathematics, "Ecole d'Eté de Saint Flour"*, 1427, 111-193, Springer Verlag, 1992.

[90] Geman D., Reynolds G., "Constrained restoration and the recovery of discontinuities", *IEEE Trans. on Pattern Anal. Machine Intell.*, 14, 367-383, 1992.

[91] Geman S., Bienenstock E., Dousart R., "Neural networks and the biais/variance dilemma", *Neural Computation*, 1991.

[92] Gill P.E., Murray W., Wright M.H., *Practical Optimization*, Academic Press, 1981.

[93] Girard D.A., "The fast Monte Carlo cross validation and C_L procedures : Comments, new results and application to image recovery problems", (with discussions), *Computational Statistics*, 10, 205-258, 1995.

[94] Girard S.C., *Building and Statistical Learning of Auto-Associative Models*, (in French), PhD thesis, University of Cergy-Pontoise, Mathematics, 1996.

[95] Girard S.C., Dinten J.M., Chalmond B., "Building and training radiographic models for flexible object identification from incomplete data", *IEE Proc. Vis. Image Signal Process.*, 143, 257-264, 1996.

[96] Girard S.C., Chalmond B., Dinten J.M., "Designing nonlinear models for flexible curves", *Curves and Surfaces with Applications in CGAD*, In A. Le Méhauté, C. Rabut, L.L. Schumaker (Eds.), Vanderbilt Univ. Press, 135-142, Nashville, 1997.

[97] Girard S.C., Chalmond B., Dinten J.M., "Position of principal component analysis among auto-associative composite models", *C. R. Acad. Sci. Paris*, t. 326, série I, Statistics, 763-768, 1998.

[98] Girard S.C., "A non linear PCA based on manifold approximation", *Computational Statistics*, 15, 145-167, 2000.

[99] Girosi F., Jones M., Poggio T., "Regularization theory and neural network architectures", *Neural Computation*, 219-269, 1995.

[100] Graffigne C., Heitz F., Perez P., Preteux F., Sigelle M., Zerubia J., "Markovian hierarchical models : a review", Invited paper in *Proc. SPIE Conf. on Neural, morphological, stochastic methods in image and signal processing*, San Diego, 1995.

[101] Green P., "Bayesian reconstruction from emission tomography data using a modified EM algorithm", *IEEE Trans. on Medical Imaging*, 9(1), 84-93, 1990.

[102] Grenander U., *General pattern theory, a mathematical study of regular structures*, Clarendon Press, 1993.

[103] Griffeath D., "Introduction to random fields", *In Denumerable Markov Chains*, Kemeny, Knapp and Snell, (Eds.), New-York, Springer-Verlag, 1976.

[104] Guyon X., *Random Fields On a Network*, Springer-Verlag, 1995.

[105] Hammersley J.M., Handscomb D.C., *Monte Carlo Methods*, Methuen and Company, London, 1964.

[106] Hajek B., "Cooling schedules for optimal annealing", *Math. Oper. Res.*, 13, 311-329, 1988.

[107] Hastie T., Stuetzle W., "Principle curves", *Journal of the American Statis. Ass.*, 84(406), 502-516, 1989.

[108] Hastie T., Tibshirani R. and Friedman J.H., *The elements of statistical learning: Data mining, inference and prediction*, Springer Verlag, New-York, 2001.

[109] Heitz F., Bouthémy P., Pérez P., "Multiscale minimization of global energy functions in visual recovery problems", *CVGIP: Image Understanding*, 59(1), 125-134, 1994.

[110] Herman G.T., *Image Reconstruction From Projections : the Fundamental of Computerized Tomography*, Academic Press, New-York, 1980.

[111] Huber P.J., "Projection pursuit", *The Annals of Statistics* 13(2), 434-525, 1985.

[112] Huber P.J., *Robust Statistics*, Wiley, 1981.

[113] Hummel R.A., Zucker S.W., Zucker B.K., "Debluring Gaussian blur", *CVGIP*, 38, 66-80, 1987.

[114] Hyvarïnen A., Oja E., "Independent Component Analysis: Algorithms and applications", *Neural Networks*, 13, 411-430, 2000.

[115] Irons B.M., Razzaque A., "Experience with the patch test for convergence of finite elements", *In : Computing Foundations of the finite element method with applications to partial differential equations*, A.K. Aziz (Ed.), 557-587, Academic Press, 1972.

[116] Isaacson D., Madsen R., *Markov Chains Theory and Applications*, Krieger, 1985.

[117] Jain A.K., Zhong Y., Lakshmaman S., "Object matching using deformable template", *IEEE Trans. on Pattern Anal. Machine Intell.*, 18(3), 267-277, 1996.

[118] Joshi S., Miller M.I., Christensen G.E., Banerjee A., Coogan T., Grenander U., "Hierachical brain mapping via a generalized Dirichlet solution for mapping brain manifolds", *Proc. of the SPIE's Int. Symp. on Optical Science, Engineering and Instrumentation : Vision and Geometry IV*, 2573, 278-289, 1995.

[119] Karuhnen J., Joutsensalo J., "Generalization of principal component analysis, optimisation problems, and neural networks", *Neural Networks*, 8(4), 549-562, 1995.

[120] Kass M., Witkin A., Terzopoulos D., "Snakes : Active contour models", *International J. of Computer Vision*, 1, 321-331, 1987.

[121] Kent J.T., Mardia K.V., "The link between kriging and thin plates", In *Probability, Statistics and Optimization*, F.P. Keller (Ed.), John Wiley, 1994.

[122] Kimeldorf G., Wabha G., "Spline functions and stochastic processes", *Sankhya : The Indian J. of Statistics*, Series A, 173-180, 1970.

[123] Kirkpatrick S., Gellat C.D., Vecchi M.P., "Optimization by simulated annealing", *Science*, 220, 671-680, 1982.

[124] Kullback S., *Information Theory and Statistics*, John Wiley, 1959.

[125] Lapeyre B., Pardoux E., Sentis R., *Monte Carlo Methods for Transport Equation and Diffusion*, (in French), Springer-Verlag, 1998.

[126] Laurent P.J., *Approximation and Optimization*, (in French), Hermann, Paris, 1972.

[127] Leitner F., Marque I., Lavallée S., Cinquin P., "Dynamic segmentation : Finding the edge with snake splines", *Curves and Surfaces* In Laurent P.J., Le Méhauté A., L.L. Schumaker (Eds.), Academic Press, Boston, 1991.

[128] Lévy P. "Chaînes doubles de Markov et fonctions aléatoires de deux variables", *C. R. Académie des Sciences*, 226, 53-55, 1948.

[129] Li S. Z., *Markov Random Field Modeling in Computer Vision*, Springer-Verlag, 1995.

[130] Lierse C., Kaciniel E., Gobel H., "Nondestructive examination of radioactive waste containers by advanced radiometric methods", *Proc. of the Int. Symposium on Computerized Tomography for Industrial Application*, Berlin, 1984.

[131] Luenberger D.G., *Introduction to Linear and Nonlinear Programming*, Addison-Wesley Publishing Company, 1972.

[132] Mallat S., "A theory for multiresolution signal decomposition: the wavelet representation", *IEEE Trans. on Pattern Anal. Machine Intell.*, 11(7), 674-693, 1989.

[133] Mardia K.V., Kent J.T., Bibby J.M., *Multivariate Analysis*, Academic Press, 1979.

[134] Meer P., Mintz D., Rosenfeld A., "Robust regression methods for computer vision", *International J. of Computer Vision*, 6, 59-70, 1991.

[135] Metropolis N., Rosenbluth A.W., Rosenbluth N.N., Teller A.H., Teller E., "Equations of state calculations by fast computational machine", *Journal of Chemical Physics*, 21, 1087-1091, 1953.

[136] Milanfar P., Karl W.C., Willsky A.S., "Reconstruction of binary polygonal objects from projections : a statistical view", *CVGIP : Graph. models and Image processing*, 56(5), 371-391, 1994.

[137] Milnor J., *Topology From the Differentiable Point of View*, The Univeristy Press of Virginia, 1965.

[138] Moghaddam B. and A. Pentland "Probabilistic visual learning for object representation". *IEEE Trans. on Pattern Analysis and Machine Intelligence*, 19(7), 696-710, 1997.

[139] Mougeot M., Azencott R., Angeniol B., "Image compression with back-propagation : improvement of the visual restoration using different cost functions", *Neural Networks*, 4, 467-476, 1991.

[140] Mumford D., Shah J., "Boundary detection by minimizing functionals", *Proc. IEEE Conf. Comput. Vision, Patt. Recog.*, San Francisco, 1985.

[141] Murase H., Nayar S., "Visual Learning and recognition of 3D objects from appearance", *International J. of Computer Vision*, 14, 5-24, 1995.

[142] Natterer F., *The Mathematics of Computerized Tomography*, John Wiley, 1986.

[143] Ord K., "Estimation methods for models of spatial interaction", *Journal of the Amer. Statist. Ass.*, 70, 120-126, 1975.

[144] Parent P., Zucker S.W., "Trace inference, curvature consistency and curve detection", *IEEE Trans. on Pattern Anal. Machine Intell.*, 11, 823-839, 1989.

[145] Pearson K., "On lines and planes of closest fit to systems of points in space", *Philos. Mag.*, 6, 559, 1901.

[146] Perez P., *Markov Random Fields and Multiresolution Image Analysis*, (in French), PhD thesis, University of Rennes, 1993.

[147] Petitjean S., "The enumerate geometry of projective algebraic surfaces and their complexity of aspect graphs", *International J. of Computer Vision*, 19, 261-287, 1996.

[148] Pilo R., Valli G., "Shape from radiological density", *CVGIP : Graph. Models and Image Processing*, 65(3), 361-381, 1997.

[149] Possolo A., "Estimation of binary random Markov field", Technical report 77, Universty of Washington, 1986.

[150] Prenter P., *Spline and Variational Methods*, Wiley, 1971.

[151] Priestley M.B., *Spectral Analysis and Time Series*, Academic Press, 1981.

[152] Ramsay J., Silverman B.W., *Functional Data Analysis*, Springer-Verlag, 1997.

[153] Rao C.R., *Linear Statistical Inference and Its Application*, John Wiley and Sons, 1973.

[154] Reinsch C., Smoothing by spline function, *Numer. Math.*, 10, 177-183, 1967.

[155] Rektorys K., (Ed.), *Survey of Applicable Mathematics*, Iliffe, London, 1969.

[156] Retraint F., *Radiographic Nondestructive Control of Industrial Pieces*, Technical report, CEA, DSYS/SCSI/98653/JMD, Grenoble, 1998.

[157] Rice J., Rosenblatt M., "Smoothing splines : regression, derivatives and deconvolution", The Annals of Statistics, 11, 141-156, 1983.

[158] Rice J.A., Silverman B.W., "Estimating the mean and the covariance structure nonparametrically when the data are curves", *JRSS B*, 53(1), 233-243, 1991.

[159] Rozanov Y. A., *Probability Theory, Random Processes and Mathematical Statistics*, Kluwer Academic Publishers, 1995.

[160] Schechter S., "Relaxation method for convex problems", *SIAM J. of Num. Anal.*, 5, 601-612, 1968.

[161] Shepard R.N., Carroll J.D., "Parametric representation of nonlinear data structures", *Int. Symp. on Multivariate Analysis*, P.R. Krishnaiah (Ed.), 561-592, Academic-Press, 1965.

[162] Shepp L.A., Vardi Y., "Maximum likelihood reconstruction in positron emission tomography", *IEEE Trans. on Medical Imaging*, 113-122, 1982.

[163] Silverman B.W., "Some aspects of the spline smoothing approach to non-parametric regression curve fitting", *J.R. Statist. Soc.* , B47, 1-52, 1985.

[164] Silverman B.W., *Density Estimation*, Chapman-Hall, 1986.

[165] Sire P., Grangeat P, Lemasson P., Melennec P., Rizo P., "NDT applications of 3D Radon transform algorithm from cone beam reconstruction", in : *Proc. Fall Meeting of the Material Research Soc.*, Boston, 1990.

[166] Stone M., "Cross-validatory choice and assesment of statistical predictions (with discussion)", *J.R. Statist. Soc.*, B36, 111-147, 1974.

[167] Strang G., Fix G.J., *An Analysis of the Finite Element Method*, Prentice-Hall, Englewood Cliffs, 1973.

[168] Terzopoulos D., "Multilevel computational processes for visual surface reconstruction", *CVGIP*, 24, 52-96, 1983.

[169] Terzopoulos D., "Regularisation of inverse visual problems involving discontinuities", *IEEE Trans. on Pattern Anal. Machine Intell.*, 8, 413-424, 1986.

[170] Terzopoulos D., Witkin A., Kass M., "Symmetry-seeking models and 3D object reconstruction", *International J. of Computer Vision*, 1(3), 211-221.

[171] Thompson A.M., Brown J.C., Kay J.W., Titterington, "A study of methods of choosing the smoothing parameter in image restoration by regularization", *IEEE Trans. on Pattern Anal. Machine Intell.*, 13, 326-339, 1991.

[172] Trouvé A., "Diffeomorphism groups and pattern matching in image analysis", *International J. of Computer Vision*, 28, 213-221, 1998.

[173] Utreras F., *Cross Validation for Spline Smoothing*, (in French), PhD thesis, University of Grenoble, INPG, Mathematics, 1979.

[174] Vaillant R., Faugeras O., "Using extremal boundaries for 3-D object modeling", *IEEE Trans. on Pattern Anal. Machine Intell.*, 14(2), 157-173, 1992.

[175] Vapnik V., *Estimation of Dependences Based on Empirical Data* Springer-Verlag, New-York, 1982.

[176] Vardi Y., Shepp L.A., Kaufman L., "A statistical model for positron emission tomography", *Journal of the Amer. Statist. Ass.*, 80, 6-37, 1985.

[177] Wabha G., "A survey of some smoothing problems and the methods of generalized cross-validation for solving them", In: Applications of Statistics, Krishnaiah (Ed.), *North-Holland Publishing Company*, 507-523, 1977.

[178] Wang J.P., *Multilevel Markov Random Fields*, (in French), PhD thesis, University of Paris-Sud, Orsay, Mathematics, 1994.

[179] Whittaker E., "On a new method of graduation", *Proc. Edinburgh Math. Soc.*, 41, 63-75, 1924.

[180] Whittle P., "On the smoothing of probability density functions", *J.R. Statist. Soc.* , B20, 334-343, 1958.

[181] Williamson F.F. "Monte-Carlo simulation of photon transport phenomena : sampling techniques", *In : Monte-Carlo simulation in the radiological sciences*, R.L. Morin (Ed.), 53-101, CRC Press, 1988.

[182] Winkler G., *Image Analysis, Random Fields and Dynamic Monte Carlo Methods : A Mathematical Introduction*, Springer-Verlag, 1995.

[183] Younes L., *Parametric Estimation for Gibbs Fields*, (in French), PhD thesis, University of Paris-Sud, Orsay, Mathematics, 1988.

[184] Younes L., "Estimation and annealing for Gibbsian fields", *Annales de l'Institut Henri Poincarré*, Vol. 2, 1988.

[185] Younes L., "Parametric inference for imperfectly observed Gibbsian fields", *Prob. Theory and Rel. Fields*, 82, 625-645, 1989.

[186] Younes L., "Maximum likelihood estimation for Gibbs fields", *In Spatial Statistics and Imaging : Proc. on AMS-IMS-SIAM Joint Summer Research Conference*, A. Possolo, (Ed.), Lecture Notes - Monograph Series, Institute of Mathematical Statistics, Hayward, California, 1991.

[187] Younes L., "Parameter estimation for imperfectly observed Gibbs fields and some comments on Chalmond's EM Gibbsian algorithm", *Lectures Notes in Statistics,"Stochastic Models, Statistical Methods and Algorithms in Image Analysis"*, 74, 240-258, Springer-Verlag, 1992.

[188] Younes L., "Computable elastic distance between shapes", *Siam. J. Appl. Math.*, 1998.

[189] Younes L., "Deformations, Warping and object comparison", Research Report, Ecole Normale Supérieure, Cachan, 2000.

Author Index

Agresti A., 152
Ahlberg J.H., 26, 57
Amit Y., 276
Ash J.M., 72
Aubert G., 19
Azencott R., 62, 80, 132, 171, 187,
    217, 227

Baldi P., 80
Ballard D.H., 256
Banerjee A., 41
Barret H., 189
Basseville M., 184
Bates R., 190
Baum L., 159
Besag J., 100, 151
Bickel P.J., 63, 234
Bienenstock E., 36
Bonin A., 211
Bookstein F.L., 39, 84
Boulard H., 80
Bouthémy P., 295
Brown C.M., 256
Brown J.C., 298
Burq C., 214

Carasso A., 202

Carroll J.D., 87
Carroll R.J., 63
Cea J., 46
Cerny V., 131
Chalmond B., 62, 82, 152, 161, 171,
    187, 192, 211, 217, 225, 227,
    253, 256, 258, 276
Charbonnier P., 248
Chen H., 82
Christensen G.E., 41, 288
Ciarlet P.G., 44
Cinquin Ph., 19, 274, 279
Clearbout J.F., 179
Cohen L., 274
Coldefy F., 172, 227, 253, 276
Comon P., 70
Coogan T., 41
Cooper D.H., 76
Cootes T., 76
Courant R., 28, 40, 46
Craven P., 38

Dacunha-Castelle D., 284
David C., 227
DeBoor C., 26, 54
Dempster A., 159

Derin H., 153
Diaconis P., 72
Dierckz P., 19
Dinten J.M., 82, 212, 216, 249, 258
do Carno M.P., 232
Dolph C.L., 284
Dosksum K.A., 63, 234
Dousart R., 36
Draper N., 54, 187
Duflo M., 19, 284

Eubank P., 19

Feller W., 125
Figueiredo M., 19
Fix G.J., 46
Friedman J.H., 19, 68, 72

Gauch J.M., 227
Gellat C.D., 131
Geman D., 19, 100, 129, 134, 190,
    194, 199
Geman S., 19, 36, 100, 129, 134, 151,
    190, 194, 248
Gill P.E., 141, 152
Girard S.C., 82, 258
Girosi F., 294
Graffigne C., 151, 152, 171, 281
Graham J., 76
Green, 40
Grenander U., 19, 41, 276
Griffeath D., 19
Guyon X., 19, 156, 197

Hajek B., 132
Hammersley J.M., 19, 125
Handscomb D.C., 19, 125
Hastie T., 19, 80
Heitz F., 281
Helliot H., 153
Herman G.T., 243
Hilbert D., 28
Hornik K., 80
Huber P.J., 32, 69
Hummel R., 202
Hyman J., 202
Hyvarinen A., 70

Irons B.M., 50
Isaacson D., 124

Jain A.K., 19, 279
Jones M., 294
Joshi S., 41, 281
Julien Ph., 187

Kamp Y., 80
Karl W.C., 253
Kass M., 256, 273
Kaufman L., 248
Kay J.W., 298
Kent J.T., 41
Kimeldorf G., 284
Kirkpatick S., 131
Kornprobst P., 19
Kullback S., 69, 184

Lévy P., 100
Labourdette C., 171
Laird N., 159
Lakshmaman S., 279
Lapeyre B., 206
Laurent P.J., 19
Lavallée S., 274
Lavayssière B., 211, 253, 276
Leitao J.M., 19
Leitner F., 274
Luenberger D.G., 141

Madsen R., 124
Mallat S., 281
Mardia K.V., 41
Marque I., 274
McDonnel M., 190
McLure D.E., 248
Meer P., 239
Metropolis N., 125
Milanfar P., 253
Miller M.I., 41, 281, 288
Milnor J., 76, 85
Mintz D., 239
Moghadem B., 296
Mougeot M., 80
Mumford D., 297
Murase H., 297
Murray W., 141, 152

Natterer F., 243
Nayar S., 297
Newton-Raphson, 152
Nilson E.N., 26, 57

Oja E., 70

Pardoux E., 206
Parent P., 227
Pearson K., 77
Pentland A., 296
Perez P., 281
Petitjean S., 270
Petrie T., 159
Piccioni M., 276
Pilo R., 256
Pizer S.M., 227
Poggio T., 294
Possolo A., 153
Préteux F., 281
Prenat M., 152
Priestley M.B., 284

Rabbitt R., 288
Ramsay J., 76, 259
Rao C.R., 69
Reinsch C., 25
Retraint F., 216
Reynolds G., 194, 199
Rice J., 38
Rice J.A., 76
Rosenblatt M., 38
Rosenfeld A., 239
Roux M., 152
Rozanov Y.A., 283
Rubin D., 159
Ruppert D., 63

Sanderson J., 202
Schechter S., 142
Sentis R., 206
Shah J., 297
Shashahani M., 72
Shepard R.N., 87
Shepp L.A., 248
Sigelle M., 281
Silverman B.W., 69, 76, 92, 259
Smith H., 54, 187

Soules G., 159
Stone M., 38
Strang G., 46
Stuetzle W., 80
Swindell W., 189

Taylor C., 76
Terzopoulos D., 39, 47, 256, 273
Thomson A.M., 298
Tibshirani R., 19
Titterington J., 298
Trouvé A., 279

Utreras F., 39

Valli G., 256
Vapnik V., 36
Vardi Y., 248
Vecchi M.P., 131

Wahba G., 38, 284
Walsh J.L., 26, 57
Wang J.P., 62, 217, 224
Weis N., 159
Whittaker E., 25
Whittle P., 34
Williamson F., 206
Willsky A.S., 253
Winkler G., 19, 125
Witkin A., 256, 273
Woodbury M.A., 284
Wright M.H., 141, 152
Wu C.F., 63

Younes L., 150, 159, 168, 269, 270

Zérubia J., 281
Zhong Y., 279
Zucker B., 202
Zucker S.W., 202, 227

Subject Index

Active contour, 31, 274
Algorithm
  AAC, 83
  axis search, 89
  block-by-block relaxation, 142,
    251, 272
  Carroll's, 64, 66, 237
  detection of valley bottom lines,
    238
  EM, 159
  fixed-point, 215
  Gibbs sampler, 129, 164, 168, 238
  Gibbsian EM
    m-ary hidden field, 164
    m-ary hidden field and Gaussian
      noise, 167
    binary hidden field, 163
    Gaussian hidden field, 170
    Gaussian hidden field and
      Gaussian noise, 251
  gradient descent, 141
  ICM, 138, 140, 142, 187, 251
  iteratively reweighted least squares,
    153
  maximum likelihood, 150, 159
  Metropolis, 126, 128
  Newton-Raphson, 152
  PPR, 70, 81
  search for parametrizing axis, 88
  simulated annealing, 88, 133, 281
    Metropolis dynamic, 131, 133
  stochastic search for parametrizing
    axis, 90
Anisotropic variance, 64, 65, *see*
    heteroscedasticity

B-spline, 52, 275
  mth order, 57
  tensor product, 58
Bias/variance (dilemma), 36
Blurring, 189, 247, 258, 270, 274, 276
Brownian motion, 284, 288

Calibration of parameters, 171, 186
Canonical analysis, 262, 265
Clique, 100
Compression, 94
  function, 80
Concave, 141, 150–152, 157, 161, 199
Confidence region, 178
Conforming condition, 46
Conical source, 253

Contraction coefficient, 135
Convex, 141, 248
Cooling diagram, 131, 133, 134
Cost function, 114, 156
Cross validation, 38, 93
Curse of dimensionality, 68
Curvature, 231, 240

Deformations, 42
    Gaussian, 259, 277
Denoising, 192
Detection
    of edges, 192
    of valley bottom lines, 233
Distribution
    Gibbs, 102, 126, 156
    marginal, 157
    posterior, 116
    prior, 115

Edge site, 191
Energy (continuous)
    1D spline, 24
    2D spline, 40, 190
    bending, 24, 25, 39, 106
    L-spline surface, 40
    Laplacian, 40
    thin plate, 40
Energy (discrete)
    3D spline curve, 276
    Bernoulli, 107, 121, 151, 154
    binomial, 112
    external, 24
    fidelity, 10, 24, 25, 115, 178
    Gaussian, 108, 153, 192, 250
    internal, 24
    matching, 280
    non-Markov, 254
    posterior, 24, 116
    potential, 49, 101
    prior, 10, 24, 115
    quadratic, 44, 143, 273
    regular curve, 234
    segmentation, 250
Entropy, 102
Equation
    Euler–Lagrange , 29
    diffusion, 202

evolution, 31, 202
Fredholm, 209
heat, 202
normal, 54, 60, 61
particle speed, 285
Poisson, 285
transport, 209
variational, 45, 274
Equilibrium distribution, see Markov
    chain
Estimation
    Bayesian, 7, 118, 271
    logistic, 171
    maximum posterior likelihood
        (MAP), 119
    maximum posterior marginal
        likelihood (MPM), 120, 168,
        238
    maximum pseudolikelihood, 151,
        196
Examples, 139, 172, 177
Extraction of image characteristics,
    see Image characteristics

Face, 42, 60, 94
Field (continuous)
    Gaussian, 277
    second-order, 277
Field (discrete)
    m-ary, 153
    hidden, 113, 140, 156
    homogeneous, 151, 154
    Markov random, 100
    multivariate, 182
    potential, 270
    random, 100
    vector, 182, 276
    virtual, 229
Finite differences, 30, 31
Flux, 214
Fourier
    series, 203, 284
    transform, 244
Function
    of Euler–Lagrange , 29
    absorption, 243
    covariance, 41, 277
    Green's, 41, 284

S-periodic, 202
sensitivity, 209, 217
smoothing, 32, 194, 200
transfer, 219

Gauss–Seidel (resolution), 49, 144
Generalization error, 92
Generalized cylinder
  Bayesian estimation, 262
  characteristic curve, 257
  deformable, 259
  radiographic projection, 257
  spline, 256
Grid
  irregular, 39
  regular, 57

Heteroscedasticity, 225
Hidden
  field of discrete 3D curves, 181
  entity, 3
  object, 269
Hierarchical approach, 205, 238, 281
Hypothesis test, 234

Image
  characteristics, 13, 179, 229, 239
  features, *see* Image characteristics
  fusion, 239
  segmentation, 168
Information
  fraction of, 86
  initial, 183
  Kullback, 69, 184
  prior, 171, 183
Interpolation, 41
Invariance, 58, 88, 270

Jensen's inequality, 69, 102, 161

Knots of approximation, 51, 53, 91

L-spline surface, *see* Energy
    (continuous)
Laplacian, 40, 202, 285
Least squares, 62
  weighted, 63
Likelihood, 7, 62, 149, 157

marginal, 161
penalized, 8
Local configuration, 171
  extreme, 186
Local specification, 100, 139, 154,
    171, 186
Luminosity gradient, 60, 228

Möbius (inversion formula), 103
Markov chain
  ergodic, 124
  homogeneous, 124
  irreducible, 125
  natural, 206
  non-homogeneous
      ergodicity, 136
  nonhomogeneous, 134
  regular, 125
  strongly ergodic, 135
  weakly ergodic, 135
  with continuous states, 206
Matching, 42
  Bayesian, 279
Matrix
  exploration, 125, 127, 131, 133
Measurement noise, 189
Microtexture, 109
Model
  auto-associative, 12
  auto-associative composite (AAC),
      82, 259
  auto-model, 110
  degradation, 117, 120, 190, 201,
      217, 228
  endogenous, 12
  exogenous, 11
  Gaussian degradation, 118
  Ising, 107, 121
  logistic, 152, 153
  PPR, 70
  projection, 261
Model-building, 177, 185
Modeling, *see* Model-building
Monte Carlo, 122, 210

Neighborhood system, 100
Neighboring sites, 87, 100
Neural network

perceptron, 80
NMR imaging, 196
Nonconforming element, 46
Nonlinear parametrization, 231

Occlusion, 256
Operator
   difference, 25, 54
   differential, 40, 284
   orthogonal projection, 266
   symmetric, 203
Outlier, 31, 239

Parameter
   regularization, 25, 106
Parametrizing axis, 80
Partial differential equation
   active contour, 31
   heat, 202
   Poisson, 285
Pattern recognition, 36, 269
Photon emission, 206, 247
Pixel site, 191
Point spread function (PSF), 201
Potential
   canonical, 101, 106, 193
   neighborhood, 101
Prediction, 12
Principal component analysis (PCA),
      76, 259
Principle
   of maximum likelihood, 7, 62, 149
   of maximum pseudolikelihood, 151
Problem
   direct, 9
   ill-posed, 9, 202
   inverse, 9
Projection index, 69, 77, 87
   optimization, 69, 88, 133
Projection model, 246
Projection Pursuit (PP), 69, 81

Quadratic error, 11, 24
   of prediction, 12
Qualitative box, see Calibration of
      parameters
   curve fitting, 172, 235

Radiographic
   acquisition, 189, 239
   film, 217
   image, 243
Radon transform, 244
Realignment of radiographic images,
      271
Regression
   linear, 62, 67
   projection pursuit (PPR), 70
Regression spline
   1D, 54, 91
   2D, 58
   3D, 60
Regular curve, 172, 193
Regularization, 10, 36
Restoration
   function, 80
   fusion of multiple films, 217
   of blurred image, 202
   of images, 190
   of scatter, 214
Revealing direction, 68
Robustness, 31

Seismic imaging, 179
Signal-dependent variance, 225
Signal-to-noise ratio, 198, 227
Simulated annealing, see Algorithm
Smoothing
   mth order, 25
   factor, 91, 274
   higher-order, 200
   parameter, 54
   robust, 31
   second-order, 25
Spatial autocorrelation, 109, 225
Spectral representation, 283
Spline
   bicubic, 57, 229
   cubic, 25
   interpolation, 26, 52
   natural cubic, 25
   robust, 31, 200, 239
   tricubic, 60, 225
   under tension, 26
Spot (Gaussian), 179, 217
Stationarity, 109, 124, 277

Stochastic
  differential equation, 286
  integral, 283
Stochastic process
  independent increments, 284
  second-order, 283
  stationary, 284
Sweep, 128, 133

Theorem
  Gauss–Markov, 63
  Karuhnen–Loève, 283
Thin plate, 40
Tradeoff, 8
Transition kernel, 207

Validation, 187
Valley bottom line, 227
Variational
  calculus, 29, 45
  derivative, 30
Voxel, 180, 254

Weighting parameter, 114

Applied Mathematical Sciences

(continued from page ii)

60. *Ghil/Childress:* Topics in Geophysical Dynamics: Atmospheric Dynamics, Dynamo Theory and Climate Dynamics.
61. *Sattinger/Weaver:* Lie Groups and Algebras with Applications to Physics, Geometry, and Mechanics.
62. *LaSalle:* The Stability and Control of Discrete Processes.
63. *Grasman:* Asymptotic Methods of Relaxation Oscillations and Applications.
64. *Hsu:* Cell-to-Cell Mapping: A Method of Global Analysis for Nonlinear Systems.
65. *Rand/Armbruster:* Perturbation Methods, Bifurcation Theory and Computer Algebra.
66. *Hlaváček/Haslinger/Necasl/Lovísek:* Solution of Variational Inequalities in Mechanics.
67. *Cercignani:* The Boltzmann Equation and Its Applications.
68. *Temam:* Infinite-Dimensional Dynamical Systems in Mechanics and Physics, 2nd ed.
69. *Golubitsky/Stewart/Schaeffer:* Singularities and Groups in Bifurcation Theory, Vol. II.
70. *Constantin/Foias/Nicolaenko/Temam:* Integral Manifolds and Inertial Manifolds for Dissipative Partial Differential Equations.
71. *Catlin:* Estimation, Control, and the Discrete Kalman Filter.
72. *Lochak/Meunier:* Multiphase Averaging for Classical Systems.
73. *Wiggins:* Global Bifurcations and Chaos.
74. *Mawhin/Willem:* Critical Point Theory and Hamiltonian Systems.
75. *Abraham/Marsden/Ratiu:* Manifolds, Tensor Analysis, and Applications, 2nd ed.
76. *Lagerstrom:* Matched Asymptotic Expansions: Ideas and Techniques.
77. *Aldous:* Probability Approximations via the Poisson Clumping Heuristic.
78. *Dacorogna:* Direct Methods in the Calculus of Variations.
79. *Hernández-Lerma:* Adaptive Markov Processes.
80. *Lawden:* Elliptic Functions and Applications.
81. *Bluman/Kumei:* Symmetries and Differential Equations.
82. *Kress:* Linear Integral Equations, 2nd ed.
83. *Bebernes/Eberly:* Mathematical Problems from Combustion Theory.
84. *Joseph:* Fluid Dynamics of Viscoelastic Fluids.
85. *Yang:* Wave Packets and Their Bifurcations in Geophysical Fluid Dynamics.
86. *Dendrinos/Sonis:* Chaos and Socio-Spatial Dynamics.
87. *Weder:* Spectral and Scattering Theory for Wave Propagation in Perturbed Stratified Media.
88. *Bogaevski/Povzner:* Algebraic Methods in Nonlinear Perturbation Theory.
89. *O'Malley:* Singular Perturbation Methods for Ordinary Differential Equations.
90. *Meyer/Hall:* Introduction to Hamiltonian Dynamical Systems and the N-body Problem.
91. *Straughan:* The Energy Method, Stability, and Nonlinear Convection.
92. *Naber:* The Geometry of Minkowski Spacetime.
93. *Colton/Kress:* Inverse Acoustic and Electromagnetic Scattering Theory, 2nd ed.
94. *Hoppensteadt:* Analysis and Simulation of Chaotic Systems, 2nd ed.
95. *Hackbusch:* Iterative Solution of Large Sparse Systems of Equations.
96. *Marchioro/Pulvirenti:* Mathematical Theory of Incompressible Nonviscous Fluids.
97. *Lasota/Mackey:* Chaos, Fractals, and Noise: Stochastic Aspects of Dynamics, 2nd ed.
98. *de Boor/Höllig/Riemenschneider:* Box Splines.
99. *Hale/Lunel:* Introduction to Functional Differential Equations.
100. *Sirovich (ed):* Trends and Perspectives in Applied Mathematics.+
101. *Nusse/Yorke:* Dynamics: Numerical Explorations, 2nd ed.
102. *Chossat/Iooss:* The Couette-Taylor Problem.
103. *Chorin:* Vorticity and Turbulence.
104. *Farkas:* Periodic Motions.
105. *Wiggins:* Normally Hyperbolic Invariant Manifolds in Dynamical Systems.
106. *Cercignani/Illner/Pulvirenti:* The Mathematical Theory of Dilute Gases.
107. *Antman:* Nonlinear Problems of Elasticity.
108. *Zeidler:* Applied Functional Analysis: Applications to Mathematical Physics.
109. *Zeidler:* Applied Functional Analysis: Main Principles and Their Applications.
110. *Diekmann/van Gils/Verduyn Lunel/Walther:* Delay Equations: Functional-, Complex-, and Nonlinear Analysis.
111. *Visintin:* Differential Models of Hysteresis.
112. *Kuznetsov:* Elements of Applied Bifurcation Theory, 2nd ed.
113. *Hislop/Sigal:* Introduction to Spectral Theory: With Applications to Schrödinger Operators.
114. *Kevorkian/Cole:* Multiple Scale and Singular Perturbation Methods.
115. *Taylor:* Partial Differential Equations I, Basic Theory.
116. *Taylor:* Partial Differential Equations II, Qualitative Studies of Linear Equations.

(continued on next page)

Applied Mathematical Sciences

(continued from previous page)

117. *Taylor:* Partial Differential Equations III, Nonlinear Equations.
118. *Godlewski/Raviart:* Numerical Approximation of Hyperbolic Systems of Conservation Laws.
119. *Wu:* Theory and Applications of Partial Functional Differential Equations.
120. *Kirsch:* An Introduction to the Mathematical Theory of Inverse Problems.
121. *Brokate/Sprekels:* Hysteresis and Phase Transitions.
122. *Gliklikh:* Global Analysis in Mathematical Physics: Geometric and Stochastic Methods.
123. *Le/Schmitt:* Global Bifurcation in Variational Inequalities: Applications to Obstacle and Unilateral Problems.
124. *Polak:* Optimization: Algorithms and Consistent Approximations.
125. *Arnold/Khesin:* Topological Methods in Hydrodynamics.
126. *Hoppensteadt/Izhikevich:* Weakly Connected Neural Networks.
127. *Isakov:* Inverse Problems for Partial Differential Equations.
128. *Li/Wiggins:* Invariant Manifolds and Fibrations for Perturbed Nonlinear Schrödinger Equations.
129. *Müller:* Analysis of Spherical Symmetries in Euclidean Spaces.
130. *Feintuch:* Robust Control Theory in Hilbert Space.
131. *Ericksen:* Introduction to the Thermodynamics of Solids, Revised ed.
132. *Ihlenburg:* Finite Element Analysis of Acoustic Scattering.
133. *Vorovich:* Nonlinear Theory of Shallow Shells.
134. *Vein/Dale:* Determinants and Their Applications in Mathematical Physics.
135. *Drew/Passman:* Theory of Multicomponent Fluids.
136. *Cioranescu/Saint Jean Paulin:* Homogenization of Reticulated Structures.
137. *Gurtin:* Configurational Forces as Basic Concepts of Continuum Physics.
138. *Haller:* Chaos Near Resonance.
139. *Sulem/Sulem:* The Nonlinear Schrödinger Equation: Self-Focusing and Wave Collapse.
140. *Cherkaev:* Variational Methods for Structural Optimization.
141. *Naber:* Topology, Geometry, and Gauge Fields: Interactions.
142. *Schmid/Henningson:* Stability and Transition in Shear Flows.
143. *Sell/You:* Dynamics of Evolutionary Equations.
144. *Nédélec:* Acoustic and Electromagnetic Equations: Integral Representations for Harmonic Problems.
145. *Newton:* The *N*-Vortex Problem: Analytical Techniques.
146. *Allaire*: Shape Optimization by the Homogenization Method.
147. *Aubert/Kornprobst:* Mathematical Problems in Image Processing: Partial Differential Equations and the Calculus of Variations.
148. *Peyret:* Spectral Methods for Incompressible Viscous Flow.
149. *Ikeda/Murota:* Imperfect Bifurcation in Structures and Materials: Engineering Use of Group-Theoretic Bifucation Theory.
150. *Skorokhod/Hoppensteadt/Salehi:* Random Perturbation Methods with Applications in Science and Engineering.
151. *Bensoussan/Frehse:* Topics on Nonlinear Partial Differential Equations and Applications.
152. *Holden/Risebro:* Front Tracking for Hyperbolic Conservation Laws.
153. *Osher/Fedkiw:* Level Set Methods and Dynamic Implicit Surfaces.
154. *Bluman/Anco:* Symmetry and Integration Methods for Differential Equations.
155. *Chalmond*: Modeling and Inverse Problems in Image Analysis.